PROBABILITY THEORY AND MATHEMATICAL STATISTICS WITH APPLICATIONS

PROBABILITY THEORY AND MATHEMATICAL STATISTICS WITH APPLICATIONS

Proceedings of the 5th Pannonian Symposium on Mathematical Statistics Visegrád, Hungary, 20-24 May, 1985

Edited by

W. GROSSMANN
University of Vienna, Austria

J. MOGYORÓDI
Eötvös Loránd University, Budapest, Hungary

I. VINCZE
Hungarian Academy of Sciences, Budapest, Hungary

W. WERTZ
Technical University of Vienna, Austria

D. REIDEL PUBLISHING COMPANY

A MEMBER OF THE KLUWER ACADEMIC PUBLISHERS GROUP

DORDRECHT / BOSTON / LANCASTER / TOKYO

Library of Congress Cataloging-in-Publication Data

Pannonian Symposium on Mathematical Statistics
(5th : 1985 : Visegrád, Hungary)
Probability theory and mathematical statistics
with applications.

Includes index.
1. Mathematical statistics—Congresses.
2. Probabilities—Congresses. I. Grossmann, Wilfried, 1948—
QA276.A1P36 1985 519.2 87-16646
ISBN 90-277-2547-0

Distributors for the U.S.A. and Canada
Kluwer Academic Publishers,
101 Philip Drive, Norwell, MA 02061, U.S.A.

Distributors for Albania, Bulgaria, People's Republic of China,
Cuba, Czechoslovakia, German Democratic Republic, Hungary,
People's Republic of Korea, Mongolia, Poland, Romania,
the U.S.S.R., Vietnam, and Yugoslavia
Kultura Hungarian Foreign Trading Company
P.O.B. 149
H-1389 Budapest 62, Hungary

Distributors for all remaining countries
Kluwer Academic Publishers Group
P.O.Box 322, 3300 AH Dordrecht, Holland

Joint edition published by D. Reidel Publishing Company
Dordrecht, Holland / Boston, U.S.A. / Lancaster, England /Tokyo, Japan

and

Akadémiai Kiadó, Budapest, Hungary

Printed in Hungary

CONTENTS

PREFACE

The papers in Part A are devoted to interesting elementary and advanced probabilistic problems, such as random filling of boxes in an increasing order from a random sample of a finite population and the corresponding stochastic model, random walk in multidimensional time which needs a profound theoretical approach, foundations of non-Kolmogorovian probability theory, an attempt to characterize the two-dimensional exponential distribution, other probabilistic characterization theorems, the rarefaction of renewal processes, etc. Algebraic and arithmetic properties of probability distribution functions are also considered in a few papers. The actual state of the extreme limit problem is summarized in one of them. Markov chains also appear in several papers. Namely, one of them considers the first passage problem for Ehrenfest's urn model, another deduces a limit distribution theorem for partial sums of random variables forming a Markov chain and taking 2Π. A further paper considers Markov chains with bound positive jumps and homogeneous second component. One of the authors deals with optimal stopping in the secretary problem and gives a remarkable generalization of the results obtained in this area. Large deviations problems are considered in the non-independent case, namely, in connection with Gibbs random fields.

The different kinds of dependence are treated in some papers: in case of exchangeable random variables a complete convergence theorem is proved and the martingale dependence is also investigated in one- and two-discrete parameter cases and also in the case when the parameter set forms a tree. In the last case, in the so-called tree-martingales, some "classical" martingale inequalities are generalized, whilst in the preceding ones regression type martingale fields are considered. A new proof of an inequality of Burkholder, Davis and Gundy is given. Some facts from theory of stochastic processes are also presented in this volume, e.g. an investigation of the largest flat interval of a stochastic process whose first passage process is a Lévy process, mimicking stochastic differential equations by simpler ones and considering the rank of multidimensional stationary processes.

Part B deals with estimation theory, testing hypotheses, statistical methods and models, regression and applications, least squares estimation, maximum likelihood estimation, its asymptotic normality, overparametrized maximum likelihood estimation of ARMA processes, estimation of contamination level, estimation with (randomly) censored data, the local structure of estimation model to the theory of unbiased estimation, estimation of cross dispersion function for stationary ergodic process, state estimation of a dynamical system perturbated by small noise, etc. The authors study the asymptotic efficiency, relative risk efficiency of estimations, the log likelihood ratios, strong consistency and so on. There are papers on goodness-of-fit tests and on testing the order of stochastic processes. The papers dealing with statistical methods and models are devoted to periodic autoregression with exogenous variables, nonlinear regression models, large non-parametric models, quadratic input linear output (QUILO) models with applications to regression, general multivariate linear models, differentiation of one-parameter families of probability distribution functions, the rejective sampling plan from a finite population and to further procedures and models. Non-parametric regression with random right censoring, QUILO models in regression and periodic autoregression are also examined. The applications are concerned with stochastic kinetics in chemical reactions, and the asymptotic normality of parameter dependent vector Gaussian processes.

Since the third symposium the Pannonian Symposia have been organized alternately in Austria and in Hungary. They offer a forum for statisticians and probabilists living in the countries of the ex-Roman Province Pannonia to present results in this field and exchange opinions and experiences about problems and research interests. We are particularly grateful to those colleagues who arrived from the countries e.g. from the Soviet Union, Poland, German Democratic Republic, German Federal Republic, France, Portugal, USA, Netherlands, Czechoslovakia and the United Kingdom.

The organization work of the Symposium was done by the staff of the Department of Probability Theory and Statistics of the Mathematical Institute of Eötvös Loránd University, Budapest. We express here our thanks to the Computing Centre of this University whose members have put a lot of effort in the organization of the Symposium.

In 1985 we commemorated the 350th anniversary of the foundation of the Eötvös Loránd University. This Symposium has also been included in the series of scientific programs of the anniversary.

Professor dr. Károly Sarkadi and dr. Pál Kosik, coauthors of a paper in this volume (whose works on mathematical statistics and its applications, as well as on probability theory have influenced the Hungarian statistical school) have in the meantime sadly passed away.

The Editors

PART A

PROBABILITY THEORY

Proc. of the 5th Pannonian Symp. on Math. Stat., Visegrád, Hungary 1985
W. Grossmann, J. Mogyoródi, I. Vincze, W. Wertz, eds

FIRST PASSAGE PROBLEMS FOR EHRENFEST MODEL

G. BARÓTI

Budapest

In Kemperman's /1961/ book the following result can be found for the discrete Ehrenfest model:
Let 2R denote the number of molecules and let $Y_{i,j}$ be the first moment the vessel V_1 has j molecules supposing that it has i molecules when the counting is started. The limit distribution of $Y_{R,R-k}$ $(R,k\to\infty, 0<c_0<\frac{k}{R}<c_1<1)$ is exponential. We have proved that this result holds for $Y_{R,0}$; $Y_{2R,0}$ and $Y_{0,0}$ too.

AMS subject classification: 60C05, 60F05

In /1907/ P. and T. Ehrenfest introduced a model representing heat exchange between two isolated bodies. This well-known model is the following.

Consider two vessels V_1 and V_2 with equal volume, separated by a membrane and containing together 2R molecules of a certain gas. At time $n (n=0,1,\ldots)$ one of the 2R molecules passes the membrane. Let X_n denote the number of molecules in V_1 after the nth time a molecule passed the membrane. It is reasonable to assume that each of the 2R molecules is equally likely to have passed, that is, given $X_{n-1}=i$, we have $X_n=i-1$ or $X_n=i+1$ with probability $\frac{i}{2R}$ resp. $1-\frac{i}{2R}$. $\{X_n\}$ is a stationary Markov chain with states $0,1,\ldots,2R$ and with transition probabilities

$$P_{i,i-1}=\frac{i}{2R}\,,\quad P_{i,i+1}=1-\frac{i}{2R} \tag{1}$$

At the 3rd 4th PSMS we examined /see Baróti/1982/,/1983// some limit distribution theorems for X_n.

Suppose that $X_0=i$ and let $Y_{i,j}$ denote the first moment $n \geq 1$ the vessel V_1 has j molecules.
Introduce the generating functions:

$$P_{ij}(z) = \sum_{n=0}^{\infty} P(X_n=j|X_0=i)z^n, \tag{2}$$

$$F_{ij}(z) = \sum_{n=1}^{\infty} P(Y_{i,j}=n)z^n, \qquad |z|<1$$

From the general theory of Markov chain it follows that all states are recurrent, $Y_{i,j}<\infty$ with probabaility 1 and

$$F_{ij}(z)=\begin{cases} \dfrac{P_{ij}(z)}{P_{jj}(z)} & \text{if} \quad i\neq j, \\ 1-\dfrac{1}{P_{jj}(z)} & \text{if} \quad i=j. \end{cases} \tag{3}$$

Some limit theorems for the first passage time $Y_{i,j}$ have been obtained by Bellman and Harris /1951/ although they have studied the corresponding continuous model in which the number of molecules crossing the membrane during a time interval of length t has a Poisson distribution with mean $\lambda t/\lambda$ is a fixed constant/. The proofs of these results for the discrete model and other results can be found in Kemperman's /1961/ book. These authors have dealt with the limit distribution of $Y_{i,j}$ if

a/ $i=R$ or $i=j$ and $R\to\infty$, $j\to\infty$ with $\frac{j}{R}\to\mu$ $(0<\mu<1)$,

b/ $i=R$ and $R\to\infty$, $j\to\infty$ with $\frac{R-j}{\sqrt{R}}\to\alpha>0$

c/ $i=R+k$, $j=R$ and $R\to\infty$, $k\to\infty$ with $k\geq\sqrt{R\log R}$,

d/ $i=j=R$ and $R\to\infty$

Now we shall prove that the limit distribution of $Y_{0,0}$, $Y_{R,0}$ and $Y_{2R,0}$ are exponential if $R\to\infty$. That is the following theorem holds.

Theorem

$$\lim_{R\to\infty} P(Y_{i,0}\cdot 4^{-R} < x) = 1-e^{-x} \quad (i=0,R,2R;\ x>0)$$

Proof of the theorem. We obtain from a partial differential equation /see Baróti /1982/ / that

$$\sum_{n=0}^{\infty}\left(\sum_{j=0}^{2R} P(X_n=j\,|\,X_0=i)x^j\right)\frac{(2R)^n}{n!}z^n = (\mathrm{sh}z + x\mathrm{ch}z)^i(\mathrm{ch}z + x\mathrm{sh}z)^{2R-i}$$

and so we are led to the following formulae

$$P_{00}(z) = \int_0^{\infty}\left(\mathrm{ch}\frac{zt}{2R}\right)^{2R} e^{-t}\,dt \tag{4}$$

$$P_{R0}(z) = \int_0^{\infty}\left(\mathrm{sh}\frac{zt}{R}\right)^{R} 2^{-R} e^{-t}\,dt \tag{5}$$

It follows from (4) and (5) that

$$P_{00}\left(e^{-\frac{S}{4R}}\right) = \sum_{k=0}^{2R}\binom{2R}{k}\frac{R}{R\cdot 4^R\left(1-e^{-\frac{S}{4R}}\right)+k\cdot 4^R e^{-\frac{S}{4R}}} =$$

$$= \frac{1}{4^R\left(1-e^{-\frac{S}{4R}}\right)} + \sum_{k=1}^{2R}\binom{2R}{k}\frac{R}{k\cdot 4^R} + \sigma(1) = \frac{1}{S} + 1 + \sigma(1)$$

$$P_{R0}\left(e^{-\frac{S}{4R}}\right) = \sum_{k=0}^{R}(-1)^k\binom{R}{k}\frac{R}{R\cdot 4^R(1-e^{-\frac{S}{4R}}) + 2k\cdot 4^R e^{-\frac{S}{4R}}} =$$

$$= \frac{1}{4^R(1-e^{-\frac{S}{4^R}})} + \sum_{k=1}^{R} (-1)^k \binom{R}{k} \frac{R}{2k \cdot 4^R} + \sigma(1) = \frac{1}{S} + \sigma(1).$$

These relations and (3) imply the statement of the Theorem if $i = 0$ or $i = R$.
Finally $Y_{2R,0} = Y_{2R,R} + Y_{R,0}$ and $\mathrm{var}(Y_{2R,R}) = O(R^2)$ /see Kemperman /1961/ / thus $Y_{2R,0} \cdot 4^{-R}$ and $Y_{R,0} \cdot 4^{-R}$ have the same limit distribution.

REFERENCES:

Baróti, G. (1982) Limit distribution for Ehrenfest's urn model; Proc. of the 3rd Pannonian Symp. on Math. Stat., Visegrád Hungary.

Baróti, G. (1983) A limit theorem for Ehrenfest urn model; Proc. of the 4th Pannonian Symp. on Math. Stat., Bad Tatzmannsdorf, Austria.

Bellman, R. and Harris, T. E. (1951) Recurrence times for Ehrenfest model; Pacific J. Math., 1, 184-188.

Ehrenfest, P. and Ehrenfest, T. (1907) Über zwei bekannte Einwände gegen das Boltzmannsche H-Theorem; Physik. Z., 8, 311-314.

Kemperman, J. H. B. (1961) The first passage problem for a stationary Markov chain; The University of Chicago Press.

György Baróti
Kandó College of Electrical
Engineering,
Budapest, H-1084
Tavaszmező u. 17.
Hungary

Proc. of the 5th Pannonian Symp. on Math. Stat., Visegrád, Hungary 1985
W. Grossmann, J. Mogyoródi, I. Vincze, W. Wertz, eds

A NEW PROOF OF THE RIGHT HAND SIDE OF THE BURKHOLDER-DAVIS-GUNDY INEQUALITY

N.L. BASSILY

Cairo

A proof of the right hand side of the Burkholder-Davis-Gundy inequality is given by using Fefferman's inequality established in the more general form in [2]. The left hand side of the inequality has been proved in another paper by the author for a large class of Young functions [14].
AMS subject classification: 60G42

1. Throughout the present note we consider a fixed probability space $(\Omega, \mathcal{F}, P)$ and a sequence of σ-fields $\mathcal{F}_0 \subset \mathcal{F}_1 \subset \dots$ such that $\sigma\left(\bigcup_{n=0}^{\infty} \mathcal{F}_n\right) = \mathcal{F}$. Let $X, Y \in L_1(\Omega, \mathcal{F}, P)$ be random variables and consider the martingales

$$X_n = E(X \mid \mathcal{F}_n), \quad Y_n = E(Y \mid \mathcal{F}_n), \quad n \geq 0,$$

where for the sake of commodity we suppose that $X_0 = Y_0 = 0$ a.s.

We say that $X \in \mathcal{H}_p$, where p is a number such that $1 \le p < +\infty$, if the quadratic variation

$$S = S(X) = \left(\sum_{i=1}^{\infty} d_i^2\right)^{1/2}$$

of the martingale $(X_n, \mathcal{F}_n)$ belongs to $L_p(\Omega, \mathcal{F}, P)$. Here

$$d_i = X_i - X_{i-1}, \quad i \geq 1, \quad d_0 = X_0 = 0 \quad \text{a.s.}$$

are the differences of the martingale $(X_n, \mathcal{F}_n)$. If $X \in \mathcal{H}_p$ then we put

$$\|X\|_{\mathcal{H}_p} = \|S\|_p .$$

It is easily seen that $\|.\|_{\mathcal{H}_p}$ defines a norm on $\mathcal{H}_p$.

We also define the space $\mathcal{K}_p$, where p is a number such that $1 \le p \le +\infty$, in the following way: if the set of the random variables γ defined by the formula

$$\Gamma_Y^{(p)} = \{\gamma : \gamma \in L_p, E(|Y-Y_{n-1}| \mid \mathcal{F}_n) \le E(\gamma \mid \mathcal{F}_n) \text{ a.s.}, \ \forall n \ge 1\}$$

is not empty, then we say that Y belongs to $\mathcal{K}_p$. In this case we set

$$\|Y\|_{\mathcal{K}_p} = \inf_{\gamma \in \Gamma_Y^{(p)}} \|\gamma\|_p .$$

It can be easily seen that $\|.\|_{\mathcal{K}_p}$ is a norm on the space $\mathcal{K}_p$. The space $\mathcal{K}_\infty$ is the well-known BMO_1-space.

S. Ishak and J. Mogyoródi [1] have proved the following generalization of the Fefferman inequality.

<u>Theorem 1</u>. Let $X \in \mathcal{H}_p$ and $Y \in \mathcal{K}_q$, where $1 \le p < +\infty$ and $q = \frac{p}{p-1}$. Then

$$\lim_{n \to +\infty} E(X_n Y_n)$$

exists and is finite. Moreover, we have

$$|\lim_{n \to +\infty} E(X_n Y_n)| \le C_p' \|X\|_{\mathcal{H}_p} \|Y\|_{\mathcal{K}_q}$$

and

$$|E(X_n Y_n)| \le C_p' \|X_n\|_{\mathcal{H}_p} \|Y_n\|_{\mathcal{K}_q} .$$

Here C_p' is a constant depending only on p.

In connection with the last inequality let us remark that together with X (resp. Y) the random variables X_n (resp. Y_n) also belong to $\mathcal{H}_p$ (resp. $\mathcal{K}_q$) and we have

$$\|X_n\|_{\mathcal{H}_p} \leq \|X\|_{\mathcal{H}_p} \quad (\text{resp. } \|Y_n\|_{\mathcal{K}_q} \leq \|Y\|_{\mathcal{K}_q}).$$

In Theorem 1 the case $p = 1$ $(q = +\infty)$ is the more interesting one.

For $1 < p < +\infty$ the inequality is easier to prove. Garsia [3] has proved the same inequality for $1 \leq p \leq 2$ and he used the BMO_2-space instead of the more natural BMO_1-space.

Let ϕ be a Young function, i.e. of the form

$$\phi(x) = \int_0^x \varphi(t)dt, \quad x \geq 0,$$

where $\varphi(t)$ is a nondecreasing, right-continuous function such that $\varphi(0) = 0$ and $\lim_{t\to+\infty} \varphi(t) = +\infty$. It is easily seen that ϕ is an increasing and convex function with $\phi(0)=0$ and $\lim_{x\uparrow+\infty} \frac{\phi(x)}{x} = +\infty$. The power of the Young function ϕ is defined by the formula

$$p = \sup_{x>0} \frac{x\varphi(x)}{\phi(x)}.$$

Put

$$\psi(x) = \varphi^{-1}(x),$$

where φ^{-1} is the generalized inverse of φ. Then $\psi(x)$ has the same properties as φ. Consequently, the function

$$\Psi(x) = \int_0^x \psi(t)dt$$

is also a Young function. Ψ is called the conjugate (complementary) function of ϕ. Its power is defined similarly.

The Orlicz space $L^{\phi}(\Omega, \mathcal{F}, P)$ is defined as the set of the random variables X on $(\Omega, \mathcal{F}, P)$ for which there exists an $a > 0$ such that

$$E(\phi(a^{-1}|X|)) \leq 1.$$

In this case the quantity

$$||X||_{\phi} = \inf(a>0: E(\phi(a^{-1}|X|)) \leq 1)$$

is a norm, which we call the Luxemburg norm of $X \in L^{\phi}(\Omega, \mathcal{F}, P)$.

For the proof of these and for further details concerning the Young functions we refer to [6] and [7].

The notion of the spaces $\mathcal{H}_{\phi}$ and $\mathcal{K}_{\Psi}$ is defined similarly as above for $\phi(x) = x^p$, $p \geq 1$ and $\Psi(x) = x^q$, $q \geq 1$.

S. Ishak and J. Mogyoródi have established their above inequality in a more general form (see [2]). Namely, for every pair (ϕ, Ψ) of complementary Young functions such that ϕ has a finite power, the following inequality holds: if $X \in \mathcal{H}_{\phi}$ and $Y \in \mathcal{K}_{\Psi}$ then the limit

$$\lim_{n \to +\infty} E(X_n Y_n)$$

exists. Moreover, we have

$$|\lim_{n \to +\infty} E(X_n Y_n)| \leq C'_{\phi} ||X||_{\mathcal{H}_{\phi}} ||Y||_{\mathcal{K}_{\Psi}}$$

and

$$|E(X_n Y_n)| \leq C'_{\phi} ||X_n||_{\mathcal{H}_{\phi}} ||Y_n||_{\mathcal{K}_{\Psi}}, \quad n \geq 1.$$

Here $C'_{\phi} > 0$ is a constant depending only on (ϕ, Ψ). The proof is based on Herz's inequality and on the right hand side of the Burkholder-Davis-Gundy inequality.

2. Conversely, supposing the validity of the Fefferman inequality we can give a proof of the right hand side of the Burkholder-Davis-Gundy inequality. This is the following: if the Young function ϕ has finite power then with some constants $c_\phi > 0$ and $C_\phi > 0$ depending only on ϕ we have

$$c_\phi E(\phi(S)) \leq E(\phi(X^*)) \leq C_\phi E(\phi(S)),$$

where X^* stands for the random variable

$$\sup_{n\geq 0} |X_n|.$$

Burkholder obtained this inequality in the L^p case for $1<p<+\infty$ (see [11]). Burkholder and Gundy [12] obtained the cases $0<p\leq 1$ for a large class of martingales. After their work, Davis [4] proved the $p = 1$ case for all martingales. Finally, Burkholder, Davis and Gundy [13] obtained the result for convex functions of the Young type with finite power.

Garsia in his book [3] also gives a proof of this inequality by using his martingale Fefferman inequality mentioned above. There are several methods of proof of the above maximal inequality. In this connection we also refer to the proof of the independent martingale difference sequence case by Marcinkiewicz and Zygmund [10].

In this paper we shall follow the method of proof of Garsia. Our proof, however, will be more homogeneous than his one. To obtain this we use the generalized Fefferman inequality of S. Ishak and J. Mogyoródi. Unfortunately, we are not able to give here the proof of the left hand side. This is proved by the author of this note in [14] for the Young functions $\phi(x^2)$, where ϕ is also a Young function.

Because of the importance of the Burkholder-Davis-Gundy inequality it seems to be interesting to give another proof of this.

First we prove the inequality

$$E(X^{*p}) \leq c_p E(S^p),$$

where $1 \leq p < +\infty$, since this contains the more interesting case of $p = 1$. Further, we shall establish the inequality

$$E(\phi(X_n^*)) \leq C_\phi E(\phi(S_n)), \quad n \geq 1.$$

3. We prove a lemma which is an adapted form of an assertion given by Garsia [3].

Lemma 1. Let $\{\varepsilon_n\}$ and $\{\Theta_n\}$ be sequences of random variables such that Θ_n is $\mathcal{F}_n$-measurable and $\sum_{n=1}^{\infty} |\varepsilon_n| \leq 1$. If for some p, where $1 \leq p < +\infty$, the variable $\Theta^* = \sup_{n\geq 1}|\Theta_n|$ is in L_q, where $q = \frac{p}{p-1}$, then the function

$$Y = \sum_{k=1}^{\infty} \Theta_k E(\varepsilon_k|\mathcal{F}_k) - E\left(\sum_{k=1}^{\infty} \Theta_k E(\varepsilon_k|\mathcal{F}_k)|\mathcal{F}_o\right)$$

is in $\mathcal{K}_q$ and we have

$$||Y||_{\mathcal{K}_q} \leq 2p||\Theta^*||_q .$$

Proof. First of all note that there is no problem with the convergence and integrability of Y since by our assumptions we have

$$E(|Y|) \leq 2 \sum_{k=1}^{\infty} E(|\Theta_k|E(|\varepsilon_k||\mathcal{F}_k)) \leq 2E(\Theta^* E\left(\sum_{k=1}^{\infty} |\varepsilon_k|\Big|\mathcal{F}_k\right)) \leq$$

$$\leq 2E(\Theta^*),$$

and we have supposed that Θ^* is in L_q.

To prove the inequality of the lemma put $Y_n = E(Y|\mathcal{F}_n)$, $n \geq 0$. Then for $n \geq 2$

$$Y_{n-1} = \sum_{i=1}^{n-1} \Theta_i E(\varepsilon_i|\mathcal{F}_i) + E\left(\sum_{i=n}^{\infty} \Theta_i \varepsilon_i|\mathcal{F}_{n-1}\right) - E\left(\sum_{k=1}^{\infty} \Theta_k E(\varepsilon_k|\mathcal{F}_k)|\mathcal{F}_o\right).$$

Introduce the notation

$$Q_n = \sum_{i=n}^{\infty} \Theta_i E(\varepsilon_i | \mathcal{F}_i), \quad n = 1, 2, \ldots .$$

Then

$$Y - Y_{n-1} = Q_n - E(Q_n | \mathcal{F}_{n-1})$$

and

$$E(|Y - Y_{n-1}| \, | \mathcal{F}_n) \leq |E(Q_n | \mathcal{F}_n)| + |E(Q_n | \mathcal{F}_{n-1})| .$$

Now

$$|E(Q_n | \mathcal{F}_n)| \leq \sum_{i=n}^{\infty} E(|\Theta_i| E(|\varepsilon_i| \, | \mathcal{F}_i) | \mathcal{F}_n) =$$

$$= \sum_{i=n}^{\infty} E(|\Theta_i| |\varepsilon_i| \, | \mathcal{F}_n) \leq \sum_{i=n}^{\infty} E(\Theta^* |\varepsilon_i| \, | \mathcal{F}_n) \leq E(\Theta^* | \mathcal{F}_n) .$$

Similarly we get

$$|E(Q_n | \mathcal{F}_{n-1})| \leq E(\Theta^* | \mathcal{F}_{n-1}) \leq \sup_{k \geq 0} E(\Theta^* | \mathcal{F}_k) = \Theta^{**} .$$

From these

$$E(|Y - Y_{n-1}| \, | \mathcal{F}_n) \leq 2E(\Theta^{**} | \mathcal{F}_n) .$$

Let first $q = +\infty$. Since by assumption Θ^* belongs to L_∞, we see that $\|\Theta^*\|_\infty < +\infty$. Consequently, $\Theta^{**} \leq \|\Theta^*\|_\infty$ and so $E(|Y - Y_{n-1}| \, | \mathcal{F}_n) \leq 2\|\Theta^*\|_\infty$. This means that $Y \in BMO_1$ and $\|Y\|_{BMO_1} \leq 2\|\Theta^*\|_\infty$. Consider now the case when $1 < q < +\infty$. Then by the inequality of Doob

$$(E((2\Theta^{**})^q))^{1/q} = 2(E((\sup_{k \geq 0} E(\Theta^* | \mathcal{F}_k))^q))^{1/q} \leq 2p(E(\Theta^{*q}))^{1/q},$$

since $\sigma\left(\bigcup_{n=0}^{\infty} \mathcal{F}_n\right) = \mathcal{F}$.

The right hand side is finite by supposition. Consequently, $Y \in \mathcal{K}_q$ and

$$\|Y\|_{\mathcal{K}_q} \leq 2p\|\Theta^*\|_q .$$

This proves the lemma.

Now we are able to formulate our

Theorem 2. For arbitrary p, where $1 \leq p < +\infty$, we have

$$E(X_n^{*p}) \leq c_p E(S_n^p), \quad n = 1,2,...$$

in the sense that both sides are at the same time infinite or finite and then the above inequality holds. Here c_p is a positive constant depending only on p and X_n^* stands for the maximal function $\max_{1\leq k\leq n} |X_k|$.

Proof. Since

$$X_n^* \leq S_n\sqrt{n} ,$$

we see that $E(S_n^p) < +\infty$ implies that $E(X_n^{*p}) < +\infty$. Conversely,

$$S_n = \left(\sum_{i=1}^{n} d_i^2\right)^{1/2} \leq 2\sqrt{n}\, X_n^* .$$

Consequently, $E(S_n^p) < +\infty$ whenever $E(X_n^{*p}) < +\infty$. Thus we can suppose that both sides of the inequality to be proved are finite.

Now to prove the asserted inequality it is a natural thing to write

$$E(X_n^{*p}) = \sum_{i=1}^{n} E(|X_i|^p \chi_{E_i})$$

where

$$E_i = \left\{X_{i-1}^* < X_n^*, |X_i| = X_n^*\right\}, \; i = 2,...,n; \; E_1 = \{X_n^* = |X_1|\} .$$

These events are disjoint and we have

$$\bigcup_{i=1}^{n} E_i = \Omega.$$

Set

$$\Theta_i = |X_i|^{p-1} \text{ sign } X_i.$$

Then we have

$$E(X_n^{*p}) = \sum_{i=1}^{n} E(|X_i|^p \chi_{E_i}) = \sum_{i=1}^{n} E(\Theta_i X_i \chi_{E_i}) =$$

$$= \sum_{i=1}^{n} E(\Theta_i X_i E(\chi_{E_i} | \mathcal{F}_i)) = \sum_{i=1}^{n} E(\Theta_i E(X_n | \mathcal{F}_i) E(\chi_{E_i} | \mathcal{F}_i)) =$$

$$= \sum_{i=1}^{n} E(X_n \Theta_i E(\chi_{E_i} | \mathcal{F}_i)) = E(X_n \sum_{i=1}^{n} \Theta_i E(\chi_{E_i} | \mathcal{F}_i)) =$$

$$= E(X_n (\sum_{i=1}^{n} \Theta_i E(\chi_{E_i} | \mathcal{F}_i) - E(\sum_{i=1}^{n} \Theta_i E(\chi_{E_i} | \mathcal{F}_i) | \mathcal{F}_o))) =$$

$$= E(X_n Y_n),$$

since

$$E(E(\sum_{i=1}^{n} \Theta_i E(\chi_{E_i} | \mathcal{F}_i) | \mathcal{F}_o) X_n) = \sum_{i=1}^{n} \sum_{k=1}^{n} E(d_k E(\Theta_i E(\chi_{E_i} | \mathcal{F}_i) | \mathcal{F}_o)) = 0.$$

Here

$$Y_n = \sum_{i=1}^{n} \Theta_i E(\chi_{E_i} | \mathcal{F}_i) - E(\sum_{i=1}^{n} \Theta_i E(\chi_{E_i} | \mathcal{F}_i) | \mathcal{F}_o)$$

and

$$Y_k = E(Y_n | \mathcal{F}_k), \quad k = 0, 1, \ldots, n$$

so that

$$Y_o = 0 \quad \text{a.s.}$$

and

$$Y_n = E(Y_n | \mathcal{F}_n).$$

The random variable Y_n satisfies the conditions of Lemma 1. Namely, with $\varepsilon_i = \chi_{E_i}$ we have $\sum_{i=1}^{n} \varepsilon_i = 1$ a.s. and the random variables Θ_i are $\mathcal{F}_i$ -measurable.

For $p = 1$ $(q=+\infty)$ we have

$$\Theta_i = \operatorname{sign} X_i, \quad \Theta_n^* = \max_{1\le i\le n} |\operatorname{sign} X_i| \le 1 \in L_\infty$$

whilst for $1 < p < +\infty$, trivially

$$\Theta_n^* = \max_{1\le i\le n} |X_i|^{p-1} = (X_n^*)^{p-1}$$

which belongs to L_q since

$$E(((X_n^*)^{p-1})^q) = E(X_n^{*p}) < +\infty.$$

Consequently, using Lemma 1 we see that $Y_n \in \mathcal{K}_q$ and

$$\|Y_n\|_{\mathcal{K}_q} \le 2p \|\Theta_n^*\|_q,$$

where

$$\|\Theta_n^*\|_q = (E(X_n^{*p}))^{1/q}.$$

So we can use the result of Theorem 1 according to which

$$E(X_n^{*p}) = E(X_n Y_n) \le c_p' \|X_n\|_{\mathcal{H}_p} \|Y_n\|_{\mathcal{K}_q}.$$

Using the preceding inequality we get

$$E(X_n^{*p}) \leq 2pc_p' \|X_n\|_{\mathcal{H}_p} (E(X_n^{*p}))^{1/q} .$$

We can suppose that $E(X_n^{*p}) > 0$ since otherwise the inequality of our assertion is trivial. Dividing by $(E(X_n^{*p}))^{1/q}$ we get

$$\|X_n^*\|_p \leq 2pc_p' \|X_n\|_{\mathcal{H}_p} ,$$

which proves the asserted inequality.

4. To generalize the case $1 < p < +\infty$ of the preceding section we shall consider in what follows a pair (ϕ, Ψ) of arbitrary Young functions. In this case Lemma 1 of the preceding section is modified in the following way.

<u>Lemma 2</u>. Suppose that the power p of ϕ is finite. Let $\{\varepsilon_n\}$ and $\{\Theta_n\}$ be random variables such that Θ_n is $\mathcal{F}_n$-measurable and

$$\sum_{n=1}^{\infty} |\varepsilon_n| \leq 1.$$

If $\Theta^* = \sup_{n\geq 1} |\Theta_n|$ is in L^Ψ then the function

$$Y = \sum_{k=1}^{\infty} \Theta_k E(\varepsilon_k | \mathcal{F}_k) - E\left(\sum_{k=1}^{\infty} \Theta_k E(\varepsilon_k | \mathcal{F}_k) | \mathcal{F}_o \right)$$

is in $\mathcal{K}_\Psi$ and we have

$$\|Y\|_{\mathcal{K}_\Psi} \leq 2p \|\Theta^*\|_\Psi .$$

For the proof of this lemma we refer to that of Lemma 1 and to Doob's generalized maximal inequality for arbitrary Young function Ψ whose conjugate Young function ϕ has finite power p (see e.g. [8] or [9]).

Theorem 3. Let (ϕ,Ψ) be a pair of conjugate Young functions such that the power p of ϕ is finite. Let further $(X_n, \mathcal{F}_n)$ be a martingale considered in section 1. Then we have

$$||X_n^*||_\phi \le 2pc'_\phi ||X_n||_{\mathcal{H}_\phi}, \quad n = 1,2,\ldots,$$

in the sense that both sides are at the same time finite or infinite and in the finite case the above inequality holds.

Proof. Since $X_n^* \le \sqrt{n}\, S_n$ and $S_n \le 2\sqrt{n}X_n^*$ and the power p of ϕ is finite we see that $||S_n||_\phi < +\infty$ implies that $||X_n^*||_\phi < +\infty$ since $||X_n^*||_\phi \le \sqrt{n}||S_n||_\phi$. In the same way $||S_n||_\phi \le 2\sqrt{n}||X_n^*||_\phi$. It follows that $||X_n^*||_\phi < +\infty$ implies that $||S_n||_\phi < +\infty$. Therefore we shall suppose that $||X_n^*||_\phi$ is finite and positive, because in the contrary case the asserted inequality is trivial.

The finiteness of the power p of ϕ implies that

$$E\left\{\phi\left(\frac{X_n^*}{||X_n^*||_\phi}\right)\right\} = 1.$$

From this

$$||X_n^*||_\phi = ||X_n^*||_\phi E\left\{\phi\left(\frac{X_n^*}{||X_n^*||_\phi}\right)\right\} = ||X_n^*||_\phi \sum_{i=1}^{n} E\left\{\phi\left(\frac{|X_i|}{||X_n^*||_\phi}\right)\chi_{E_i}\right\},$$

where we use the notations of the proof of Theorem 2, i.e.

$$E_1 = \{|X_1| = X_n^*\},\ E_i = \{X_{i-1}^* < X_n^*,\ |X_i| = X_n^*\},\ i = 2,\ldots,n.$$

Let the random variables Θ_i, $i = 1,\ldots,n$, be defined by the following way: on the set $\{|X_i| \neq 0\}$ put

$$\Theta_i = \frac{\phi\left(\frac{|X_i|}{||X_n^*||_\phi}\right)}{\frac{|X_i|}{||X_n^*||_\phi}} \operatorname{sign} X_i$$

and let $\Theta_i = 0$ on the set $\{|X_i| = 0\}$. Then, similarly as in the proof of Theorem 2

$$\|X_n^*\|_\phi = \|X_n^*\|_\phi \sum_{i=1}^n E(\Theta_i X_i \chi_{E_i} / \|X_n^*\|_\phi) = \sum_{i=1}^n E(\Theta_i X_i E(\chi_{E_i} |\mathcal{F}_i)) =$$

$$= E\left(X_n\left(\sum_{i=1}^n \Theta_i E\left(\chi_{E_i} |\mathcal{F}_i\right) - E \sum_{i=1}^n \Theta_i E\left(\chi_{E_i} \mathcal{F}_i\right) |\mathcal{F}_o\right)\right) =$$

$$= E(X_n Y_n),$$

since

$$E\left(X_n E\left(\sum_{i=1}^n \Theta_i E\left(\chi_{E_i} |\mathcal{F}_i |\mathcal{F}_o\right)\right)\right) = \sum_{i=1}^n \sum_{k=1}^n E\left(d_k E\left(\Theta_i E\left(\chi_{E_i} |\mathcal{F}_i\right) |\mathcal{F}_o\right)\right) = 0.$$

Here

$$Y_n = \sum_{i=1}^n \Theta_i E\left(\chi_{E_i} |\mathcal{F}_i\right) - E\left(\sum_{i=1}^n \Theta_i E\left(\chi_{E_i} |\mathcal{F}_i\right) |\mathcal{F}_o\right)$$

and

$$Y_k = E(Y_n|\mathcal{F}_k), \quad k = 0,1,\ldots,n$$

so that $Y_o = 0$ a.s. and $Y_n = E(Y_n|\mathcal{F}_n)$ a.s.

The random variable Y_n satisfies the conditions of Lemma 2. Namely, with $\varepsilon_i = \chi_{E_i}$ we have $\sum_{i=1}^n \varepsilon_i = 1$ and the random variables Θ_i are $\mathcal{F}_i$-measurable. Further, on the set where $X_n^* \neq 0$ we have

$$\Theta_n^* = \max_{1\leq k\leq n} \frac{\phi\left(\dfrac{|X_i|}{\|X_n^*\|_\phi}\right)}{\dfrac{|X_i|}{\|X_n^*\|_\phi}} = \frac{\phi\left(\dfrac{X_n^*}{\|X_n^*\|_\phi}\right)}{\dfrac{X_n^*}{\|X_n^*\|_\phi}},$$

while on the set $\{X_n^* = 0\}$ we define $\Theta_n^* = 0$. This form of Θ_n^* follows from the fact that $\phi(x)/x$ is increasing as x increases.

We show that $\theta_n^* \in L^{\Psi}$. In fact, using the inequality

$$\Psi\left(\frac{\phi(x)}{x}\right) \leq \phi(x)$$

we get

$$E(\Psi(\Theta_n^*)) = E\left(\Psi\left(\frac{\phi\left(\frac{X_n^*}{||X_n^*||_\phi}\right)}{\frac{X_n^*}{||X_n^*||_\phi}}\right)\right) \leq E\left(\phi\left(\frac{X_n^*}{||X_n^*||_\phi}\right)\right) = 1.$$

Consequently, $\Theta_n^* \in L^{\Psi}$ and

$$||\Theta_n^*||_\Psi \leq 1,$$

further, by Lemma 2

$$||Y_n||_{\mathcal{K}_\Psi} \leq 2p||\Theta_n^*||_\Psi \leq 2p.$$

Now by the generalized Fefferman-Garsia inequality formulated in section 1

$$||X_n^*||_\phi = E(X_nY_n) \leq c_\phi' ||X_n||_{\mathcal{H}_\phi} ||Y_n||_{\mathcal{K}_\Psi} \leq 2pc_\phi' ||X_n||_{\mathcal{H}_\phi},$$

which proves our assertion.

REFERENCES

1. S. Ishak and J. Mogyoródi: On the generalization of the Fefferman-Garsia inequality. In: Stochastic differential systems. Proceedings of the 3rd IFIP-WG Working Conference, Visegrád, Hungary, Sept. 15-20. 1980. Lecture Notes in Control and Information Sciences, 36 (1981), 85-97.
2. S. Ishak and J. Mogyoródi: On the $\mathcal{P}_\phi$-spaces and the generalization of Herz's and Fefferman's inequalities.

I., II., III. Studia Scientiarum Mathematicarum Hungarica, 17 (1982), 229-234., 18 (1983), 205-210, 18 (1983), 211-219.

3. A.M. Garsia: Martingale inequalities. Seminar Notes in Recent Progress. Benjamin, Reading, Mass., 1973.
4. B. Davis: On the integrability of the martingale square function. Israel Journal of Mathematics. 8 (1970), 187-190.
5. D.L. Burkholder: Distribution function inequalities for martingales. Annals of Probability. 1 (1973), 19-42.
6. J. Neveu: Discrete parameter martingales. North-Holland, Amsterdam, 1975.
7. M.A. Krasnoselskii and Ya.B. Rutickii: Convex functions and Orlicz spaces. (Transl.: L.F. Boron, P. Noordhoff Ltd. Groningen, 1961.)
8. J. Mogyoródi and T.F. Móri: Necessary and sufficient condition for the maximal inequality of convex Young functions. Acta Scientiarum Mathematicarum, Szeged, 45 (1983), 325-332.
9. J. Mogyoródi: On an inequality of Marcinkiewicz and Zygmund. Publicationes Mathematicae, Debrecen, 26 (1979), 267-274.
10. J. Marcinkiewicz and A. Zygmund: Quelques théorèmes sur les fonctions indépendantes. Studia Mathematica, 7 (1938), 104-120.
11. D.L. Burkholder: Martingale transforms. Annals of Mathematical Statistics. 37 (1966), 1494-1504.
12. D.L. Burkholder and R.F. Gundy: Extrapolation and interpolation for convex functions of operators on martingales. Acta Mathematica. 124 (1970), 249-304.
13. D.L. Burkholder, B. Davis and R.F. Gundy: Integral inequalities for convex functions of operators on martingales. Proceedings Sixth Berkeley Symp. Math. Statist. Prob. 2 (1972), 223-240.
14. N.L. Bassily: A proof of the left hand side of the Burkholder-Davis-Gundy inequality. Annales Univ. Sci. Budapest, Sectio Mathematica XXX (1987) (Accepted for publication).

N.L. Bassily
Ain Shams University
Faculty of Science
Department of Mathematics.
Cairo
EGYPT

Proc. of the 5th Pannonian Symp. on Math. Stat., Visegrád, Hungary 1985
W. Grossmann, J. Mogyoródi, I. Vincze, W. Wertz, eds

ON THE LENGTH OF THE LONGEST FLAT INTERVAL

E. CSÁKI, A. FÖLDES

Budapest

Let $X(t)$ be a stochastic process, $M(t) = \sup_{0\le s\le t} X(s)$, $\nu(t) = \inf\{s: X(s) = M(t),\ 0\le s\le t\}$ and denote by $\varphi(t)$ the length of the longest interval in $(0,t]$, where $M(t)$ is constant (length of the longest flat interval of $M(s)$, $0\le s\le t$). The joint distribution of $(M(t),\ \nu(t),\ \varphi(t))$ is given in terms of multiple Laplace transform in the case when the first passage process associated with $X(t)$ is a Lévy process. Some examples are discussed.
AMS subject classification: 60J25, 60E99

1. INTRODUCTION

Let $\{W(t),\ t\ge 0\}$ be a standard Wiener process and $L(t)$ be its local time at 0, i.e.,

$$L(t) = \lim_{\varepsilon\downarrow 0} \frac{1}{2\varepsilon}\, \text{meas}\{s: |W(s)|<2\varepsilon,\ 0\le s\le t\} \tag{1.1}$$

where meas $\{\cdot\}$ stands for the Lebesque measure.

In Csáki-Erdős-Révész (1985) the length of the largest excursion (length of the longest zero-free subinterval of $(0,t]$) of the Wiener process was investigated and certain strong limit theorems were proved. The method - used there - was based on random walk approach and on certain recursion formulae for this case.

Since $L(t)$ increases only when the Wiener process is zero, the length of the largest excursion can also be interpreted as the length of the largest flat interval of $L(s)$ $(0\le s\le t)$ i.e. the length of the largest interval (in $(0,t]$), where $L(s)$ is constant.

The aim of this paper is to derive certain distributions and joint distributions directly for the Wiener process (one of which refers to the length of the largest excursion) and extend the results for more general processes with continuous time.

In this paper we consider a stochastic process $\{X(t), t \geq 0\}$ with $X(0) = 0$ and define its maximum

$$M(t) = \sup_{0 \leq s \leq t} X(s), \tag{1.2}$$

the first time when the maximum is achieved:

$$\nu(t) = \inf\{s: X(s) = M(t),\ 0 \leq s \leq t\} \tag{1.3}$$

and the length of the longest flat interval of $\{M(s),\ 0 \leq s \leq t\}$:

$$\varphi(t) = \sup\{\alpha : M(s+\alpha) = M(s) \text{ for some } 0 \leq s \leq t-\alpha\}. \tag{1.4}$$

Define also the first passage process

$$\tau_a = \tau(a) = \inf\{s: X(s) \geq a,\ 0 \leq s < \infty\} \quad (a \geq 0) \tag{1.5}$$

Obviously, this is the inverse of $M(.)$. By convention, $\tau_a = \infty$ if $\sup_{0 \leq s < \infty} X(t) < a$.

Our basic assumptions are as follows:

(i) $X(t)$ has no positive jumps, i.e. $\{M(t), t \geq 0\}$ is continuous a.s.

(ii) $\{\tau_a, a \geq 0\}$ is a non-decreasing Lévy process (subordinator), with zero drift, i.e. τ_a has stationary independent increments and hence (cf. Gihman - Skorohod, 1969) admits the representation

$$E(e^{-\lambda \tau_a}) = \exp\{-a\int_0^\infty (1-e^{-\lambda x})n(dx)\}, \quad a > 0, \quad \lambda > 0, \tag{1.6}$$

where n(.) is a measure (so-called Lévy measure) for which

$$\int_0^\infty (1-e^{-x})n(dx) < \infty \tag{1.8}$$

For some basic properties, distributions, joint distributions of Lévy processes we refer to Horowitz (1972), Bingham (1975) and the references therein.

In section 2 we derive the joint distribution of $(M(t), \nu(t), \varphi(t))$ in the form of multiple Laplace transform, where instead of the fixed time t we use a random time T exponentially distributed with parameter λ and independent of the process $\{X(t), t\geq 0\}$.

In section 3 some examples will be given.

2. MAIN RESULT

The basic result is summarized in the following multiple Laplace transform type theorem (the notations are the same as in Sect. 1).

Theorem: Let T be an exponentially distributed random variable with parameter λ; $P(T>z) = e^{-\lambda z}$ which is independent from the process $X(t)$. Then

$$\frac{1}{\lambda} E\left(e^{-\mu\nu(T)-\theta M(T)};\ \varphi(T)<v\right) =$$

$$= \frac{\int_0^v e^{-\lambda u}\, n(u,\infty)du}{\theta+\int_0^v (1-e^{-(\lambda+\mu)x})n(dx)+n(v,\infty)}. \qquad \begin{matrix} 0\le\mu<+\infty \\ 0\le\theta<+\infty \\ 0\le v<+\infty \\ 0<\lambda<+\infty \end{matrix} \tag{2.1}$$

To prove the above result we need two lemmas.

Lemma 1. For $a > 0$, $0 < v < \infty$, $t - v < s \le t$

$$P(M(t)\in da,\ \nu(t)\in ds,\ \varphi(t)<v) = P(\tau_a\in ds,\ \varphi(\tau_a)<v)h(t-s)da \tag{2.2}$$

where

$$h(u) = n(u,\infty) = \int_u^\infty n(dx). \tag{2.3}$$

Proof: Expressing the left hand side in terms of the inverse process τ_a and using that τ_a is a process with stationary independent increments

$$\begin{aligned} &P(M(t)\in da,\ \nu(t)\in ds,\ \varphi(t)<v) = \\ &P(\tau_a\in ds,\ \varphi(\tau_a)<v,\ \tau_{a+da}-\tau_a>t-s) = \\ &= P(\tau_a\in ds,\ \varphi(\tau_a)<v)\ P(\tau_{da}>t-s). \end{aligned} \tag{2.4}$$

Therefore it is enough to show that

$$P(\tau_\varepsilon>u) \sim \varepsilon n(u,\infty) \quad \text{as} \quad \varepsilon\downarrow 0. \tag{2.5}$$

Taking the Laplace transform of $\frac{1}{\varepsilon}P(\tau_\varepsilon>u)$ and using (1.6) we have

$$\begin{aligned} &\int_0^\infty \frac{1}{\varepsilon}P(\tau_\varepsilon>u)e^{-\lambda u}du = \int_0^\infty\int_u^\infty \frac{1}{\varepsilon}P(\tau_\varepsilon\in dy)e^{-\lambda u}du = \\ &= \int_0^\infty \frac{1}{\varepsilon}P(\tau_\varepsilon\in dy)\int_0^y e^{-\lambda u}du = \int_0^\infty \frac{1}{\varepsilon}P(\tau_\varepsilon\in dy)\frac{1-e^{-\lambda y}}{\lambda} = \\ &= \frac{1}{\lambda\varepsilon}\left(1-e^{-\varepsilon\int_0^\infty(1-e^{-\lambda x})n(dx)}\right). \end{aligned} \tag{2.6}$$

According to the elementary limit relation $1-e^{-z} = z+o(z)$ when $z\to 0$, we have from (2.6) that

$$\lim_{\varepsilon\to 0}\int_0^\infty \frac{1}{\varepsilon}P(\tau_\varepsilon>u)e^{-\lambda u}du = \frac{1}{\lambda}\int_0^\infty(1-e^{-\lambda x})n(dx) \tag{2.7}$$

as $\int_0^\infty(1-e^{-\lambda x})n(dx) < \infty$.

But observing that the Laplace transform of $n(u,\infty)$ is

$$\int_0^\infty n(u,\infty)e^{-\lambda u}\,du = \int_0^\infty \int_u^\infty n(dz)e^{-\lambda u}\,du =$$

$$= \int_0^\infty n(dz)\int_0^z e^{-\lambda u}\,du = \int_0^\infty n(dz)\,\frac{1-e^{-\lambda z}}{\lambda} \tag{2.8}$$

we arrive to the conclusion that

$$\lim_{\varepsilon\to 0} \mathcal{L}\left(\frac{1}{\varepsilon}P(\tau_\varepsilon > u)\right) = \mathcal{L}\,(n(u,\infty)) \tag{2.9}$$

which implies that $\lim_{\varepsilon\to 0}\frac{1}{\varepsilon}P(\tau_\varepsilon>u) = n(u,\infty)$, hence (2.5) follows, implying Lemma 1.

<u>Lemma 2.</u> For $a>0$, $\lambda>0$, $v>0$

$$E(e^{-\lambda\tau_a}, \varphi(\tau_a)<v) = \exp\{-a[\int_0^v(1-e^{-\lambda x})n(dx)+n(v,\infty)]\}. \tag{2.10}$$

<u>Proof:</u> Let $A=[v,+\infty)$ and its indicator function be $\chi_A(\cdot)$. Let

$$\tau_a(A) = \sum_{z<a}[\tau(z+0)-\tau(z-0)]\chi_A(\tau(z+0)-\tau(z-0))$$

and (2.11)

$$\tau_a(\bar{A}) = \tau_a - \tau_a(A)$$

Then clearly $\tau_a(A)$ contains all the jumps of τ_a which are greater than v. Then

$$E(e^{-\lambda\tau_a}, \varphi(\tau_a)<v) = E(e^{-\lambda(\tau_a(A)+\tau_a(\bar{A}))}, \tau_a(A) = 0) =$$

$$E(e^{-\lambda\tau_a(\bar{A})}, \tau_a(A)=0)=E(e^{-\lambda\tau_a(\bar{A})})\cdot P(\tau_a(A)=0),$$

being $\tau_a(A)$ and $\tau_a(\bar{A})$ independent. Now from Gihman - Skorohod (1969, p. 272(5))

$$E\left[e^{-\lambda\tau_a(\bar{A})}\right] = e^{-a\int_0^v (1-e^{-\lambda x})n(dx)} .$$

Moreover

$$P(\tau_a(A) = 0) = P\text{ (no jumps are greater than v)} = e^{-an(v,\infty)},$$

proving the lemma.

Proof of the theorem:

By Lemma 1 and 2

$$\frac{1}{\lambda} E\left(e^{-\mu\nu(T)-\theta M(T)},\ \varphi(T)<v\right) =$$

$$= \int_{a=0}^{\infty} \int\int_{0\le s\le t<\infty} e^{-\theta a-\mu s-\lambda t} P(M(t)\in da,\ \nu(t)\in ds,\ \varphi(t)<v)dt =$$

$$= \int_{a=0}^{\infty} e^{-\theta a} \int\int_{0\le s\le t<\infty} e^{-(\lambda+\mu)s} P(\tau_a\in ds,\ \varphi(\tau_a)<v)e^{-\lambda(t-s)}h(t-s)dt\ da$$

$$= \int_{a=0}^{\infty} e^{-\theta a}\left(\int_{s=0}^{\infty} e^{-(\lambda+\mu)s} P(\tau_a\in ds,\ \varphi(\tau_a)<v)\int_0^v e^{-\lambda u}h(u)du\right)da$$

$$= \int_{a=0}^{\infty} e^{-\theta a}\left[E\left(e^{-(\lambda+\mu)\tau_a},\ \varphi(\tau_a)<v\right)\int_0^v e^{-\lambda u}h(u)du\right]da$$

$$= \int_{a=0}^{\infty}\left\{e^{-\theta a}\, e^{-a\left(\int_0^v(1-e^{-(\lambda+\mu)x})n(dx)+n(v,\infty)\right)} \int_0^v e^{-\lambda u} h(u)du\right\}da$$

$$= \frac{\int_0^v e^{-\lambda u}h(u)du}{\theta+\int_0^v(1-e^{-(\lambda+\mu)x})n(dx)+n(v,\infty)}$$

as it was stated.

Remarks.

1. By putting $v=+\infty$ in (2.1), we get the double Laplace transform of the joint distribution of $(\nu(T), M(T))$ given by Pecherskii - Rogozin (1969) (see also Bingham, 1975).

2. By putting $\mu=\theta=0$ in (2.1), the distribution of $\varphi(T)$ is obtained. An equivalent formula for this distribution was given by Krylov - Juškevič (1965) and by Horowitz (1972).

3. The process $\varphi(\tau_a)$ as a function of a is the largest jump of the Lévy process $\{\tau_a, a\geq 0\}$ and hence (cf. Resnick-Rubinovitch, 1973) is an extremal process.

4. The case when τ_a has independent but non-stationary increments is more complicated. For $X(t) = |W(t)|$ $t \geq 0$ (reflected Wiener process), see Csáki-Földes-Révész (1985). The authors want to return to this problem in general case in a subsequent paper.

3. EXAMPLES

3.1. Wiener process

If $X(t) = W(t)$, the standard Wiener process, then (cf. Ito - Mc Kean, 1965)

$$E(e^{-\lambda\tau_a}) = e^{-a\sqrt{2\lambda}} = \exp\left\{-a\int_0^\infty (1-e^{-\lambda x})\frac{dx}{\sqrt{2\pi x^3}}\right\},$$

hence

$$n(dx) = \frac{dx}{\sqrt{2\pi x^3}},$$

and

$$h(u) = n(u,\infty) = \sqrt{\frac{2}{\pi u}}.$$

We have furthermore

$$\int_0^v (1-e^{-\lambda x}) \frac{dx}{\sqrt{2\pi x^3}} + \sqrt{\frac{2}{\pi v}} = {}_1F_1\left(-\frac{1}{2}, \frac{1}{2}, -\lambda v\right)\sqrt{\frac{2}{\pi v}}, \tag{3.1}$$

where

$${}_1F_1(a,b,y) = 1 + \frac{ay}{b} + \frac{(a)_2}{(b)_2}.\frac{y^2}{2!} + \ldots + \frac{(a)_n}{(b)_n} \cdot \frac{y^n}{n!} + \ldots$$

with $(a)_n = a(a+1)\ldots(a+n-1)$ is the confluent hypergeometric function (Slater, 1960).
/(3.1) can be seen, using the well-known identity

$${}_1F_1\left(-\frac{1}{2}, \frac{1}{2}, y\right) = e^y - y\int_0^1 \frac{e^{yz}}{\sqrt{z}}\, dz. /$$

Moreover,

$$\int_0^v e^{-\lambda u}\, n(u,\infty)du = \int_0^v e^{-\lambda u}\sqrt{\frac{2}{\pi u}}\, du = 2\sqrt{\frac{2v}{\pi}}\, {}_1F_1\left(\frac{1}{2}, \frac{3}{2}, -\lambda v\right).$$

Hence

$$E\left(e^{-\mu\nu(T)-\theta M(T)}, \varphi(T)<v\right) = \frac{2\lambda\sqrt{\frac{2v}{\pi}}\, {}_1F_1(\frac{1}{2}, \frac{3}{2}, -\lambda v)}{\theta + \sqrt{\frac{2}{\pi v}}\, {}_1F_1(-\frac{1}{2}, \frac{1}{2}, -(\lambda+\mu)v)}.$$

By a theorem of Lévy, $M(t) \overset{D}{=} L(t)$ for a standard Wiener process, hence for $X(t) = L(t)$ $(t\geq 0)$, we have also

$$E\left(e^{-\mu\nu^*(T)-\theta L(T)}; \varphi^*(T)<v\right) =$$

$$= \frac{2\lambda\sqrt{\frac{2v}{\pi}}\, {}_1F_1(\frac{1}{2}, \frac{3}{2}, -\lambda v)}{\theta + \sqrt{\frac{2}{\pi v}}\, {}_1F_1(-\frac{1}{2}, \frac{1}{2}, -(\lambda+\mu)v)},$$

where in this case $\nu^*(t)$ is the last zero of $W(s)$ before t and $\varphi^*(t)$ is the length of the largest excursion of $W(s)$ in $(0,t]$.

3.2 Stable processes

Now let $X(t)$ be a stable process having no positive jump, so called spectrally negative stable process. Suppose that $X(t)$ is centered and has exponent $1<\alpha<2$. Then its characteristic function can be written as

$$E\left[e^{isX(t)}\right] = \exp\left\{-ct|s|^{\alpha}(1+i \operatorname{sgn} s \tan(\tfrac{1}{2}\pi\alpha))\right\}$$

/where c is some positive scale factor/
and for the first passage process we have (cf. Bingham, 1975)

$$E(e^{-\lambda\tau(a)}) = e^{-a\eta(\lambda)}$$

where $\eta(\lambda) = \left(\frac{\lambda}{q}\right)^{\frac{1}{\alpha}}$ and $q=-c\sec(\frac{1}{2}\pi\alpha)>0$. The Lévy measure of $\tau(a)$ is given by

$$n(dx) = \frac{dx}{\alpha\Gamma\left(1-\frac{1}{\alpha}\right)x^{1+\frac{1}{\alpha}}q^{\frac{1}{\alpha}}}$$

In this case

$$h(u) = n(u,\infty) = \frac{1}{\Gamma(1-\frac{1}{\alpha})(qu)^{\frac{1}{\alpha}}}\,.$$

Moreover we get by simple calculation, that

$$\int_0^v(1-e^{-\lambda x})n(dx) + h(v) = \frac{{}_1F_1(-\frac{1}{\alpha},1-\frac{1}{\alpha},-\lambda v)}{(vq)^{\frac{1}{\alpha}}\cdot\Gamma(1-\frac{1}{\alpha})}$$

and

$$\int_0^v e^{-\lambda u}\, n(u,\infty) = \frac{v^{1-\frac{1}{\alpha}}\ {}_1F_1(1-\frac{1}{\alpha},\ 2-\frac{1}{\alpha},\ -\lambda v)}{(1-\frac{1}{\alpha})\Gamma(1-\frac{1}{\alpha})q^{\frac{1}{\alpha}}},$$

implying that

$$E(e^{-\mu\nu(T)-\theta M(T)},\ \varphi(T)<v) =$$

$$= \frac{v\ {}_1F_1(1-\frac{1}{\alpha},\ 2-\frac{1}{\alpha},\ -\lambda v)}{(1-\frac{1}{\alpha})\left[(v\cdot q)^{\frac{1}{\alpha}}\ \Gamma(1-\frac{1}{\alpha})\cdot\theta + {}_1F_1(-\frac{1}{\alpha},\ 1-\frac{1}{\alpha},\ -(\lambda+\mu)v)\right]}$$

REFERENCES

Bingham, N.H. (1975) Fluctuation theory in continuous time, *Adv. in Appl. Probab.* 7, 705-766.

Csáki E., Erdős, P. and Révész, P. (1985) On the length of the longest excursion, *Z. Wahrsch. Verw. Gebiete* 68, 365-382.

Csáki, E., Földes, A. and Révész, P. (1985) On the maximum of a Wiener process and its location, Preprint no. 29/1985, Math. Inst. Hung. Acad. Sci.

Gihman, I.I. and Skorohod, A.V. (1969) *Introduction to the Theory of Random Processes*, W.B. Saunders, Philadelphia.

Horowitz, J. (1972) Semilinear Markov Processes, subordinators and renewal theory, *Z. Wahrsch. Verw Gebiete* 24, 167-193.

Ito, K. and McKean, H.P. (1965) *Diffusion Processes and their Sample Paths*, Springer, Berlin.

Krylov, N.V. and Juškevič, A.A. (1965) Markov random sets, *Trans. Moscow Math. Soc.* 13, 114-135.

Pecherskii, E.A. and Rogozin, B.A. (1969) On joint distributions of random variables associated with fluctuations of a process with independent increments, *Theor. Probability Appl.* 14, 410-423.

Resnick, S.I. and Rubinovitch, M. (1973) The structure of extremal processes, *Adv. in Appl. Probab.* 5, 287-307.

Slater, L.J. (1960) *Confluent Hypergeometric Functions*, University Press, Cambridge.

E.Csáki, A.Földes
Mathematical Institute of the
Hungarian Academy of Sciences
1053 Budapest
Reáltanoda u. 13-15.

Proc. of the 5th Pannonian Symp.
on Math. Stat., Visegrád, Hungary 1985
W. Grossmann, J. Mogyoródi, I. Vincze, W. Wertz, eds

ON A FULL PROTOCOL PROBLEM

G. ELEK, K. GRILL

Budapest Vienna

This paper deals with the following problem: From an urn containing numbers 1 to n one successively draws k ($k \leq n$) numbers. Immediately upon drawing each number one has to assign to it its rank within the final ordered sample. Let p_{nk} be the probability that, following an optimal strategy, one can successfully do so with all k numbers drawn.
The aim of this paper is to investigate the p_{nk} sequence, especially with respect to its limiting behaviour.
AMS subject classification: 60C05

1. INTRODUCTION

The problem we consider has been inspired by a Hungarian first grade mathematics text (1) which states as follows:

Write the numbers from 1 to 10 on small leaflets of paper and put them in a hat. A number is drawn and one has to write it down in one of four boxes prepared, e.g.

		6	

Three more numbers are drawn successively, and each has to be put in one of the empty boxes, but in such a way that it is bigger than all numbers to its left and smaller than those to the right, e.g.

3		6	

or

	3	6	

but not

		6	3

The goal is to place all numbers into the boxes.

We, of course, shall make matters more general by putting n instead of 10 and k instead of 4, and ask for the probability p_{nk} that one can actually fill up all k boxes in the way indicated above.

2. BASIC PROPERTIES OF THE p_{nk} SEQUENCE

We shall now derive some inequalities and a recursion formula for the p_{nk}.

<u>Proposition 1.</u> The following recursion formula holds:

$$p_{nk} = \frac{1}{n} \sum_{i=1}^{n} \operatorname*{Max}_{j} \frac{\binom{i-1}{j-1}\binom{n-i}{k-j}}{\binom{n-1}{k-1}} p_{i-1;j-1}\, p_{n-i;k-j} \tag{2.1}$$

<u>Proof.</u> Suppose that the first number drawn is i. Of course we have to put it in some box with index, say, j. Now, in order to be able to fill up the remaining boxes, first of all it has to be guaranteed that j is the real rank of i in the final sample. This has probability

$$\frac{\binom{i-1}{j-1}\binom{n-i}{k-j}}{\binom{n-1}{k-1}} \tag{2.2}$$

Then, under this condition, the remaining task amounts to independently filling up the j-1 boxes to the left of i and the k-j ones to the right. This can be done with probability

$$p_{i-1;j-1}\, p_{n-i;k-j} \tag{2.3}$$

Now, putting (2.2) and (2.3) together, maximizing with respect to j (the only choice one has), and averaging over the possible values of i (which, of course, is uniformly distributed over {1,2,...,n} yields (2.1).)

Proposition 1 enables us to get any of the p_{nk} we like along with the optimal strategy leading up to them (which can easily be extracted from the proof of proposition 1), so we get, e.g., for our initial problem

$$p_{10;4} = \frac{551}{1260}$$

Before going on with listing the properties of the p_{nk}, let us introduce some new quantities for the sake of notational convenience. Namely, define

$$q_{nk} := \binom{n}{k} p_{nk} \tag{2.4}$$

In these terms, (2.1) now reads:

$$q_{nk} = \frac{1}{k} \sum_{i=1}^{n} \operatorname*{Max}_{j} q_{i-1;j-1}\, q_{n-i;k-j} \tag{2.5}$$

Now we have:

<u>Proposition 2.</u> The following inequalities hold:

$$p_{n+1;k} \leq p_{nk} \tag{2.6}$$

$$q_{n+1;k} \geq \left(\frac{n+1}{n}\right)^k q_{nk} \tag{2.7}$$

$$q_{n+n';k+k'} \geq q_{nk}\, q_{n'k'} \tag{2.8}$$

$$q_{n'+n'';k} \leq \sum_{k'+k''=k} q_{n'k'}\, q_{n''k''} \tag{2.9}$$

$$q_{n+1;k'+k''+1} \leq \sum_{n'+n''=n} q_{n'k'}\, q_{n''k''}\,. \tag{2.10}$$

We only shall prove (2.7) and sketch the proof of (2.10). The remaining statements are proved in a similar way as (2.10).

So, let us first consider this statement. To this end, let us suppose that we are told the value of the number with rank $k'+1$. If this value is $n'+1$ then the maximum probability that we are able to fill up all boxes is $p_{n'k'}p_{n''k''}$. Multiplying this by the probability that $n'+1$ is the number with rank $k'+1$ and summing over n' one gets the maximum probability that all $k'+k''+1$ boxes can be filled when one uses the information about the number with rank $k'+1$; of course this cannot be less then the probability one gets without using this information, which amounts to $p_{n+1;k'+k''+1}$. Multiplying both sides of the so obtained inequality by appropriate binomial coefficients one finally gets (2.10).

Now we shall prove (2.7). We proceed by induction. Since $q_{n0} = 1$ and $q_{n1} = n$, (2.7) holds for $k = 1,2$. Now, in order to prove (2.7) for some n and k, let us first define $j(i)$ as the value of j that maximizes $q_{i-1;j-1}\, q_{n-i;k-j}$. We clearly have

$$\begin{aligned}\operatorname*{Max}_{j}\, &(q_{i-1;j-1}\, q_{n-i+1,k-j}) \geq \\ &\geq \frac{i-1}{n}\, q_{i-1;j(i-1)-1}\, q_{n+1-i;k-j(i-1)} + \\ &+ \frac{n+1-i}{n}\, q_{i-1;j(i)-1}\, q_{n+1-i;k-j(i)}\end{aligned} \tag{2.11}$$

Putting (2.11) into (2.5), using the induction hypotheses on the first factor $q_{..}$ of the first right-side term and on the second $q_{..}$ of the second right-side term of (2.11) we get:

$$\begin{aligned} q_{n+1;k} &\geq \frac{1}{k} \sum_{i=1}^{n} \frac{n+1}{n}\, q_{i-1;j(i)-1}\, q_{n-i;k-j(i)} \times \\ &\times \left[\frac{i}{n+1}\left(\frac{i}{i-1}\right)^{j(i)-1} + \frac{n+1-i}{n+1}\left(\frac{n+1-i}{n-i}\right)^{k-j(i)}\right]\end{aligned} \tag{2.12}$$

Now, using the facts that the function $\left(\frac{x}{x-1}\right)^{x}$ is decreasing and that the geometric mean is smaller than the arithmetic mean, we get that the expression in the square brackets is larger than $\left(\frac{n+1}{n}\right)^{k-1}$ which in turn implies (2.7).

3. LIMITING BEHAVIOUR OF p_{nk} FOR FIXED k

In this section we are concerned with the behaviour of the p_{nk} as n tends to infinity and k remains fixed. Our inequality (2.6) readily implies that

$$\pi_k := \lim_{n\to\infty} p_{nk}$$

exists, and from (2.7) it follows that these quantities are positive. One also easily verifies that one can pass to the limit in (2.1) to obtain

$$\pi_k = \int_0^1 \operatorname*{Max}_j \left[\pi_{j-1}\ \pi_{k-j}\ x^{j-1}(1-x)^{k-j}\binom{k-1}{j-1}\right] dx$$

which also is the recursion one gets if the numbers drawn from an urn in our original setting are replaced by i.i.d. U(0,1) random variables, as one might probably guess. We can also pass to the limit in Proposition 2 to get

<u>Proposition 3:</u> The following assertions hold:

$$\pi_{k+k'} \geq \binom{k+k'}{k} \frac{k^k\ k'^{k'}}{(k+k')^{k+k'}}\ \pi_k \pi_{k'} \tag{3.1}$$

$$\pi_{k+k'+1} \leq \pi_k \pi_{k'}\ . \tag{3.2}$$

For a proof, just pass to the limit in (2.8) resp. (2.10). Now, observe that (3.2) implies that log π_{k-1} is a subadditive function, and, since for any subadditive function f the limit of f(x)/x exists as x tends to infinity (cf. Hille (2), pp. 380 ff.), we have already proved:

<u>Proposition 4.</u> The limit

$$\rho := \lim_{k\to\infty} \pi_k^{1/k} \tag{3.3}$$

exists.

Theorem 13.1.3. in Hille (2) implies even more, namely that $\pi_k \geq \rho^{k+1}$.

Applying the same method to the quantities $(k/e)^k \pi_k / k!$ and using inequality (3.1) yields

$$\pi_k \leq k!(e/k)^k \rho^k = \sqrt{2\pi k}\ \rho^k(1+o(1)) \tag{3.4}$$

and, fact, we strongly believe that this is the right asymptotic formula (up to the constant factor, which will be different from $\sqrt{2\pi}$).

4. ASYMPTOTIC BEHAVIOUR AS $n/k \to x$

We now let n and k grow simultaneously in such a way that their ratio approaches a value $x \geq 1$. We have:

Proposition 5. Let n(k) be a sequence of integers such that $n(k)/k \to x \geq 1$ as $k \to \infty$. Then

$$\gamma(x) := \lim_{k\to\infty} q_{n(k)\ k}^{1/k} \text{ and} \tag{4.1}$$

$$\rho(x) := \lim_{k\to\infty} p_{n(k)\ k}^{1/k} \tag{4.2}$$

exist, and do not depend on the special choice of n(k).

Proof. We first prove (4.1) for $n(k) = [xk]$. By (2.8) one gets that $-\log(q_{n(k);k})$ is a subadditive function. As in the proof of Proposition 4 we may conclude that $\gamma(x)$ exists. Now another application of (2.8) yields that $\log(q_{n(k),k})$ is a concave function and hence continuous in the interior of its domain. Continuity in $x = 1$ follows from (2.6) and (2.7). The fact that γ is continuous now implies. That the limit in (4.1) is the same for all sequences n(k) that fulfill the condition stated. (4.2) now follows from (4.1) by some simple calculations. It should be noted that

$$\rho = \rho(\infty) := \lim_{x\to\infty} \rho(x)$$

5. SOME PROBLEMS

We finally would like to state some open problems. First, we would like to know something about the $\rho(.)$ and $\gamma(.)$ functions or at least the exact value of ρ. All that we know now is that $x \leq \gamma(x) < ex$ where both equalities are strict for $x \neq 1$.

Then, there are some modifications to the original problem. In the first one we do not ask for the maximum probability of putting all numbers in right, but we want to maximize the expected number of boxes that we can fill until there occurs a number that we cannot put in a box without violating the monotonicity condition we imposed.

The second modification is the following: now we may "shift" the numbers already placed but have to pay a certain amount of money for each time we move a number. The question now is how to fill up all boxes with paying as little as possible.

REFERENCES

1. Matematika feladatgyüjtemény Általános iskola I. osztály. Szerkesztette Oravecz Pálné, Tankönyvkiadó, Budapest, 1985. 71. oldal (Mathematics textbook for 1st grade pupils in the Hungarian elementary schools (in Hungarian))
2. Hille, E.(1972) Methods in classical and functional analysis, Addison-Wesley, Reading, Massachusetts

Karl Grill
Institut für Statistik und
Wahrscheinlichkeitstheorie
TU Wien
Karlsplatz 13
1040 Wien

Gábor Elek
Eötvös Loránd University
1119 Budapest
Fehérvári ut 123.

Proc. of the 5th Pannonian Symp. on Math. Stat., Visegrád, Hungary 1985
W. Grossmann, J. Mogyoródi, I. Vincze, W. Wertz, eds

ON THE CONVERGENCE OF REGRESSION TYPE MARTINGALE FIELDS

I. FAŻEKAS

Debrecen

In this paper a genaralization of two-parameter martingales is studied. A two-parameter process is called a regression martingale if it satisfies certain autoregressive type stochastic difference equations. It is proved that Cairoli's martingale convergence theorems are true for homogeneous regression martingale fields. The proof is based on the investigation of a two-parameter matrix-valued process corresponding to the regression martingale field.
AMS subject classification: 60G42, 60G60

1. INTRODUCTION

In this paper we deal with a regression type generalization of the notion of a two-parameter margingale. The introduction of the idea of a regression type martingale /Definition 1/ was motivated by the concepts of an autoregressive field /see e.g. Terdik, 1977/ and an L-Markovian field /Rozanov, 1967/.

The aim of this paper is to prove convergence theorems for the regression martingale field ξ_{ij}. Convergence theorems for two-parameter martingales have been proved by Cairoli /1970/. We show that the analoques of Cairoli's theorems are true for homogeneous regression martingale fields as well /Theorem 3/.

The method of the proofs is similar to that used by Fazekas /1986/. By the transformation described in Section 2, we reduce our problem to the study of a matrix-valued process X_{ij}. In Section 3, we prove convergence theorems for X_{ij}. The bahaviour of X_{ij} can be characterized by that of the so--called accompanying martingale Y_{ij} of X_{ij}. Therefore, well-known results on martingales can be applied. More precisely, we use Cairoli's theorem and a two-parameter version of Burkholder's inequality /due to Métraux, 1978/.

2. DEFINITIONS AND PRELIMINARY REMARKS

Let N^2 denote the two-dimensional positive integer lattice points with the coordinatewise ordering. If $(i,j)\in N^2$, then $(i,j)\to\infty$ means that $i\to\infty$ and $j\to\infty$.

Let $(\Omega,\mathcal{A},P)$ be a probability space, $\{\xi_{ij};\ (i,j)\in N^2\}$ a sequence of random variables /r.v.'s/ defined on Ω and $\{\mathcal{F}_{ij};\ (i,j)\in N^2\}$ an increasing sequence of σ-subalgebras of $\mathcal{A}$. Suppose that the σ-algebras $\mathcal{F}_{ij}$ satisfy the so-called condition /F4/ introduced by Cairoli and Walsh /1975/:

$$E\{.|\mathcal{F}_{ij}\} = E\{E\{.\ |\mathcal{F}_{i\infty}\}|\mathcal{F}_{\infty j}\} = E\{E\{.\ |\mathcal{F}_{\infty j}\}|\mathcal{F}_{i\infty}\} \qquad /F4/$$

for every $(i,j)\in N^2$, where $\mathcal{F}_{i\infty}$ and $\mathcal{F}_{\infty j}$ are the following generated σ-algebras: $\mathcal{F}_{i\infty}=\sigma\{\mathcal{F}_{ij}:\ j=1,2,\dots\}$ and $\mathcal{F}_{\infty j}=\sigma\{\mathcal{F}_{ij}:\ i=1,2,\dots\}$.

Definition 1. The process $\{\xi_{ij},\mathcal{F}_{ij};\ (i,j)\in N^2\}$ is called a regression type martingale field if

/i/ ξ_{ij} is $\mathcal{F}_{ij}$-measurable and integrable for every $(i,j)\in N^2$;

/ii/ $E\{\xi_{ij}|\mathcal{F}_{i-1,\infty}\} = a_1(i)\xi_{i-1,j}+\dots+a_m(i)\xi_{i-m,j}$

and /1/

$$E\{\xi_{ij}|\mathcal{F}_{\infty,j-1}\}= b_1(j)\xi_{i,j-1}+\dots+b_m(j)\xi_{i,j-m}$$

for every $i>m$ and $j>m$, where m is a fixed positve integer and $a_k(i)$, $b_k(j)$ are non-negative non-random coefficients for which $\sum_{k=1}^{m} a_k(i) = 1$ and $\sum_{k=1}^{m} b_k(j)=1$.

If the coefficients $a_k(i)$ and $b_k(j)$ do not depend on i and j, respectively, then ξ_{ij} is called a homogeneous regression martingale field.

In order to study the convergence properties of ξ_{ij} we introduce the following matrices:

$$X_{ij} = \begin{pmatrix} \xi_{i,j-m+1} & \cdots & \xi_{i,j} \\ \vdots & & \\ \xi_{i-m+1,j-m+1} & \cdots & \xi_{i-m+1,j} \end{pmatrix},$$

$$A_i = \begin{pmatrix} a_1(i) & \cdots & a_m(i) \\ 1 & & 0 \quad \vdots \\ & \ddots & \vdots \\ 0 & & \ddots \quad \vdots \\ & 1 & 0 \end{pmatrix}, \quad B_j = \begin{pmatrix} 0 & & b_m(j) \\ 1 & \ddots \quad 0 & \vdots \\ & \ddots & \vdots \\ 0 & \ddots & \vdots \\ & 1 & b_1(j) \end{pmatrix}$$

From hypothesis /F4/ and equations /1/, one can easily get the following.

<u>Proposition 1.</u> The process $\{X_{ij}, \mathfrak{F}_{ij};\ (i,j)\in N^2\}$ satisfies equation

$$E\{X_{i+t,j+s} | \mathfrak{F}_{ij}\} = A(i+t,i)\ X_{ij}\ B(j,j+s) \qquad /2/$$

for every $i>m$, $j>m$, where $A(i+t,i)=A_{i+t}\ A_{i+t-1}\cdots A_{i+1}$ and $B(j,j+s) = B_{j+1}\ B_{j+2}\ \cdots\ B_{j+s}$.

Generalizing this property, we get the following notion.

<u>Definition 2.</u> A matrix-valued process $\{X_{ij}, \mathfrak{F}_{ij}; (i,j)\in N^2\}$ is called an AB-martingale field if

/i/ X_{ij} is $\mathfrak{F}_{ij}$-measurable and integrable for every $(i,j)\in N^2$;

/ii/ equation /2/ is satisfied for every (i,j), $(s,t)\in N^2$, where A_i and B_j are given non-random matrices. /All matrices considered are of type mxm./

Let us introduce the martingale difference

$$\Delta_{i,j} = X_{ij} - E\{X_{ij}|\ _{i-1,j}\} - E\{X_{ij}|\ _{i,j-1}\} + E\{X_{ij}|\ _{i-1,j-1}\} =$$
$$= X_{ij} - A_i X_{i-1,j} - X_{i,j-1} B_j + A_i X_{i-1,j-1} B_j$$

for i,j > 1 and $\Delta_{11}=X_{11}$, $\Delta_{i1} = X_{i1} - A_i X_{i-1,1}$, $\Delta_{1j} = X_{1j} - X_{1,j-1}B_j$.

We remark that in the case of the regression martingale field ξ_{ij}

$$\Delta_{ij} = \begin{pmatrix} 0 & \dots & 0 & \delta_{ij} \\ & & & 0 \\ & & & \cdot \\ & 0 & & \cdot \\ & & & \cdot \\ & & & 0 \end{pmatrix},$$

where

$$\delta_{ij} = \xi_{ij} - E\{\xi_{ij}|\mathcal{F}_{i-1,j}\} - E\{\xi_{ij}|\mathcal{F}_{i,j-1}\} + E\{\xi_{ij}|\mathcal{F}_{i-1,j-1}\}.$$

Proposition 2. For the AB-martingale field X_{ij}, we have the representation:

$$X_{ij} = \sum_{k=1}^{i} \sum_{l=1}^{j} A(i,k)\, \Delta_{kl}\, B(l,j)$$

for $(i,j) \in N^2$, where $A(k,k) = B(l,l) = I$ /the identity matrix/.

Proof. By induction.

3. CONVERGENCE OF AB-MARTINGALE FIELDS

In this section, we shall prove convergence theorems for AB-martingale fields under the following conditions /which are satisfied in the case of homogeneous regression martingale fields/.

Let us suppose that $A(i+t,i) \to A(\infty,i)$ as $t \to \infty$ for every i and $B(j,j+s) \to B(j,\infty)$ as $s \to \infty$ for every j and these convergences are "fast" in the following sense:

$$||A(i+t,i) - A(\infty,i)|| < c_t$$

and /3/

$$||B(j,j+s) - B(j,\infty)|| < d_s ,$$

where $\sum_{t=1}^{\infty} c_t < \infty$ and $\sum_{s=1}^{\infty} d_s < \infty$. /Throughout this paper $||.||$ denotes the norm of a matrix ./

For the limit matrices $A(\infty,k)$, we assume that there exists a positive number c such that

$$||A(\infty,k)\Delta_{kj}|| \geq c||\Delta_{kj}||$$

and /4/

$$||A(\infty,k)\sum_{l=1}^{j}\Delta_{kl}\,B(l,\infty)|| \geq c||\sum_{l=1}^{j}\Delta_{kl}\,B(l,\infty)||$$

for every j and k. Suppose that analogous conditions hold for $B(l,\infty)$. We understand /4/ to include these conditions too.

The behaviour of X_{ij} is closely related to a certain martingale Y_{ij}. This martingale is called the accompanying martingale of X_{ij}, and it is defined by the equation

$$Y_{ij} = \sum_{k=1}^{i}\sum_{l=1}^{j} A(\infty,k)\,\Delta_{kl}\,B(l,\infty), \qquad (i,j)\ N^2.$$

<u>Lemma 1.</u> If $\sup_{i,j} E\,\varphi(||X_{ij}||) < c < \infty$, where $\varphi: R^+ \to R^+$ is a convex non-decreasing function, then for the accompanying martingale $\sup_{i,j} E\ (||Y_{ij}||) < c$ as well.

<u>Proof.</u> Let $Y_{ij}^{ts} = \sum_{k=1}^{i}\sum_{l=1}^{j} A(t,k)\,\Delta_{kl}\,B(l,s)$,

where t and s are fixed. It is easy to see that

$\{Y_{ij}^{ts}, \mathfrak{F}_{ij},\ 1\leq i\leq t,\ 1\leq j\leq s\}$ is a martingale.

Since $\varphi(||Y_{ij}^{ts}||)$ is a real submartingale, we have

$$E\,\varphi(||Y_{ij}^{ts}||) \leq E\varphi(||Y_{ts}^{ts}||) = E\,\varphi(||X_{ts}||) < c$$

for every $(i,j)\leq(t,s)$. On the other hand, $Y_{ij}^{ts} \to Y_{ij}$ as $(t,s)\to\infty$. Thus by Fatou's lemma $E\varphi(||Y_{ij}||<c$ for every $(i,j)\in N^2$.

<u>Theorem 1.</u> Let $\{X_{ij},\ _{ij}; (i,j)\in N^2\}$ be an AB-martingale

field for which conditions /F4/, /3/ and /4/ are satisfied. If

$$\sup_{k,l} E||X_{kl}||\log^+(||X_{kl}||)<\infty,$$

then X_{ij} converges a.s. and in L_1 as $(i,j)\to\infty$.

Proof. First, we prove the a.s. convergence. By Lemma 1 and Cairoli's theorem /Cairoli, 1970; see also Fazekas, 1983/, $Y_{ij}\to Y$ a.s. as $(i,j)\to\infty$. We show that $X_{ij}\to Y$ a.s. We mention that Y_{ij} converges if at least one of the parameters tends to infinity. Therefore, by /4/, $\Delta_{kl}\to 0$ a.s. and $\{\Delta_{kl};(k,l)\in N^2\}$ is bounded. Proposition 2, the definition of Y_{ij} and condition /3/ imply that

$$||X_{ij}-Y_{ij}||\leqq \sum_{k=1}^{i}\sum_{l=1}^{j} c_{i-k}||\Delta_{kl}||d_{j-l}+$$

$$+\sum_{k=1}^{i} c_{i-k}||\sum_{l=1}^{j} \Delta_{kl}\, B(l,\infty)|| +$$

$$+\sum_{l=1}^{j} d_{j-l}||\sum_{k=1}^{i} A(\infty,k)\Delta_{kl}||=U+V+W.$$

We estimate these three expressions separately.

$$U = \sum_{u=0}^{i-1}\sum_{s=0}^{j-1} c_u||\Delta_{i-u,j-s}||d_s \leqq \sum_{u=0}^{v-1}\sum_{s=0}^{z-1} c_u||\Delta_{i-u,j-s}||d_s +$$

$$+\sum_{u=v}^{i-1} c_u \sum_{s=0}^{j-1} ||\Delta_{i-u,j-s}||d_s +\sum_{s=z}^{j-1} d_s \sum_{u=1}^{i-1} c_u||\Delta_{i-u,j-s}||.$$

Let $\varepsilon>0$ be arbitrary. According to /3/, one can fix v and z such that the last two terms of the above expression are less than ε for every $i>v$ and $j>z$. If v and z are fixed, then the first term tends to zero as $(i,j)\to\infty$. Therefore, $U\to 0$ as $(i,j)\to\infty$.

Now, we turn to V.

$$V = \sum_{u=0}^{i-1} c_u||\sum_{l=1}^{j} \Delta_{i-u,l}\, B(l,\infty)|| \leqq$$

$$\leqq \sum_{u=0}^{n} c_u||\sum_{l=1}^{j}\Delta_{i-u,l}B(l,\infty)||+\sum_{u=n+1}^{i-1} c_u||\sum_{l=1}^{j} \Delta_{i-u,l}B(l,\infty)||. \qquad /5/$$

We remark that the convergence of Y_{ij} and /4/ imply that $\|\sum_{l=1}^{j} \Delta_{il} B(l,\infty)\| \to 0$ a.s. as $i\to\infty$ for every j. Let $\varepsilon>0$ be arbitrary. By /3/, one can choose n such that the last term in /5/ is less than ε for every $i>n$ and $j>1$. If n is fixed, then the first term of /5/ tends to zero if $i\to\infty$.

The convergence of W^* can be proved similarly. Therefore $X_{ij}\to Y$ a.s. as $(i,j)\to\infty$.

Finally, the assumptions of the theorem imply the uniform integrability of X_{ij}, so X_{ij} converges in L_1, too.

Theorem 2. Suppose that for the AB-martingale field $\{X_{ij}, \mathcal{F}_{ij}\}$ conditions /F4/, /3/ and /4/ hold and

$$\begin{aligned} &\|A(i,u)\| < K < \infty \qquad (i>u), \\ &\|B(v,j)\| < K < \infty \qquad (j>v). \end{aligned} \tag{/6/}$$

If $\sup_{i,j} E\|X_{i,j}\|^{\alpha} < \infty$, where $\alpha>1$, then X_{ij} converges in L_α as $(i,j)\to\infty$.

Proof. By Lemma 1 $\sup_{i,j} E\|Y_{ij}\|^{\alpha}<\infty$. Therefore, according to Theorem 4.6 of Fazekas /1983/,

$$Y_{ij} \to Y_\infty \tag{/7/}$$

a.s. and in L_α if i or j tends to infinity.

The main step of our proof is the following sequence of inequalities:

$$W_{kl}^{ij} = E\|\sum_{u=k}^{i} \sum_{v=1}^{j} A(i,u)\Delta_{u,v} B(v,j)\|^{\alpha} \leqq$$

$$\leqq C_1 E\left(\sum_{u=k}^{i} \sum_{v=1}^{j} \|\Delta_{u,v}\|^2\right)^{\frac{\alpha}{2}} \leqq$$

$$\leqq C_2 \, E \left|\left| \sum_{u=k}^{i} \sum_{v=l}^{j} A(\infty,u) \, \Delta_{u,v} \, B(v,\infty) \right|\right|^{\alpha} = \qquad /8/$$

$$= C_2 \left|\left| Y_{ij} - Y_{kj} - Y_{il} + Y_{kl} \right|\right|^{\alpha}$$

for $i \geqq k$, $j \geqq l$. These inequalities are consequences of Burkholder's inequalities /the two-parameter version of Burkholder's inequalities is due to Métraux, 1978/. In the first inequality, we applied also condition /6/ and in the second one condition /4/.

Consequently, $W^{ij}_{kl} \to 0$ if at least one of the parameters k,l tends to infinity.

We show that $X_{ij} \to Y_{\infty}$.

$$||X_{ij} - Y_{\infty}||_{L_\alpha} \leqq ||Y_{kl} - Y_{\infty}||_{L_\alpha} + \left|\left| \sum_{u=1}^{k} \sum_{v=1}^{l} A(i,u) \Delta_{uv} B(v,j) - \right.\right.$$

$$\left.\left. - Y_{kl} \right|\right|_{L_\alpha} + \left|\left| \sum_{\substack{(u,v) \leqq (i,j) \\ (u,v) \not\leqq (k,l)}} \sum A(i,u) \Delta_{uv} B(v,j) \right|\right|_{L_\alpha}.$$

Let $\varepsilon > 0$ be arbitrary. By /7/ and /8/, one can fix k and l such that the first and the third terms in the above expression are less than ε. If k and l are fixed, then the second term tends to zero as $(i,j) \to \infty$.

4. CONVERGENCE OF HOMOGENEOUS REGRESSION MARTINGALE FIELDS

We prove that Cairoli's martingale convergence theorems are true for homogeneous regression martingale fields.

Theorem 3. Let $(\xi_{ij}, \mathfrak{F}_{ij}, i,j = 1,2,\ldots)$ be a homogeneous regression martingale field, and suppose that the σ-algebras $\mathfrak{F}_{ij}$ satisfy condition /F4/. Moreover, assume that the greatest common divisor of those numbers k for which $a_k > 0$ /resp. $b_k > 0$/ is equal to one.

/a/ If $\sup_{i,j} E|\xi_{ij}|\log^+|\xi_{ij}|<\infty$, then ξ_{ij} converges a.s. and in L_1 as $(i,j)\to\infty$.

/b/ Let $\alpha>1$. If $\sup_{i,j} E|\xi_{ij}|^{\alpha} < \infty$, then ξ_{ij} converges in L_{α} /and a.s./ as $(i,j)\to\infty$.

Proof. In Section 2, we have constructed the AB-martingale X_{ij}, the matrices A_i, B_j, and the martingale difference Δ_{ij} corresponding to ξ_{ij}. Because of the conditions of our theorem, $A=A_i$ is the transition matrix of a non-decomposable acyclic Markov chain. The elements of the matrices $A(i+t,i) = A^t$ converge exponentially fast to the elements of the matrix $A(\infty,i) = A(\infty) = (a_{kj})$ as $t\to\infty$, where $a_{kj} = p_j$ $(k,j = 1,\dots,m)$ and $p_j = \sum_{k=j}^{m} a_k \Big/ \sum_{k=1}^{m} k\, a_k$ is the unique stationary distribution of the Markov chain /see e.g. Seneta, 1981/. Similar assertions are true for the matrices B_j. Therefore, conditions /3/, /4/ and /6/ hold.

Now, it is easy to see that Theorems 1 and 2 imply the desired result.

Question. For what kind of autoregressive type schemes, more general than those defined in /1/, can martingale convergence theorems be proved?

REFERENCES

Cairoli, R. /1970/. Une inégalité pour maringales à indices multiples et ses applications, Séminaire de Probabilités IV, Université de Strasbourg, Lecture Notes in Math., Vol. 124, 1-27, Springer-Verlag, Berlin.

Cairoli, R. and Walsh J. B./1975/. Stochastic integrals in the plane, Acta Math., 134, 111-183.

Fazekas, I. /1983/. Convergence of vector-valued martingales with multidimensional indices, Publ. Math. Debrecen, 30, 157-164.

Fazekas, I. /1986/. On the convergence of linear margingales, Publ. Math. Debrecen /submitted for publication/.

Métraux, Ch. /1978/. Quelques inégalités pour martingales à paramètre bidimensionnel. Séminaire de Probabilités XII, Université de Strasbourg, Lecture Notes in Math., Vol. 649, 170-179, Springer-Verlag, Berlin.

Rozanov, Yu. A. /1967/. On the Gaussian homogeneous fields with given conditional distributions, Teor. Veroyatnost. i Primenen., 12, 433-443.

Seneta, E. /1981/. Non-negative Matrices and Markov Chains, Springer-Verlag, New York.

Terdik, G. /1977/. The moving average representation of an autoregressive field and the maximum likelihood estimate of its coefficients /in Hungarian/, Alkalmaz. Mat. Lapok, 3, 379-388.

I. Fazekas
Kossuth Lajos University
Department of Mathematics
4010 Debrecen, Pf. 12.
Hungary

*Proc. of the 5th Pannonian Symp.
on Math. Stat., Visegrád, Hungary 1985
W. Grossmann, J. Mogyoródi, I. Vincze, W. Wertz, eds*

TREE-MARTINGALES

S. FRIDLI, F. SCHIPP

Budapest

In this paper we generalize some of the classical results of the martingale theory. We deal with martingales whose index set is a tree.

AMS subject classification: 60G48, 60G46

1. PRELIMINARIES AND RESULTS

Let $(X, \mathcal{A}, P)$ be a probability space. Further let $(\mathcal{A}_n, n \in \mathbb{N})$ $(\mathbb{N} := \{0,1,2,\ldots\})$ be a non-decreasing sequence of sub-σ-fields of $\mathcal{A}$. Denote by $||f||_p$ $(1 \le p \le \infty)$ the $L^p := L^p(X, \mathcal{A}, P)$-norm of the function $f \in L^p$, and by f_n the conditional expectation of $f \in L^1$ relative to $\mathcal{A}_n$. By the well-known martingale maximal inequality [4] the maximal function $f^* := \sup_n |f_n|$ $(f \in L^1)$ is of weak type (1,1), i.e.

$$y\, P(f^* > y) \le \int_{\{f^* > y\}} |f|\, dP \le ||f||_1 \quad (y > 0,\ f \in L^1) . \tag{1}$$

This immediately gives the Doob's inequality [4]

$$||f^*||_p \le q ||f||_p \quad (1 < p \le \infty,\ p^{-1} + q^{-1} = 1,\ f \in L^p) . \tag{2}$$

If $a_n \in L^\infty(X, \mathcal{A}_{n-1}, P)$ and $||a_n||_\infty \le 1$ $(n \in \mathbb{N})$, then the sequence $(\sum_{k=1}^{n} a_k (f_k - f_{k-1}),\ n \in \mathbb{N} \setminus \{0\})$ $(f \in L^1)$ is a martingale, which is called the transform of the martingale $(f_n, n \in \mathbb{N})$. The generalization of the martingale maximal inequality for martingale transforms is due to D. L. Burkholder [1].

This states that the operator

$$T_a f := \sup_n \left| \sum_{k=1}^{n} a_k(f_k - f_{k-1}) \right| \qquad (f \in L^1)$$

is of weak type (1,1), i.e.

$$y\, P(T_a f > y) \le 18 \|f\|_1 \qquad (y > 0,\ f \in L^1) \quad . \tag{3}$$

It is known that there is a close connection among the maximal function, the martingale transform and the quadratic variation, which is defined by

$$Qf := \left(\sum_{n=0}^{\infty} |f_n - f_{n-1}|^2 \right)^{1/2} \qquad (f \in L^1)$$

(where $f_{-1} := 0$).

D. L. Burkholder generalized the Paley inequality for martingales. By this generalization the L^p-norm of Qf is equivalent to the L^p-norm of f ($1<p<\infty$) (briefly $\|Qf\|_p \sim \|f\|_p$), more precisely

$$c_p \|f\|_p \le \|Qf\|_p \le C_p \|f\|_p \qquad (1<p<\infty,\ f \in L^p)$$

(where c_p and C_p are constants depending only on p). This inequality fails to hold for p=1. However $\|Qf\|_p \sim \|f^*\|_p$ ($1 \le p < \infty$). The operator Q is not bounded from L^1 to L^1, but it is of weak type (1,1), moreover [2]

$$y\, P(Qf>y,\ f^* \le y) \le 2\|f\|_1 \qquad (y>0,\ f \in L^1) \tag{4}$$

The purpose of our paper is to generalize some of the above mentioned theorems for martingales whose index set is a tree.

Let T be a countable ordered set having the following properties:

i. every non-empty subset of T has a minimal element,

ii. for every $s, t \in T$ there exists the envelope
$s \vee t := \sup\{s,t\} \in T$, (5)

iii. for every $s \in T$ the set $\{t \in T : s \le t\}$ is linearly ordered,

iv. for every $s \in T$ the set $\{t \in T : t \le s\}$ is finite.

Let $(\mathcal{A}_t, t \in T)$ be such a non-decreasing sequence of sub-σ-fields of $\mathcal{A}$, which can be linearly ordered and $\mathcal{A} = \bigcup_{t \in T} \mathcal{A}_t$. The concept of the conditional expectation operator is generalized as follows:

$E_t : L^1 \to L^1$ is said to be a quasi conditional expectation operator (briefly QCE) relative to $\mathcal{A}_t$ if there exists $\varphi_t \in L^\infty$ such that

$$E_t f = \varphi_t \, E(f \, \bar{\varphi}_t | \mathcal{A}_t) \qquad (f \in L^1) .$$

It is obvious that any conditional expectation operator is a QCE.

The sequence $(\mathcal{A}_t, E_t, t \in T)$ is called a tree-basis if:

i. E_t is a QCE relative to $\mathcal{A}_t$.

ii. if $E_t f = \varphi_t \, E(f \, \bar{\varphi}_t | \mathcal{A}_t)$, then

$$E_s \, f = \varphi_\ell E(f \, \bar{\varphi}_\ell | \mathcal{A}_s) \qquad (\ell \le s) ,$$

iii. $E_s E_\ell f = 0$ $(f \in L^1)$ if s and ℓ are not comparable $(\ell, s, t \in T)$.

Throughout this paper C denotes an absolute but not always the same constant and similarly C_p denotes an only on p depending constant, which is not always the same. Further we suppose that the sequence $(\varphi_t, t \in T)$ is uniformly bounded, i.e. $\|\varphi_t\|_\infty < C$ $(t \in T)$.

Since $(\mathcal{A}_t, t \in T)$ can be linearly ordered and $|E_t f| \le C E(|f| \, |\mathcal{A}_t)$, therefore by (1) E^* is of type (p,p) $(1<p\le\infty)$ and it is of weak type $(1,1)$, where $E^* f := \sup_{t \in T} |E_t f|$ $(f \in L^1)$.
We shall prove a stronger estimation for $2 \le p \le \infty$ in Theorem 1.
We say that $(f_t, t \in T)$ $(f_t \in L^1)$ is a tree-martingale, if $\ell < s$ implies $E_\ell f_s = f_\ell$ $(\ell, s \in T)$.

Example. Let $(m_n, n \in \mathbb{N})$ be a sequence of natural numbers whose terms are at least 2. Introduce the notation $M_o := 1$,

$$M_{n+1} := \prod_{k=0}^{n} m_k \qquad (n \in \mathbb{N}) .$$

Define T as follows

$$T := \{[k M_n, (k+1)M_n) : k, n \in \mathbb{N}\}$$

and introduce in T the ordering generated by the inclusion. Then T satisfies all the conditions of (5).

The functions

$$r_n(x) := \exp \frac{2\pi i x_n}{m_n} \qquad (x \in [0,1))$$

(where $x = \sum_{k=0}^{\infty} \frac{x_k}{M_{k+1}}$, $0 \le x_k < m_k$, $x_k \in \mathbb{N}$, $x \in [0,1))$ - are the so called generalized Rademacher functions. Let $\mathcal{A}_t$ be the σ-algebra generated by $\{r_o,\dots,r_{n-1}\}$, if there exists $k \in \mathbb{N}$ such that $t = [kM_n, (k+1) M_n)$ $(n \in \mathbb{N})$.

Thus $(\mathcal{A}_t, t \in T)$ is non-decreasing and it can be linearly ordered. The so called Vilenkin-system [6] $(\Psi_n, n \in \mathbb{N})$ is defined as follows

$$\Psi_n := \prod_{k=0}^{\infty} r_k^{n_k} \qquad (n \in \mathbb{N})$$

(where $n = \sum_{k=0}^{\infty} n_k M_k$, $0 \le n_k < m_k$, $n_k \in \mathbb{N}$).

Let $E_t f := \Psi_{kM_n} E(f\bar{\Psi}_{kM_n} | \mathcal{A}_t)$ $(f \in L^1)$, if

$$t = [kM_n, (k+1) M_n) \quad (k, n \in \mathbb{N}, t \in T) .$$

Thus it is not hard to check that $(\mathcal{A}_t, E_t, t \in T)$ is a tree-basis and $(E_t f, t \in T)$ $(f \in L^1)$ is a tree-martingale.

The martingale maximal inequality (1) can be written in the following equivalent form:

$$y \sum_{n} \text{mes } D_n^y \leq \int_{\{f^* > y\}} |f| dP \qquad (y>0) \tag{6}$$

where $D_n^y := \{x \in X: |f_n| > y,\ |f_k| \leq y,\ k<n\}$ $(n \in \mathbb{N})$. (6) can be generalized as follows.

<u>Theorem 1</u>. Let $f \in L^p$ $(2 \leq p \leq \infty)$ and

$$D_t^y := \{x \in X: |E_t f| > y,\ |E_\ell f| \leq y,\ \ell < t\} \quad (y>0).$$

Then $y^p \sum_{t\ T} \text{mes } D_t^y \leq C_p \,\|\chi_{\{E^* f > y\}} f\|_p$.

Theorem 1. is a simple consequence of the more general Theorem 2. To formalize Theorem 2. we need the concept of the $L^p(\ell^q)$ $(1 \leq p,\ q \leq \infty)$ spaces. The sequence $g := (g_r,\ r \in \mathbb{N})$ $(g_r \in L^1)$ belongs to the space $L^p(\ell^q)$ $(1 \leq p \leq \infty,\ 1 \leq q < \infty)$ resp. to $L^p(\ell^\infty)$ $(1 \leq p \leq \infty)$ if and only if

$$\|(\sum_{r \in \mathbb{N}} |g_r|^q)^{1/q}\|_p < \infty$$

resp. $\|\sup_{r \in \mathbb{N}} |g_r|\|_p < \infty$. The (p,q)-norm is defined by

$$\|f\|_{(p,q)} := \|(\sum_{r \in \mathbb{N}} |g_r|^q)^{1/q}\|_p \qquad (1 \leq q < \infty)$$

$$\|f\|_{(p,\infty)} := \|\sup_{r \in \mathbb{N}} |g_r|\|_p \qquad (1 \leq p \leq \infty)$$

<u>Theorem 2</u>. Let $(K_t,\ t \in T)$ be a sequence of subsets of X such that

i. $K_t \cap K_\ell = \emptyset$ if $t<\ell$ or $\ell<t$

ii. K_t is $\mathcal{A}_t$-measurable $(\ell,\ t \in T)$.

Further let $P_t f := \chi_{K_t} E_t f \quad (t\in T,\ f\in L^1)$.

Then for the operator $Pf := (P_t f,\ t\in T)$ the following inequality is true

$$||Pf||_{(p,p)} \le C||\chi_K f||_p \qquad (2\le p\le\infty,\ f\in L^p)$$

where $K := \bigcup_{t\in T} K_t$.

(χ_H denotes the characteristic function of $H\subset X$.)

The proof of this theorem is based on a modified Riesz--Thorin-type interpolation theorem. This states that if a linear operator $\mathcal{U}$ is bounded from L^{p_i} to $L^{q_i}(\ell^{r_i})$ $(i=1,2,\ 1\le p_i,\ q_i,\ r_i\le\infty)$ then it is bounded from L^p to $L^q(\ell^r)$ (where $p^{-1}:=\lambda p_1^{-1} + (1-\lambda)p_2^{-1}$, $q^{-1}:=\lambda q_1^{-1} + (1-\lambda)q_2^{-1}$, $r^{-1}:=\lambda r_1^{-1} + (1-\lambda)r_2^{-1}$.)

Denote by T^* the set of the minimal elements of T and by t^+ $(t\in T)$ the minimum of $\{s\in T: t\le s,\ t\ne s\}$. The existence and unicity of t^+ evidently follows from the definition of T. Further the operator

$$Qf := \sup_{t\in T^*} \Big(\sum_{t\le s} |f_{s^+} - f_s|^2\Big)^{1/2} \qquad (f\in L^1)$$

will be considered.

Theorem 3. The operator Q is of type (p,p) $(2<p<\infty)$.

Remark. Let us turn to the above mentioned example. It is proved in [5], that if $m_n = 2$ $(n\in \mathbb{N})$, then Q has type (p,p) $(1<p<\infty)$. It is not hard to see, that the proof can be easily generalized for the so called bounded case (i.e. for the case $\limsup m<\infty$), but it can't be transferred for the unbounded case. The above theorem holds not only for bounded but for unbounded case as well.

To prove this theorem the generalization of a lemma of D. L. Burkholder [2] is needed.

Lemma. Let $(X, \mathcal{A}, P)$ be a probability space and $(\mathcal{A}_n, n \in \mathbb{N})$ be a non-decreasing sequence of sub-σ-fields of $\mathcal{A}$. Then for any $\lambda > 0$ and $f \in L^{2p}$ $(1 \leq p < \infty)$

$$||Q_{\mu-1} f||_{2p} \leq C_p \lambda^{1/2} ||f||_p^{1/2}$$

where $\mu(x) := \min\{n \in \mathbb{N} : |f_n(x)| > \lambda\}$ $(x \in X)$.

2. PROOFS.

Proof of Theorem 2. It follows from the assumption for K_t and for E_t $(t \in T)$, that the operators $P'_t f := \frac{1}{|\varphi_t|^2} P_t f$ $(t \in T,\ f \in L^1)$ are orthogonal projections, i.e. $P'_t P'_\ell = \delta_{t,\ell} P'_t$ $(t, \ell \in T)$ (where $\delta_{t,\ell} = 0$ if $t \neq \ell$ and $\delta_{t,t} = 1$).

Thus we have by $C ||P'_t f||_2 > ||P_t f||_2$ that

$$C||\chi_K f||_2 \geq C|| \sum_{t \in T} P'_t f||_2 \geq ||Pf||_{(2,2)} \quad (f \in L^2)$$

On the other hand, since $||E_t f||_\infty \leq ||\varphi_t||_\infty^2 ||f||_\infty$ $(t \in T,\ f \in L^\infty)$ therefore $||Pf||_{(\infty,\infty)} = \text{ess sup} \sup_{t \in T} |\chi_{K_t} E_t f| \leq C||\chi_K f||_\infty$.

Applying the above mentioned modified Riesz-Thorin-type interpolation theorem the proof of Theorem 2. is obtained.

Proof of Lemma. The case $p=1$ is proved in [2]. Let $p>1$. Applying the equality

$$Q^2_{n-1} f + f^2_{n-1} = 2f_n f_{n-1} - 2g_n \qquad (n \geq 1)$$

where $g_n = \sum_{k=1}^{n} f_{k-1}(f_k - f_{k-1})$ we have

$$||Q_{\mu-1}||_{2p} \leq (2||f_\mu f_{\mu-1}||_p + 2||g_\mu||_p)^{1/2} .$$

Since $f_{\mu-1} < \lambda$, therefore $||f_\mu f_{\mu-1}||_p \leq \lambda ||f||_p$. On the other hand g_μ is a transform of the martingale $(\chi_{\{\mu \geq k\}} f_k,\ k \in \mathbb{N})$. It is known [1] that the operator of the martingale transform is of strong type (p,p) $(1<p<\infty)$. Thus $\|g_\mu\|_p \le C_p\,\lambda\|f_\mu\|_p$. The Doob's inequality [4] immediately gives (7).

Proof of Theorem 3. It is enough to prove (see [5]) that the operators

$$Q_t\, f := \sup_{\substack{s\in T^*\\ s\le t}} \Big(\sum_{s\le \ell<t} |E_{\ell^+} f - E_\ell f|^2 \Big)^{1/2} \qquad (f\in L^1)$$

are of restricted weak type (p,p) for any $2<p<\infty$, i.e. for every $\mathcal{A}$-measurable set H of X

$$y^p\, P(Q_t\, \chi_H > y) < C_p \|\chi_H\|_p^p \qquad (y>0)\ .$$

Further f will denote the characteristic function of an $\mathcal{A}$-measurable set. Let $\varepsilon_t^y := \frac{1}{y}\,\chi_{B_t^y}\,\chi_{C_t^y}$ (y>0) where

$$B_t^y := \bigcup_{s\le t} \{x\in X : |E_s f(x)| > y\}\quad ,$$

$$C_t^y := \bigcap_{s\le t} \{x\in X : |E_s f(x)| \le ey\}\ .$$

Thus ε_t^y is $\mathcal{A}_t$-measurable and $\int_0^\infty \varepsilon_t^y\, dy = 1$.

Applying Lemma 2. of [5] we have as in [5] that

$$Q_t\, f \le \int_0^\infty \sup_{s\le t} \frac{2}{y}\,\chi_{D_s^y} \Big(\sum_{s\le r\le t} |\chi_{C_T^y} (E_{r^+} f - E_r f)|^2\Big)^{1/2} dy =$$

$$=: \int_0^\infty \sup_{s\le t} R_s^y\, f\, dy,$$

where $D_s^y := B_s^y \setminus \bigcup_{r\le s} B_r^y$, consequently

$$\chi_{\{E^* f<\lambda\}}\, Q_t f \le \int_0^\lambda \sup_{s\le t} R_s^y\, dy\ .$$

Since D_s^y is $\mathcal{A}_s$-measurable, therefore using Lemma we obtain

$$||R_s^y f||_p \le C_p\, y^{-\frac{1}{2}} ||\chi_{D_s^y}||_p^{\frac{1}{2}} \quad .$$

First let p>4 and q:=2p-4. Thus by Theorem 1. we have

$$\lambda P(Q_t f>\lambda,\ E^* f\le\lambda)^{\frac{1}{q}} \le ||\chi_{\{E^*f\le\lambda\}} Q_t f||_q \le$$

$$\le \int_0^\lambda (\sum_{s\le t} ||R_s^y f||_q^q)^{\frac{1}{q}} \le C_p \int_0^\lambda y^{-\frac{1}{2}} (\sum_{s\le t} \text{mes } D_s^y)^{\frac{1}{q}} \le$$

$$\le C_p \int_0^\lambda y^{-\frac{1}{2}-\frac{2}{q}}\, dy (\int |f|)^{\frac{1}{q}} \le C_p\, \lambda^{\frac{1}{2}-\frac{2}{q}} ||f||_q \ .$$

From this we have

$$\lambda^p\ P(Q_t f>\lambda,\ E^*f<\lambda) \le C_p\ ||f||_p^p \quad .$$

Combining this and the maximal inequality we get the desired inequality. Let now 2<p<4. An integral transformation gives

$$\chi_{\{E^*f\le\lambda^p\}}\ Q_t f < \lambda \int_0^{\lambda^{p-1}} \sup_{s\le t} R_s^{\lambda t}\, dt.$$

Since $E^*f\le C$, therefore

$$\chi_{\{E^*f\le\lambda^p\}}\ Q_t f < \lambda \int_0^{\lambda_1} \sup_{s\le t} R_s^{\lambda t}\, dt,$$

where $\lambda_1 := \min\{\lambda^{p-1},\ C\,\lambda^{-1}\} \le C$.

It is easy to see that there exist q>4 and s>0 such that

$$\frac{q}{2} + 2 - sq = p \quad \text{and} \quad \frac{1}{2} + \frac{2}{q} + \frac{s}{q} < 1 \ .$$

Thus $t^{-\frac{1}{p-1}} \geq \lambda^{-1}$ Lemma and Theorem 1. imply that

$$||\chi_{\{E^*f\leq\lambda^p\}} Q_t f||_q \leq \lambda \int_0^{\lambda_1} (\sum_{s\leq t} ||R_s^{\lambda t}||_q^q)^{\frac{1}{q}} \, dt \leq$$

$$\leq C_p \lambda^{1/2} \int_0^{\lambda_1} t^{-\frac{1}{2}} (\sum_{s\leq t} \text{mes } D_s^{\lambda t}) \, dt <$$

$$< C_p \lambda^{\frac{1}{2}-\frac{2}{q}+s} \int_0^{c} t^{-\frac{1}{2}-\frac{2}{q}-\frac{s}{p-1}} \, dt \, (\int f)^{\frac{1}{q}} \leq$$

$$\leq C_p \lambda^{\frac{1}{2}-\frac{2}{q}+s} (\int f)^{\frac{1}{q}} .$$

From this follows that

$$\lambda^p P(Q_t f>\lambda, E^*f\leq\lambda^p) \leq C_p ||f||_p^p .$$

Applying the maximal inequality we get the proof of Theorem 3.

REFERENCES

1. D. L. Burkholder, Martingale transforms, Ann. Math. Statist. 37 (1966), 1494-1504.
2. D. L. Burkholder, Distribution function inequalities for martingales, Annals of Prob. 1 (1973), 29-42.
3. B. Davis, On the integrability of the martingale square function, Israel J. Math., 8 (1970), 187-190.
4. J. Neveu, Discrete-parameter martingales, Math. Library, North-Holland (Amsterdam-Oxford-New York, 1975).
5. F. Schipp, On a Paley-type inequality, Acta Sci. Math., 45 (1983), 357-364.

6. N. Ja. Vilenkin, On a class of complete orthonormal systems, Izv. Akad. Nauk. SSSR, Ser. Mat. 11 (1947).

S. Fridli - F. Schipp
L. Eötvös University
Institute of Mathematics
Budapest, Muzeum krt. 6-8.
H-1088.

Proc. of the 5th Pannonian Symp. on Math. Stat., Visegrád, Hungary 1985
W. Grossmann, J. Mogyoródi, I. Vincze, W. Wertz, eds

SOME REMARKS ON RANDOM WALKS IN MULTIDIMENSIONAL TIME

J. GALAMBOS, I. KÁTAI

Philadelphia Budapest

Let X_i be integer valued, aperiodic, independent random variables with distribution function $F(x)$, $S_n = X_1+\ldots+X_n$, $Q(k) := \sum_{n=1}^{\infty} d_r(n)P(S_n=k)$, $d_r(n)$ is the number of solutions of $n = n_1 n_2 \ldots n_r$, $n_i \geq 1$ integers. It is proved that $Q(k) = Q_1(k) + o(1)$ $(k\to\infty)$ under the condition $\int_{|x|\geq z} x^2 dF(x) = O(z^{-\delta})$, where $\delta>0$ is a suitable constant, $Q_1(k)$ is a quantity that depends only on the mean value and variance of X_i and on r, furthermore $Q_1(k) > c_1 (\log k)^{r-1}$ with a suitable constant $c_1 > 0$.
AMS subject classification: 60K05

1. INTRODUCTION.

Let K_r be the set of r-tuples $\underline{n} = (n_1, n_2, \ldots, n_r)$, $r \geq 1$, where the components n_j are positive integers. Let $\{X, X_{\underline{n}}, \underline{n} \in K_r\}$ be a family of independent and identically distributed random variables with positive mean $EX = \mu<\infty$ and finite variance $VX = \sigma^2<\infty$. In a recent work, M. MAEJIMA and T. MORI [1] have shown that if X is integer valued, aperiodic and $E|X|^3<\infty$, then for r=2 or 3,

$$\sum_{\underline{n}\in K_r} P(S_{\underline{n}}=k) \sim \frac{1}{(r-1)!\mu}(\log k)^{r-1} \quad \text{as} \quad k\to+\infty$$

where

$$S_{\underline{n}} = \sum X_{\underline{j}}$$

the summation being extended over all members $\underline{j} = (j_1, j_2, \ldots, j_r)$ of K_r that satisfy $j_t \leq n_t$ for all $1 \leq t \leq r$.

Let $d_r(n)$ denote the number of solutions of $n = n_1 n_2 \ldots n_r$ in positive integers n_j. Then

$$Q(k) := \sum_{\underline{n} \in K_r} P(S_{\underline{n}} = k) = \sum_{n=1}^{\infty} d_r(n) P(S_n = k),$$

where $S_n = X_1 + X_2 + \ldots + X_n$ is the ordinary partial sum of n independent copies X_j of X. The limitation $r = 2$ or 3 in the result of Maejima and Mori is due to the lack of good asymptotic results for the sum of $d_r(n)$ over n. In a recent paper [2] by a new method of proof we established the following extension of the result of Maejima and Mori.

Theorem 1. *Let X be integer valued, aperiodic and* $E|X|^3 < \infty$. *Then there is a polynomial*

$$s(u) = \frac{1}{(r-1)!\mu} u^{r-1} + b_{r-2} u^{r-2} + \ldots + b_0$$

such that, as $k \to +\infty$,

$$Q(k) = s(\log k\, \mu^{-1}) + R_k + O(1), \tag{1.1}$$

where $R_k \to 0$ *for* $r = 2$ *or* 3. *Furthermore, for arbitrary* r,

$$\frac{1}{N} \sum_{k=N+1}^{2N} |R_k| = O(1) \qquad as \qquad N \to +\infty$$

implying, as $N \to +\infty$

$$\limsup \frac{1}{N} \sum_{k=N+1}^{2N} |Q(k) - s(\log k\, \mu^{-1})| < +\infty.$$

In the course of the proof we have quoted a method of explicitly computing the coefficients b_j of s(u).

The proof has been based on some quite new results for the remainder term of the divisor function, and on the following form of the central limit theorem: /see PETROV [4], p. 207/

$$P(S_n=k) = \phi_n(k;\mu,\sigma^2) + \frac{\beta}{6\sigma^2 n^{1/2}} (\xi^3_{n,k} - 3\xi_{n,k})\phi_n(k;\mu,\sigma^2) +$$

$$+ \frac{c_{n,k}}{n(1+|\xi_{n,k}|^3)}$$

where $\beta = E(X-\mu)^3$,

$$\phi_n(k) = \phi_n(k;\mu,\sigma^2) = \frac{1}{(2\pi n)^{1/2}\sigma} \exp\left(-\frac{1}{2}\xi^2_{n,k}\right), \tag{1.2}$$

$$\xi_{n,k} = \frac{n\mu-k}{\sigma n^{1/2}}$$

and $c_{n,k}$ are constants such that $c_n = \sup_k |c_{n,k}| \to 0$ as $n \to +\infty$.

In particular, we proved in [2] that

$$Q(k) = Q_1(k) + O(1) \quad \text{as} \quad k \to +\infty, \tag{1.3}$$

where

$$Q_1(k) := \sum_{n=1}^{\infty} d_r(n)\phi_n(k). \tag{1.4}$$

By using a more careful estimation we are able to prove that

$$Q(k) = Q_1(k) + o(1) \quad \text{as} \quad k \to +\infty. \tag{1.5}$$

It is worth to mention that $Q_1(k)$ does not depend on the distribution of X. Furthermore, we can get easily that

$$Q_1(k) > c_1(\log k)^{r-1} \tag{1.6}$$

with a suitable positive constant c_1 that may depend on r, σ, μ. To prove (1.6) we have to take into consideration the next Lemma 1 which is wellknown in number theory.

Lemma 1. *We have*

$$\frac{1}{\sqrt{N}} \sum_{n=N}^{N+\sqrt{N}} d_r(n) > c_2 (\log N)^{r-1},$$

with a suitable constant $c_2 = c_2(r) > 0$.

Now, since the terms $d_r(n)\phi_n(k)$ in $Q_1(k)$ are positive, and $\phi_n(k) > c_3 k^{-1/2}$ as n runs over $\left[\frac{k}{\mu} - \left(\frac{k}{\mu}\right)^{1/2}, \frac{k}{\mu} + \left(\frac{k}{\mu}\right)^{1/2}\right]$, from Lemma 1 (1.6) immediately follows.

The question, whether for the validity of Theorem 1 the finiteness of the third moment $E|X|^3$ is necessary seems to be open even in the case $r = 2$. In the next sections we shall give a negative answer to this, and prove Lemma 1.

2. FORMULATION AND PROOF OF THE THEOREM

Theorem 2. *Let us assume that* X *takes on integer values, aperiodic* $0 < \mu < +\infty$, *and*

$$\int_{|x| \geq z} x^2 dF(x) = O(z^{-\delta}) \tag{2.1}$$

with a suitable $0 < \delta < 1$, *where* F *denotes the distribution function of* X. *Then*

$$Q(k) = Q_1(k) + o(1) \qquad (k \to \infty), \tag{2.2}$$

consequently the assertion stated in Theorem 1 holds.

Proof. Let

$$T(k) = \sum_{n=1}^{\infty} d_r(n)H(n,k), \tag{2.3}$$

$$H(n,k) = |P(S_n=k)-\phi_n(k)|. \tag{2.4}$$

It is obvious that

$$|Q(k)-Q_1(k)| \leq T(k). \tag{2.5}$$

We shall estimate $T(k)$ by using a result due to I.A. IBRAGIMOV and Yu. V. LINNIK /see [3], Theorem 4.5.3., p. 169/ which we state now as

Lemma 2. *If* (2.1) *holds, then*

$$\max_k H(n,k) = O(n^{-\frac{1}{2}-\delta/2}) \quad \textit{as} \quad n\to\infty. \tag{2.6}$$

It is wellknown that $d_r(n) = O(n^{\varepsilon})$ for an arbitrary but fixed constant $\varepsilon>0$. Let δ^* be so chosen that $0<\delta^*<\delta/2-\varepsilon$ and

$$N_1 = \frac{k}{\mu}+k^{1/2+\delta^*}, \; N_2 = \frac{k}{\mu}-k^{1/2+\delta^*}, \; J = [N_1,N_2].$$

Then, by using (2.6) we get that

$$\sum_{n\in J} d_r(n)H(n,k) \ll k^{\varepsilon}\cdot k^{-1/2-\delta/2}(N_2-N_1) \ll$$

$$\ll k^{\varepsilon+\delta^*-\delta/2} = o(1) \quad \text{as} \quad k\to\infty.$$

To estimate the contribution of the terms $d_r(n)H(n,k)$ for $n\notin J$, first we consider the sums

$$L_1 := \sum_{n>N_1} d_r(n)P(S_n=k), \; L_2 := \sum_{n<N_2} d_r(n)P(S_n=k). \tag{2.7}$$

Let

$$R(k,M) := \sum_{n=M+1}^{\infty} P(S_n=k), \tag{2.8}$$

$$r(v) := \sum_{\ell=1}^{\infty} P(S_\ell=-v), \tag{2.9}$$

$$Q(k,M) := \sum_{n\le M} P(S_n=k). \tag{2.10}$$

Lemma 3. *Let* X *be integer valued, aperiodic with* $0<\mu<+\infty$ *and* $\sigma<+\infty$. *Then the series defined by* (2.8), (2.9), (2.10) *are convergent, furthermore*

$$r(v) = O(v^{-1}) \qquad (v\ge 1) \tag{2.11}$$

$$R(k,M) \le c_1 P(S_M\le k+L) + \frac{c_2}{L}, \qquad 1<L<\infty, \tag{2.12}$$

$$Q(k,M) \le c_3 \frac{M}{(k-M\mu)^2} \tag{2.13}$$

where c_1, c_2, c_3 *are suitable positive constants that may depend on* μ, σ, *and* L *is an arbitrary real number*, $1<L<\infty$.

We shall prove Lemma 3 in section 4. Now we shall use it to deduce that $L_1 \to 0$, $L_2 \to 0$ as $k\to+\infty$.

Since $d_r(n) = O(k^{\varepsilon})$ uniformly as $k\le N_2$, from (2.13) with $M = N_2$ we get immediately that $L_2 \to 0$ as $k\to\infty$. To estimate L_1 we split the sum into subsums

$$L_1 = \Sigma_0 + \Sigma_1 + \Sigma_2 + \dots$$

where in Σ_0 we sum over $n\in[N_1,k^5]$ and in Σ_j for $j\ge 1$ we sum over $n\in(M_j,M_{j+1}]$, $M_j = 2^{j-1}\cdot k^5$.

To estimate Σ_0 we observe that $d_r(n) = O(k^{\varepsilon})$, and so $\Sigma_0 = O(k^{\varepsilon}R(k,N_1))$. Let us choose $M = N_1$, $L = \frac{1}{2}k^{\frac{1}{2}+\delta^*}$ in (2.12) and apply the Chebyshev-inequality. We have immediately that $\Sigma_0 = o(1)$ as $k\to\infty$.

Let us now consider Σ_j, $j \geq 1$. Since $d_r(n) = O(M_j^{\varepsilon})$, we have

$$\Sigma_j = O(M_j R(k,M_j)),$$

and by (2.12) and the Chebyshev-inequality that

$$\Sigma_j \ll \frac{M_j^{1+\varepsilon}}{(M_j\mu-k-L)^2} + \frac{M_j^{\varepsilon}}{L}.$$

Let $L = \frac{1}{2}M_j\mu$. Then $\Sigma_j \ll M_j^{-1+\varepsilon}$, consequently $\sum_j \Sigma_j = o(1)$ and so $L_1 = o(1)$ as $k \to +\infty$.

The proof of the relation

$$\sum_{n \notin J} d_r(n)\phi_n(k) = o(1) \qquad (k \to \infty)$$

was done in [2] and so we omit it.

By this (2.2) has been proved.

3. Proof of Lemma 1. Let $n \in [N, N+\sqrt{N}]$ and consider only those solutions of $n = n_1 n_2 \dots n_r$ for which $1 \leq n_j \leq N^S$, $S = \frac{1}{2(r-1)}$, $j = 1,\dots,r-1$, hold. Fixing $n_1,\dots,n_{r-1}$, the variable n_r takes the integer values from the interval

$$\left[\frac{N}{n_1 n_2 \dots n_{r-1}}, \frac{N+\sqrt{N}}{n_1 n_2 \dots n_{r-1}}\right],$$

the number of which is

$$\frac{\sqrt{N}}{n_1 n_2 \dots n_{r-1}} + O(1).$$

Summing over n_j $(j = 1,2,\dots,r-1)$ we get that

$$\sum_{n=N}^{N+\sqrt{N}} d_r(n) \geq \sqrt{N} \sum \frac{1}{n_1 n_2 \dots n_{r-1}} + O(\sqrt{N}) \geq$$

$$\geq \sqrt{N}\left(\sum_{m<N^S} \frac{1}{m}\right)^{r-1} + O(\sqrt{N}) > c\sqrt{N}(\log N)^{r-1}$$

4. Proof of Lemma 3. The convergence of the series (2.8), (2.9) is a consequence of a theorem of CHUNG and POLLARD [5] asserting that $r(v)\to 0$ $r(-v)\to 1/\mu$ as $v\to+\infty$.

To prove (2.11) let E_ℓ, $F_{m,\ell}$ denote the events

$$E_\ell = \left\{S_\ell = -v,\ S_t \neq -v \text{ for } t<\ell\right\},\ F_{m,\ell} = \left\{X_{m+1}+\ldots+X_\ell = 0\right\}.$$

It is obvious that

$$\left\{S_\ell = -v\right\} = E_\ell + \sum_{m<\ell} E_m \cdot F_{m,\ell},$$

and consequently that

$$P(S_\ell = -v) \leq P(E_\ell) + \sum_{m<\ell} P(E_m)P(S_{\ell-m} = 0). \tag{4.1}$$

Hence upon summing up for ℓ, we get

$$r(v) \leq \sum_{\ell=1}^{\infty} P(E_\ell) + \sum_{m=1}^{\infty} P(E_m) \sum_{n=0}^{\infty} P(S_n=0)$$

$$= (1+r(0)) \sum_{\ell=1}^{\infty} P(E_\ell).$$

Since the events E_ℓ are disjoint, we get that $r(v) \leq$ $\leq (1+r(0))P(\mathcal{F})$, where

$$\mathcal{F} = \{S_\ell = -v \text{ for at least one } \ell\}.$$

Let $T_0 = 0$, $1 \leq T_1 < T_2 < \ldots$ be a sequence of integers, and the events $\mathcal{F}_0, \mathcal{F}_1, \ldots$ be defined as

$$\mathcal{F}_j = \left\{\min_{\ell=T_j,\ldots,T_{j+1}} (S_\ell - \ell\mu) \leq -v - T_j\mu\right\}.$$

It is clear that $\mathcal{F} \subseteq \Sigma \mathcal{F}_j$. Furthermore, by using the Kolmogorov inequality, by choosing $\eta_0 = S_{T_j} - T_j\mu$, $\eta_1 = X_{T_j+1} - \mu$, ..., $\eta_{T_{j+1}-T_j} = X_{T_{j+1}} - \mu$, we get immediately that

$$P(\mathcal{F}_j) \le \frac{\Sigma V\eta_\ell^2}{(v+T_j\mu)^2} \le \sigma^2 \frac{T_{j+1}}{(v+T_j\mu)^2} .$$

Let now $T_j = 2^j v$. Then

$$P(\mathcal{F}) \le \sum_{j=0}^{\infty} P(\mathcal{F}_j) = O(1/v),$$

i.e. (2.11) is true.

To prove (2.12) we need only to observe that

$$R(k,M) = \sum_{H=-\infty}^{\infty} P(S_M=H) \sum_{\ell=1}^{\infty} P(S_\ell=k-H) = \sum_{H=-\infty}^{\infty} P(S_M=H)r(H-k)$$

$$= \sum_{H\le k+L} P(S_M=H)r(H-k) + \sum_{H>k+L} P(S_M=H)r(H-k)$$

Since $r(H-k)=O(1)$ in the first sum and $r(H-k)=C\left(\frac{1}{L}\right)$ in the second one, (2.12) is true.

To prove (2.13) we start from (4.1). Summing up for $\ell=1,2,\ldots, M$, and observing that $v(0)<+\infty$, we get

$$Q(k,M) \le c \sum_{\ell=1}^{M} P(E_\ell) = c\, P\left(\sum_{\ell=1}^{M} E_\ell\right).$$

Taking into consideration that

$$\sum_{\ell=1}^{M} E_\ell = \left\{\max_{j=1,\ldots,M} (S_j - j\mu) \ge k - M\mu\right\},$$

the Kolmogorov inequality immediately implies (2.13). ∎

REFERENCES

1. *M. Maejima and T. Mori*, Some renewal theorems for random walks in multidimensional time, Math. Proc. Camb. Phil. Soc. 95 (1984), 149-154
2. *J. Galambos and I. Kátai*, A note on random walks in multidimensional time, Math. Proc. Camb. Phil. Soc. 99(1986), 163-170.
3. *I. A. Ibragimov and Yu. V. Linnik*, Independent and stationary dependent random variables, Moscow 1965, Nauka
4. *V. V. Petrov*, Sums of independent random variables, Berlin 1975, Springer
5. *K. L. Chung and H. Pollard*, An extension of renewal theory, Proc. Amer. Math. Soc. 3 (1952), 303-309

J. Galambos
Department of Mathematics
Temple University
Philadelphia
PA 19122, USA

I. Kátai
Institute of Mathematics
L. Eötvös University
Budapest 1088,
Hungary

Proc. of the 5th Pannonian Symp. on Math. Stat., Visegrád, Hungary 1985
W. Grossmann, J. Mogyoródi, I. Vincze, W. Wertz, eds

ON THE SUMS OF RANDOM VARIABLES REDUCED MOD 2Π

B. GYIRES

Debrecen

In this paper we are concerned with the limit distribution of random variables $Z_n = X_1 + \ldots + X_n$, reduces mod 2Π, where random variables $\{X_n\}$ are defined on a finite Markov chain. Among other we show if random variables $\{X_n\}$ are identically distributed with matrix valued distribution function $F(x)$ of infinite type then the powers of the Fourier coefficients of $F(x)$ converge on quasi weak sense to the Fourier coefficients of the matrix valued uniform distribution.
AMS subject classification:60F05.

I. ON THE QUASI WEAK CONVERGENCE

Let $0<m\leq 1$. We say function $\mathfrak{F}(x)$ defined on the real line to be a distribution function, if non-decreasing, right continuous,

$$\mathfrak{F}(-\infty) = \mathfrak{F}(-0) = 0, \quad \mathfrak{F}(2\Pi) = \mathfrak{F}(\infty) = m.$$

Let the set of these functions be denoted by $A(m)$. $A(1)$ is the set of the probability distribution functions.

A distribution functions is called of the finite type if its range consists of a finite number of values. In the opposite case it is called of the infinite type.

The Fourier coefficients

$$\varphi(k) = \int_0^{2\Pi} e^{ikx}\, d\mathfrak{F}(x) \quad (k=0,\pm 1, \pm 2, \ldots) \tag{1.1}$$

of $\mathfrak{F}\in A(m)$ satisfy obviously the conditions

$$\varphi(-k) = \overline{\varphi}(k) \quad (k = 1,2, \ldots).\ \varphi(0) = m. \tag{1.2}$$

It is well-known the uniqueness theorem, i.e. if two trigonometric series converge everywhere to the same sum, then the two series are identical, that is, the corresponding coefficients of both series are the same.

In the following we use on several occasions the so called theorem of the triginometric moment problem, which can be formulated as follows ([7], 108). Let $\{\varphi(k)\}$ be a sequence of complex constants satisfying conditions (1.2). Then necessary and sufficient conditions in order that an $\mathfrak{F} \in A(m)$ exists for which the equations (1.1) hold is that the Toeplitz matrices

$$T_n(\varphi) = \begin{bmatrix} \varphi(0) & \varphi(-1) & \dots & \varphi(-n) \\ \varphi(1) & \varphi(0) & \dots & \varphi(-n+1) \\ . & & \dots & . \\ \varphi(n) & \varphi(n-1) & \dots & \varphi(0) \end{bmatrix} \quad (n=0,1, \dots) \tag{1.3}$$

let be positive definite, or non-negative semidefinite.

In order that it can be expressed us easier, we shall use that notation

$$\varphi(t) = \int_0^{2\Pi} e^{itx} \, d\mathfrak{F}(x), \; t \in R$$

for the Fourier transform of $\mathfrak{F} \in A(m)$.

Definition 1.1. Let $\gamma_n(t)$ be the Fourier transform of $G_n \in A(m_n)$ $(n = 1,2, \dots)$, and let $\gamma(t)$ be the Fourier transform of $G \in A(m)$. We say that a sequence $\{G_n\}$ converges quasy weakly to G, $G_n \Rrightarrow G$, if

$$\lim_{n\to\infty} \gamma_n(k) = \gamma(k) \quad (k = 0, \pm 1 \pm 2, \dots). \tag{1.4}$$

It is obvious if $\{G_n\}$ converges weakly to G, then one converges quasi weakly too, but not conversely.

Theorem 1.1. Let $\gamma_n(t)$ be the Fourier transform of $G_n \in A(m_n)$ $(n=1,2, \ldots)$. Suppose that the conditions (1.4) are satisfied. Then such a $G \in A(m)$ exists that $G_n \Rightarrow G$.

Proof. It is evident that conditions (1.2) are satisfied. Since the determinant of a matrix is a continuous function of its elements, using notation (1.3) we get,

$$\operatorname{Det} T_n(\gamma) = \lim_{j\to\infty} \operatorname{Det} T_n(\gamma_j) \geq 0,$$

which gives us the proof of the Theorem by the help of the theorem of trigonometric moment problem.

Theorem 1.2. Let $G \in A(m)$ be given. Then such a sequence $G_n \in A(m)$ $(n=1,2,\ldots)$ exists for which $G_n \Rightarrow G$.

Proof. Let

$$\varepsilon_j > 0 \quad (j= 1,2, \ldots), \quad \varepsilon_j \to 0, \quad j\to\infty.$$

Let $\gamma(t)$ be the Fourier transform of G, and let

$$\gamma_j(t) = \frac{\gamma(t) + m\varepsilon_j}{1 + \varepsilon_j} \qquad (j = 1,2, \ldots).$$

Obviously

$$\gamma_j(k) \to \gamma(k) \qquad (k=0,\pm 1 \ldots), \ j\to\infty.$$

Thus it is necessary to show only that $\gamma_j(k)$ $(k=0,\pm 1, \ldots)$ are the Fourier coefficients of certain $G_j \in A(m)$.

It is clear that

$$\gamma_j(k) = \gamma_j(-k) \quad (k=1,2, \ldots), \qquad \gamma_j(0)= m.$$

On the other hand

$$\operatorname{Det} T_n(\gamma_j) = \left(\frac{1}{1+\varepsilon_j}\right)^{n+1} \left[\operatorname{Det} T_n(\gamma) + \sum_{\alpha=0}^{n} \sum_{\beta=0}^{n} T_{\alpha\beta} \right],$$

where

$$(T_{\alpha\beta})_{\alpha,\beta=0}^{n} = \operatorname{Adj} T_n(\gamma).$$

$\operatorname{Adj} T_n(\gamma)$ is positive definite or non-negative definite since $T_n(\gamma)$ has the similar property. Thus

$$\text{Det } T_n(\gamma_j) \geq 0 \quad (n=0,1,\ldots),$$

wich gives us the proof of the Theorem by the theorem of trigonometric moment problem.

Theorem 1.3. Let $\mathfrak{F} \in A(m)$ be a distribution functions of infinite type. Then

$$|\varphi(k)| < m \qquad (k = \pm 1, \pm 2, \ldots),$$

where $\varphi(k)$ is defined by (1.1).

Proof. Suppose that the statement does not hold, i.e. there is such an integer k that

$$|\varphi(k)|^2 = \left(\int_0^{2\Pi} \cos(kx)\, d\mathfrak{F}(x)\right)^2 + \left(\int_0^{2\Pi} \sin(kx)\, d\mathfrak{F}(x)\right)^2 = m^2. \tag{1.5}$$

Let $\frac{1}{m}\mathfrak{F} = G$, then

$$\left(\int_0^{2\Pi} \sin(kx)\,dG(x)\right)^2 = 1 - \left(\int_0^{2\Pi} \cos(kx)\, dG(x)\right)^2 =$$

$$= \left(1 + \int_0^{2\Pi} \cos(kx)\, dG(x)\right)\left(1 - \int_0^{2\Pi} \cos(kx)\, dG(x)\right) =$$

$$= \int_0^{2\Pi} [1 + \cos(kx)]\, dG(x) \int_0^{2\Pi} [1 - \cos(kx)]\, dG(x).$$

Using the well-known trigonometric identities, we get

$$\left(\int_0^{2\Pi} \sin\frac{kx}{2} \cos\frac{kx}{2}\, dG(x)\right)^2 =$$

$$= \left(\int_0^{2\Pi} \sin^2\frac{kx}{2}\, dG(x)\right)^2 \left(\int_0^{2\Pi} \cos^2\frac{kx}{2}\, dG(x)\right)^2. \tag{1.6}$$

It is obvious that (1.5) and (1.6) are equivalent. By the Schwarz' inequality the equality (1.6) holds if and only if

$$\sin \frac{kx}{2} = \lambda \cos \frac{kx}{2} \tag{1.7}$$

with $\lambda \neq 0$ constant is satisfied in the values of the range of $\mathfrak{F}(x)$. But the equality (1.7) is satisfied only by a finite number of x in the interval $[0,2\pi]$, contradicting that is a distribution function of infinite type.

II. MOD 2π REDUCED SUMS OF RANDOM VARIABLES DEFINED ON A FINITE MARKOV CHAIN

1. In this chapter we are concerned with finite matrices the elements of these are functions defined on $[0,2\pi)$. Such a matrix will be called matrix valued function. Under the integral of a matrix valued function we mean the matrix formed from the integrals of the elements of the matrix valued function. Similarly under the limit of a matrix valued sequence we mean the matrix formed from the limits of the sequence of corresponding elements of the matrix valued sequence, and such a matrix valued sequence is convergent if all these sequences are convergent.

2. If we speak on random variables, we mean that they are defined on a same probability space $(\Omega, \mathcal{A}, P)$.

Suppose now that random variables

$$D_{hj}^{(n)} \qquad (n = 1,2, \ldots; \; h,j = 1,\ldots,p) \tag{2.1}$$

with integer $p \geq 1$ are independent. Let $\{Y_n\}$ be a Markov chain with states 1, ..., p , and with transition matrices

$$S_n = (p_{hj}^{(n)})_{h,j=1}^{p} \qquad (n = 1,2,\ldots).$$

Suppose that random variables (2.1) and $\{Y_n\}$ are independent. Let

$$X_n = D_{Y_{n-1}Y_n}^{(n)}, \qquad \text{i.e. } X_n = D_{hj}^{(n)}, \text{ if } Y_{n-1} = h, \; Y_n = j.$$

Moreover let

$$\mathfrak{F}^{(n)}_{hj}(x) = P(X_n<X, Y_n = j|Y_{n-1}=h), \; X\in R,$$

and suppose that

$$\mathfrak{F}^{(n)}_{hj}(-0) = 0, \quad \mathfrak{F}^{(n)}_{hj}\cdot(2\Pi+0) = p^{(n)}_{hj}.$$

We say the p x p matrix valued function

$$\mathfrak{F}_n(X) = \begin{cases} (0), & X\leq 0 \\ (\mathfrak{F}^{(n)}_{hj}(X))^p_{h,j=1}, & X\in[0,2\Pi) \\ S_n, & X\geq 2\Pi \end{cases} \tag{2.2}$$

to be a matrix valued distribution function, where (0) is the zero matrix of p x p type.

Denote by M(S) the set of the matrix valued distribution functions defined on the same transition matrix S.

Let $R_n = X_1 + \ldots + X_n$, and let

$$G_n(X) = (G^{(n)}_{hj}(X))^p_{h,j=1},$$

with elements

$$G^{(n)}_{hj}(X) = P(R_n<X, Y_n = j|Y_0 = h), \; X\in R.$$

Thus ([4], formula (4))

$$G_n(X) = \int_{-\infty}^{\infty} G_{n-1}(X-Y)d\mathfrak{F}_n(y), \; X\in R \tag{2.3}$$

holds.

Let now $Z_n \equiv R_n \pmod{2\Pi}$, and let

$$H_n(X) = (H^{(n)}_{hj}(X))^p_{h,j=1}, \quad X\in R$$

be the matrix valued distribution function of random variable Z_n with elements

$$H_{hj}^{(n)}(X) = P(Z_n < X, Y_n = j \mid Y_0 = h).$$

Theorem 2.1. If $H_0(X)=E$, where E is the unit matrix of type pxp, then

$$H_n(x) = \begin{cases} (0), & x \leq 0 \\ \int_0^x H_{n-1}(x-y)\,d\mathfrak{F}_n(y) + \\ + \int_x^{2\Pi} H_{n-1}(2\Pi+x-y)\,d\mathfrak{F}_n(y), & x \in [0,2\Pi] \\ S_1 \ldots S_n, & x \geq 2\Pi. \end{cases} \qquad (2.4)$$

Proof. Since the Theorem holds for n=1, suppose that $H_{n-1} \in M(s,\ldots s_{n-1})$. Let $H^*(x)$ be the matrix valued distribution function of $Z_{n-1}+X_n$. We obtain

$$H^*(x) = \int_{-\infty}^{\infty} H_{m-1}(x-y)\,d\mathfrak{F}_n(y) =$$

$$= \begin{cases} (0), & x \leq 0 \\ \int_0^{2\Pi} H_{m-1}(x-y)\,d\mathfrak{F}_n(y), & x \in [0,4\Pi], \\ S_1 \ldots S_n, & x \geq 4\Pi. \end{cases}$$

by (2.3). Thus

$$H_n(x) = H^*(x) + H^*(2\Pi+x), \qquad x \in [0,2\Pi],$$

which is the statement of the middle part of (2.4). The first part of one is trivial. Since $\mathfrak{F}_n \in M(S_n)$, $H_{n-1} \in M(S_1 \ldots S_{n-1})$, we get the last part of (2.4) after a short consideration.

In the following we shall suppose on several occasion that $\mathfrak{F}(x)$ and $H_n(x)$ are matrix valued periodic function with 2Π, and they have the values of the middle part of (2.2) and

(2.4), respectively, in the interval $[0,2\Pi)$. Denote by $N(S_n)$ and $N(S_1 \dots S_n)$ the set of matrix valued distribution function with these properties.

Supposing that $\mathfrak{F} \in N(S_n)$, $H_{n-1} \in N(S_1 \dots S_{n-1})$, formula (2.4) can be written in the more simple form

$$H_n(x) = \int_0^{2\Pi} H_{m-1}(x-y)\, d\mathfrak{F}_n(y), \quad x \varepsilon [0,2\Pi]$$

Theorem 2.2. (Cf. [8], Theorem 1.5.) If $\mathfrak{F}_n \in N(S_n)$, $H_n \in N(S_1 \dots S_n)$, and if

$$\gamma_n(t) = \int_0^{2\Pi} e^{itx}\, dH_n(x), \quad \varphi_n(t) = \int_0^{2\Pi} e^{itx}\, d\mathfrak{F}_n(x),$$

then

$$\gamma_n(k) = \gamma_{n-1}(k)\, \varphi_n(k) \quad (k=0,\pm 1,\pm 2,\dots). \tag{2.5}$$

Proof. It is known the identity ([6]; p. 44. The identity over the formula 3.3.2.)

$$\gamma_n(t) = \int_0^{2\Pi} \left[\int_{-x}^{2\Pi-x} e^{ity}\, dH_{n-1}(y) \right] e^{itx}\, d\mathfrak{F}_n(x). \tag{2.6}$$

Since e^{ity} is a periodic function with period 2Π if and only if t is an integer, we have

$$\int_{-x}^{2\Pi-x} e^{ity} dH_{n-1}(y) = \int_0^{2\Pi} e^{ity} dH_{n-1}(y),$$

if t is an integer. Taking this into consideration we get (2.5) by formula (2.6).

From Theorem 2.2 we get immediately the following result.

Corollary 2.1. If $\mathfrak{F}_j \in N(S_j)$ $(j=1,\dots, n)$, and if $H_n \in N(S_1 \dots S_n)$ is defined by the middle part of (2.4), then

$$\gamma_n(k) = \varphi_1(k) \ldots \varphi_n(k) \quad (k=0, \pm 1, \pm 2, \ldots),$$

where $\gamma_n(t)$ and $\varphi_j(t)$ is the Fourier transform of the matrix valued distribution functions H_n and $\mathfrak{F}_j$, respectively.

The last connection can be regarded as the measure of the dependence among the random variables $X_1, \ldots, X_n$, if $X_1 + \ldots + X_n$ reduced modulo 2π.

Let $\varphi(t)$ be the Fourier transform of $\mathfrak{F} \in M(S)$. It is ovbious that the elements of sequence $\{\varphi(k)\}$ are the matrix valued Fourier coefficients of satisfying conditions

$$\varphi(0) = S, \quad \varphi(-k) = \bar{\varphi}(k) \quad (k = 1, 2, \ldots) \tag{2.7}$$

where $\bar{\varphi}(k)$ is the matrix formed form the conjugate of the elements of $\varphi(k)$.

Obviously matrices (2.7) are matrix valued Fourier coefficients of $\mathfrak{F} \in M(S)$ if and only if the theorem of the trigonometric moment problem is satisfied for all sequences formed from the corresponding elements of (2 7).

The adequat definition of Definition 1.1 is the following.

Definition 2.1. Let $G \in M(S)$ with Fourier transform $\varphi(t)$, and $\mathfrak{F}_n \in M(S_n)$ with Fourier transform $\varphi_n(t)$ $(n = 1, 2, \ldots)$ be given. We say that $\{\mathfrak{F}_n\}$ converges quasi weakly to G, $\mathfrak{F}_n \Rrightarrow G$, if

$$\lim_{n\to\infty} \varphi_n(k) = \varphi(k) \quad (k=0, \pm 1, \pm 2, \ldots),$$

The theorem as follows is a consequence of Theorem 1.1.

Theorem 2.3. Let $\mathfrak{F}_n \in M(S_n)$ with Fourier transform $\varphi_n(t)$ $(n = 1, 2, \ldots)$ be given. If

$$\varphi_n(k) \to \varphi(k), \; n\to\infty \quad (k=0, \pm 1, \ldots),$$

then there exists such a $\mathfrak{F} \in M(B)$ with $\varphi(0) = S$ for which $\mathfrak{F}_n \Rrightarrow \mathfrak{F}$.

Similarly the adequat of Theorem 1.2. holds too, i.e. it is valid the following.

In his paper [2] the author investigated this question in the case of $p=1$, and if the random variables are distributed with distribution functions of finite type. (For more detail cf. [5]; R. 2.1. Combination powers on a compact group, p. 153.)

Theorem 2.6. Let $p \geq 1$ be an integer. Assume that random variables

$$D_{hj}^{(n)} \quad (n = 1,2, \ldots; \quad h,j = 1,\ldots, p) \tag{2.8}$$

are independent. Let the homogeneous Markov chain $\{Y_n\}$ with the states $1,\ldots,p$, and with transition matrix S be given. Suppose that random variables (2.8) and $\{Y_n\}$ are independent. Assume that random variables

$X_n = D_{Y_{n-1}Y_n}$ $(n = 1,2, \ldots)$ have common matrix valued distribution function with the property that at least one element of all rows is a distribution function of infinite type. Let $H_n(x)$ be the matrix valued distribution function of $Z_n \equiv X_1 + \ldots + X_n$ reduced modulo 2π. Then

$$H_n(x) \Rightarrow G(x) = \begin{cases} (0), & x \leq 0 \\ \dfrac{1}{2\pi} Px, & x \in [0,2\pi) \\ P, & x \geq 2\pi \end{cases}$$

holds if and only if the moduli of the roots of stochastic matrix S less than 1, except if the root equal to one, i.e. if and only if $\lim_{n\to\infty} S^n = =$ exists.

Proof.

Let $\varphi(t)$ be the Fourier transform of the matrix valued distribution function of $\mathfrak{F}$. By Theorem 1.3. we get that the sum of moduli of the rows' elements of $\varphi(k)$ $(k=\pm1,\pm2, \ldots)$ are less than 1, hence ([1]) the modulus of all roots of (k) are less then 1. Thereby we have

Theorem 2.4. Let $\mathfrak{F} \in M(S)$ be given. Then such a sequence $\mathfrak{F}_n \in M(S)$ $(n = 1, 2, \ldots)$ exist for which $\mathfrak{F}_n \Rightarrow \mathfrak{F}$.

Proof. Let $\varphi(t)$ be the Fourier transform of $\mathfrak{F}$, and let the sequence $\{\varepsilon_j\}$ with positive elements, and with $\varepsilon_n \to 0$, $n \to \infty$ be given. It is evident that $\varphi_j(t) \to \varphi(t)$, $j \to \infty$, if $\varphi_j(t)$ is defined by

$$\varphi_j(t) = \frac{1}{1+\varepsilon_j} \left[\varphi(t) + \varepsilon_j S \right] \quad (j = 1, 2, \ldots).$$

Using Theorem 1.2 it can show that the sequences of the corresponding elements of $\{\varphi_j(k)\}_{k=-\infty}^{\infty}$ satisfy the conditions of the theorem of trigonometric moment problem, i.e. $\varphi_i(t)$ is the Fourier transform of a certain $\mathfrak{F}_j \in M(S)$.

From Theorem 2.3. we obtain the following result.

Theorem 2.5. Let $\mathfrak{F}_n \in M(S_n)$ with Fourier transform $\varphi_n(t)$ $(n=1, 2, \ldots)$ be given. Let

$$\mathfrak{F}(x) = \begin{cases} (0), & x \leq 0 \\ \frac{1}{2\pi} Sx, & x \in [0, 2\pi) \\ S, & x \geq 2\pi. \end{cases}$$

The satisfaction of the relations

$$\lim_{n\to\infty} \varphi_n(k) = 0, \quad (k = \pm 1, \pm 2, \ldots). \quad \lim_{n\to\infty} \varphi_n(0) = S$$

is the necessary and sufficient condition for $\mathfrak{F}_n \Rightarrow \mathfrak{F}$.

3. In this section we deal with the limit distribution of the sums reduced module 2π of the identically distributed random variables defined on a homogeneous Markov chain, supposing that each row of the common matrix valued distribution function has at least one distribution function of infinite type.

$$\lim_{n\to\infty}[\varphi(k)]^n = (0) \qquad (k = \pm 1, \pm 2, \dots).$$

which gives us the statement of the Theorem in view of Theorem 2.5. On the other hand under the assumption $\lim_{n\to\infty} [\varphi(0)]^n = P$.

In the case of p=1 Theorem 2.6. says the following.

Corollary 2.2. Let the radom variables $\{X_n\}$ be identically distributed independent random variables with common distribution function of infinite type. Let $G_n(x)$ be the distribution function of $Z_n \equiv X_1 + \dots + X_n$ reduced modulo 2π. Then $G_m(x) \Rrightarrow G(x)$, where $G(x)$ is the distribution function of the uniformly distributed random variable on $(0,2\pi)$.

REFERENCES

1. Brauer, A. Limits for the characteristic roots of a matrix. Duke Math. Journal 13 (1946), 187-195.
2. Gyires, B. On the limit of distributions of sums of independent random variables. Transactions of the 1st Congress of Hungarian Mathematicians. Budapest, 1952. 741-748. (Hungarian)
3. Gyires, B. On the limit of matrices satisfyng the Cauchy's function equation. Acta Universitatis Debreceniensis de Lajos Kossuth nominatae. I. (1956), 136-144. (Hungarian)
4. Gyires, B. Eine Verallgemeinerung des zentralen Grenzwertsatzes. Acta Math. Acad. Sci. Hung. XIII (1962), 69-80.
5. Heyer, H. Probability measure on locally compact groups. Springer Verlag (Berlin, Heidelberg, New York), 1977.
6. Lukács, E. Characteristic functions. Charles Griffin & Co. Limited, London, 1960.
7. Riesz, F. and Sz. Nagy B. Vorlesungen über Funktionalanalysis. VEB Deutscher Verlag der Wissenschaft. Berlin, 1956.
8. Zygmund, A. Triagonometric series. Second edition. Vol. I. & II. combined. Cambridge Univ. Press, 1968.

Béla Gyires
Kossuth Lajos University
Department of Mathematics
4010 Debrecen Pf. 12
Hungary

Proc. of the 5th Pannonian Symp. on Math. Stat., Visegrád, Hungary 1985
W. Grossmann, J. Mogyoródi, I. Vincze, W. Wertz, eds

MIMICKING COMPLICATED STOCHASTIC DIFFERENTIAL EQUATIONS BY SIMPLER ONES

I. GYÖNGY

Budapest*

Let $\xi(t)$ be a stochastic process with Ito differential

$$d\xi(t) = \delta(t,\omega)\, dW(t) + \beta(t,\omega)\, dt$$

where $(W(t), \mathfrak{F}_t)$ is a Wiener process, δ and β are bounded $\mathfrak{F}_t$-adapted processes and $\delta\delta^T$ is uniformly positive definite. We investigate the possibility of replacing the process $\xi(t)$ by a diffusion process which mimics a particular aspect of its behaviour, for instance, its one-dimensional marginal distributions, its Green measures, or its hitting probabilities.
AMS subject classification: 60H10

1. INTRODUCTION

One of the basic requirement of a good statistical model is simplicity: the model should be as simple as possible and still capture the essential features of interest in the probabilistic situation being considered. The solutions of stochastic differential equations form an important class of models for random phenomena evolving in time. In their general form they can represent very complicated behaviour indeed. But suppose we are not interested in a full description of the system, but only in some single aspect of it. Then one might hope that in many cases it is possible to construct a simpler model which still describes this aspect of the behaviour of the original process. Our aim is to illustrate this possibility in the case when the original model is an Ito process.

*This paper was written while the author was visiting the J. W. Goethe-University in Frankfurt as a fellow of the Alexander von Humboldt Foundation.

2. NOTATIONS AND RESULTS

Let $\xi(t)$ be a d-dimensional Ito process starting from zero, i.e., let $\xi(t)$ be a stochastic process of the form

$$\xi(t) = \int_0^t \delta(s,\omega)dW(s) + \int_0^t \beta(s,\omega)ds \ , \tag{2.1}$$

where $(W(t), \mathcal{F}_t)$ is a k-dimensional Wiener process on a probability space $(\Omega, \mathcal{F}, P)$ with filtration $(\mathcal{F}_t)_{t\geq 0}$, and δ, β are measurable $\mathcal{F}_t$-adapted stochastic processes taking values in $\mathbb{R}^{d\times k}$ and $\mathbb{R}^d$ respectively. For simplicity we assume that the coefficients δ and β are bounded.
It is well known that $\xi(t)$ can be represented as a solution of a stochastic differential equation with non-random coefficients depending on the time and the paths of the process:

(M1) <u>Representation of Ito processes as processes of diffusion type</u>

<u>Theorem 1.</u> (Th. 7.17 from [6]). There exist measurable functionals $\bar{\delta} = \bar{\delta}(t,\xi(\cdot))$ $\bar{\beta} = \bar{\beta}(t,\xi(\cdot))$ and a Wiener process $\bar{W}$ such that the process ξ satisfies the equation

$$\xi(t) = \int_0^t \bar{\delta}(s,\xi)d\bar{W}(s) + \int_0^t \bar{\beta}(s,\xi)ds \quad . \tag{2.2}$$

Moreover the coefficients $\bar{\delta}$ and $\bar{\beta}$ have a simple probabilistic representation

$$\bar{\delta}(s,\xi) = (E(\delta\delta^T(s)|\xi_u \quad 0\leq u\leq s))^{1/2}$$

$$\bar{\beta}(s,\xi) = E(\beta(s)|\xi_u \quad 0\leq u\leq s)$$

$\lambda \times P$ almost all $(s,\omega) \in [0,\infty)\times\Omega$, where λ is the Lebesgue measure on $\mathbb{R}$.

This representation plays an important role in some statistical problems (such as filtering, interpolation and extrapolation) of random processes (see [6]). Equation (2.2) is simpler in a certain sense than the original equation (2.1), but the coefficients are still very complicated functionals of the paths. N. V. Krylov had the idea that the equation could be much more radically simplified if we wish to study not the whole process but merely some feature of it. For instance, suppose we are interested in

(M2) The expected time the Ito process spends in the vicinity of various points in the state space

Since this property intuitively depends only on the position of the process in the state space, one might hope to find a stochastic process $x(t)$ satisfying a "simple" stochastic differential equation whose coefficients depend only on the position, such that the processes $x(t)$ and $\xi(t)$ spend, on the average, the same amount of time in each Borel set of the state space. To formulate Krylov's result more precisely we introduce the notion of the Green measure, defined by

$$\mu(\Gamma) := E\int_0^\infty \chi_\Gamma(\xi(t))\exp(-\int_0^t \gamma(s,\omega)ds)dt,$$

which measures the expected time $\xi(t)$ spends in the Borel set Γ, where the non-negative process γ is a discounting factor and χ_Γ is the indicator function of the set Γ.

Theorem 2 (Krylov [2]). Suppose that $\delta\delta^T \geq \lambda I$ and $\gamma \geq \lambda$, where δ^T is the transpose of the matrix δ, I is the identity matrix and λ is a positive constant. Then there exist coefficients σ, b and c depending only the state variable, such that the stochastic differential equation

$$dx(t) = \sigma(x(t))dW(t) + b(x(t))dt$$

$$x(0) = 0$$

admits a solution $x(t)$ (on a suitable probability space $(\bar{\Omega}, \mathcal{F}, \bar{P})$ with suitable Wiener process $(\bar{w}(t), \mathcal{F}_t)$), which has the same Green measure with discounting factor $c(x(t))$ as $\xi(t)$ has with discounting factor $\gamma(t)$.

Moreover the coefficients have the representation

$$\sigma(x) = d\mu_{\delta\delta} / d_{\mu}(x)$$

$$b(x) = d\mu_{\beta}/d_{\mu}(x)$$

$$c(x) = d\mu_{\gamma}/d_{\mu}(x)$$

for almost all $x \in \mathbb{R}^d$ with respect to the Lebesgue measure, where for a stochastic process α the measure μ_α is defined by

$$\mu_\alpha(\Gamma) = E \int_0^\infty \chi_\Gamma(\xi(t))\alpha(t)\exp\left(-\int_0^t \gamma(s,\omega)ds\right)dt.$$

For the proof we refer to [2].

This result inspires us to consider the possibility of mimicking other features of ξ. Following a suggestion of Krylov from [2] we consider.

(M3) The one-dimensional marginal distributions of the Ito process, i.e., $P[\xi(t) \in dx]$

Since this property intuitively depends only on the time t and the position x, we might hope to find a process x(t) satisfying a stochastic differential equation whose coefficients depend only on t and x and mimicking this feature. It turns out that this is indeed possible:

Theorem 3. Suppose that $\delta\delta^T \geq \lambda I$ with positive constant λ. Then there exist non-random coefficients σ and b, defined on the time-state space, such that the stochastic equation

$$dx(t) = \sigma(t,x(t))dW(t) + b(t,x(t))dt$$

$$x(0) = 0$$

has a solution x(t) (on a suitable probability space $(\bar\Omega, \bar{\mathcal{F}}, \bar P)$ with suitable Wiener process), which has the same one-dimensional marginal distributions as $\xi(t)$, i.e., $P[x(t) \in \Gamma] = P[\xi(t) \in \Gamma]$ for every Borel set Γ, $t \geq 0$. Moreover the coefficients in this equation have a simple probabilistic interpretation

$$\sigma(t,x) = (E(\delta\delta^T(t)|\xi(t) = x))^{1/2}$$

$$b(t,x) = E(\beta(t)|\xi(t) = x)$$

Proof. We give a hint of the proof. The technical details can be found in [1]. First we derive the equation for the evolution of the one-dimensional marginal distributions of ξ. Since ξ is not a Markov process, one cannot use the Kolmogorov equations, but making use of the fact that the Itô formula is valid not only for diffusion processes but also for Ito processes, one can derive the corresponding equation for $P_t(dx):=P[\xi(t)\in dx]$ as follows:
(For the sake of notational simplicity we consider the special case in which $\beta:=0$ and all the processes are scalar-valued.)
Choose a smooth function $u \in C_0^\infty([0,\infty) \times \mathbb{R})$ and apply Itô's formula to $u(t,\xi(t))$, where

$$\xi(t) = \int_0^t \delta(s,\omega)dW(s)$$

$$du(t,\xi(t)) = \frac{\partial u}{\partial t}(t,\xi(t))dt + \frac{\partial u}{\partial x}(t,\xi(t)) + \\ + \frac{1}{2}\frac{\partial^2}{\partial x^2}u(t,\xi(t))\delta^2(t)dt \quad .$$

Taking expectations we get

$$Eu(T,\xi(T))-u(0,0) = E\int_0^T \frac{\partial}{\partial t}u(t,\xi(t)) + \\ + \frac{1}{2}\delta^2(t)\frac{\partial^2}{\partial x^2}u(t,\xi(t))dt = \\ = \int_0^T E\left\{\frac{\partial}{\partial t} + \frac{1}{2}E(\delta^2(t)|\xi(t))\frac{\partial^2}{\partial x^2}\right\}u(t,\xi(t))dt$$

for every $T\geq 0$. This gives us the equation

$$u(0,0)= -\int_0^\infty\int_{-\infty}^\infty(\frac{\partial}{\partial t} + \frac{1}{2}\sigma^2(t,x)\frac{\partial^2}{\partial x^2})u(t,x)P_t(dx)dt \qquad (2.3)$$

for $P_t(dx)=P[\xi(t)\in dx]$, where $\sigma(t,x) := (E(\delta^2(t)|\xi(t) = x))^{1/2}$ and the test functions u are from $C_0^\infty([0,\infty)\times\mathbb{R})$.

Now consider any solution $x(t)$ of the stochastic differential equation

$$dx(t) = \sigma(t,x(t))dW(t)$$
$$x(0) = 0 \tag{2.4}$$

on any probability space $(\bar{\Omega}, \bar{\mathcal{F}}, \bar{P})$. It is clear that the one-dimensional distributions $\bar{P}_t(dx):=\bar{P}[x(t) \in dx]$ satisfy the same equation (2.3). If the coefficient σ happens to be smooth, then the equation (2.4) has a unique solution $x(t)$ and the equation (2.3) uniquely determines the distributions $P_t(dx)$. Consequently $P[\xi(t) \in dx] = \bar{P}[x(t) \in dx]$ for every $t\geq 0$, and our theorem is proved in this case. But unfortunately the uniqueness theorem we need is not known in the general case of measurable coefficient σ. We circumvent this difficulty by using a regularization method from [2] for the equations (2.3) and (2.4):

We choose a sequence of smooth coefficients $\{\sigma_n(t,x)\}_{n=1}^{\infty}$ and a sequence of Itô processes $\{\xi_n(t)\}_{n=1}^{\infty}$ such that

(i) $\xi_n \to \xi$ (as $n\to\infty$) weakly in $C[0,T]$ for every $T\geq 0$,

(ii) $\sigma_n(t,x)\to\sigma(t,x)$ for almost all (t,x) w.r. to the Lebesgue measure,

(iii) $P[\xi_n(t) \in dx] = P[x_n(t) \in dx]$ for every $t\geq 0$, where $x_n(t)$ is the unique solution of the stochastic differential equation

$$dx_n(t) = \sigma_n(t,x_n(t))dW(t)$$

with appropriate initial condition $x_n(0) = \eta_n$ being independent of the Wiener process W and converging to 0 as $n\to\infty$.
Using the method of the proof of Theorem II.6.1 from [3] one can show that (ii) and $\eta_n\to 0$ imply that for a subsequence n_k the processes X_{n_k} converge in distribution to a solution X of the equation (2.4). Hence our theorem follows immediately from (i) and (iii).

The next aspect of the behaviour of the Ito process we are interested in is

(M4) <u>The hitting probability of the Ito process on the boundary of a given domain</u>

Let D be a bounded domain in $\mathbb{R}^d$ containing the origin and having smooth boundary ∂D. Let us consider the hitting probability of our Ito process ξ on the boundary, i.e., consider the probability distribution of $\xi(\tau)$, where τ is the first hitting time of ξ on ∂D. We would like to mimic this distribution by a process $x(T)$ satisfying a stochastic differential equation with non-random coefficients. Our next theorem says that this is indeed possible.

<u>Theorem 4</u>. Suppose that $\delta\delta^T \geq \lambda I$ with a positive constant λ. Then there exist coefficients $\sigma=\sigma(x)$ and $b=b(x)$ such that the stochastic differential equation

$$dx(t) = \sigma(x(t))d\,W(t) + b(x(t))dt$$

$$x(0) = 0$$

has a solution $x(t)$ on a suitable probability space $(\bar{\Omega}, \bar{\mathcal{F}}, \bar{P})$, with the following properties:

$$\text{(i)}\quad E\int_0^{\bar{\tau}} \chi_\Gamma(x(t))dt = E\int_0^{\tau} \chi_\Gamma(\xi(t))dt$$

for every Borel set $\Gamma \subset D$

$$\text{(ii)}\quad \bar{P}[x(\bar{\tau}) \in S] = P[\xi(\tau)\in S]$$

for every Borel set $S \subset \partial D$, where $\bar{\tau}$ is the first hitting time of the process $x(t)$ on the boundary ∂D.

<u>Proof</u>. We give only a sketch of the proof here:

For every integer n consider the Green measure μ_n of ξ with discounting factor $C_n(\xi(t)) = n\,\mathcal{K}(\xi(t))$, where $\mathcal{K}$ is a function vanishing on $\bar{D}$ and positive on $\mathbb{R}^d\setminus\bar{D}$. Then by Theorem

2 one can show the existence of a process x_n satisfying a stochastic differential equation

$$dx_n(t) = \sigma_n(x(t))d\,W(t) + b_n(x(t))dt$$
$$x_n(0) = 0$$

with non-random coefficients σ_n and b_n such that for every n

$$E\int_0^\infty (f+n\chi\varphi)(\xi(t))\exp(-\int_0^t c_n(\xi(s))ds)dt = \tag{2.5}$$
$$= E\int_0^\infty (f+n\chi\varphi)(x_n(t))\exp(-\int_0^t c_n(x_n(s))ds)dt$$

for every bounded continuous functions f and φ. One can also prove that the sequences $\sigma_n(x)$ and $b_n(x)$ converge almost everywhere with respect to the Lebesgue measure, say to σ and b respectively. Consequently, for a subsequence $n_k\to\infty$, the processes x_{n_k} converge to a solution, say x, of the equation

$$dx(t) = \sigma(x(t))d\,W(t) + b(x(t))dt$$
$$x(0) = 0.$$

Taking the limit $n_k\to\infty$ in both sides of the equation (2.5) we get

$$E\int_0^\tau f(\xi(t))dt + \varphi(\xi(\tau))) = E(\int_0^{\bar\tau} f(x(t))dt + \varphi(x(\bar\tau))).$$

Hence the assertion of Theorem 4 is obvious.

3. APPLICATIONS

Results of the type we have been discussing make it possible to reduce investigations of the Ito processes to those of diffusion processes in many cases. To illustrate this we consider the following example (see also [2]):

A very deep result of Krylov states that for arbitrary d-dimensional Ito process

$$\xi(t) = \int_0^t \delta(s,\omega)dW(s) + \int_0^t \beta(s,\omega)ds$$

with coefficients satisfying the conditions $\delta\delta^T \geq \lambda I$ and $|\delta| + |\beta| \leq K$ with positive constant λ and K, the following estimate holds

$$E\int_0^\infty e^{-\gamma t}|f(t,x+\xi(t))|dt \leq N||f||_{L_p(\mathbb{R}^{d+1})} \tag{3.1}$$

for every Borel function $f:\mathbb{R}^{d+1} \to \mathbb{R}$, vector $x \in \mathbb{R}^d$ and positive constant γ, where $p \geq d+1$ and N is a constant depending only on the bounds K, λ, on the discounting factor γ and p, d. (See Theorem II.4.4 in [3].)

This estimate plays an essential role in the stochastic control theory (see [3]) and in other fields. Its original proof is extremely difficult (see [4], [5]). Now using Theorem 3 the proof can be reduced to the case when $\xi(t)$ is a diffusion process. More precisely, instead of the given Ito process ξ we take the diffusion process $x_n(t)$ from the proof of Theorem 3. For this process x_n the estimate

$$E\int_0^\infty e^{-\gamma t}|f(t,x+x_n(t))|dt \leq N||f||_{L_p(\mathbb{R}^{d+1})} \tag{3.2}$$

is known from the theory of partial differential equations. The diffusion and drift coefficients of x_n have the same bounds as δ and β. Consequently, taking the limit $n\to\infty$ in (3.2) we get (3.1).

ACKNOWLEDGEMENTS

The author wishes to express his thanks to N.V. Krylov in Moscow for helpful advice and to B. Ferebee in Frankfurt for his help in preparing this paper.

REFERENCES

1. I. Gyöngy, Mimicking the one-dimensional marginal distributions of processes having Ito differential, to appear.
2. N. V. Krylov, Once more about the connection between elliptic operators and Ito's stochastic equations, Steklov Seminar 1984, edited by N. V. Krylov et al., Optimization Software, Inc. Publications Division, New York, 214-229, 1985.
3. N. V. Krylov, Controlled diffusion processes, Springer-Verlag, 1980.
4. N. V. Krylov, An inequality in the theory of stochastic integrals, Theor. Prob. Appl. 16, 438-448, 1971.
5. N. V. Krylov, Some estimates of the probability density of a stochastic integral, Math. USSR Izvestija 8, 233-254, 1974.
6. R. S. Liptser, A. N. Shiryayev, Statistics of random processes 1, Springer-Verlag, 1977.

István Gyöngy
J. W. Goethe-University
Dept. of Mathematics
6000 Frankfurt am Main
Robert-Mayer-Strasse 6-10.
F.R.G.

Proc. of the 5th Pannonian Symp. on Math. Stat., Visegrád, Hungary 1985
W. Grossmann, J. Mogyoródi, I. Vincze, W. Wertz, eds

LARGE DEVIATIONS THEOREM FOR GIBBS RANDOM FIELDS

M. JANŽURA

Prague

Large deviations theorem, the usual version of which describes asymptotic behaviour of the probability that the empirical distribution of some i.i.d. sample belongs to a given convex set, is formulated in case when some special (but general enough) type of dependence occurs. Thus, the underlying probability measure is given by a Gibbs random field (as defined in frame of statistical mechanics), and the empirical distributions are substituted by a suitably defined vector statistics which determine the random field in a sufficient way.
AMS subject classification: 60G60, 60F10

1. INTRODUCTION

The asymptotic behaviour of

$$n^{-1} \log \Pr(n^{-1} S_n \in J),$$

where $S_n = Y_1 + \dots + Y_n$ is the n'th partial sum of i.i.d. random variables, and J is an infinite interval, say $J = [\lambda, \infty)$, was first considered by Cramér (1938) who proved existence of the limit under some additional assumptions on the common distribution of the Y_i. Later, this result was generalized by Chernoff (1952) in sense of omitting these assumptions. The initial work of Cramér and Chernoff has been continued and extended in several ways by various authors. Latest very general results were obtained by Bahadur and Zabell (1979).

Sanov (1957) studied the asymptotic behaviour of

$$n^{-1} \log \Pr(\hat{F}_n \in J),$$

where $\hat{F}_n$ is the empirical distribution obtained from the i.i.d.

sample $Y_1, \ldots, Y_n$, and J is a set of probability distributions. This approach has been also followed by many authors, lately by Csiszár (1984) who emphasized the connection between the large deviations problem and the basic information theory notions.

In the present paper the "statistical mechanics approach" is introduced to study the dependent sample case. We suppose the random variables under consideration to constitute so called Gibbs random field. This assumption involves certain kind of homogeneity and rather weak dependence. A short exposition of Gibbs random fields theory (basic definitions and results only) is contained in Section 2, while Section 3 is devoted to the large deviations theorem. In spite of using most of the original terminology and notation developed in frame of statistical mechanics, we do not, naturally, insist on the physical meaning of notions. And, eventually, it is worth emphasising that the notion of Gibbs random field is wide enough to involve e. g. both the independent and the Markovian (of any order) cases (cf. Averintsev (1972)).

2. GIBBS RANDOM FIELDS

Considering a finite state space X with the σ-algebra of all its subsets $\mathfrak{F} = \exp X$, by *random field* (r. f.) we mean a probability measure on the product measurable space $(X^T, \mathfrak{F}^T)$ where the parameter set T is given by the d-dimensional ($d \geq 1$) lattice Z^d. Providing the state space X is fixed, we denote by M the set of all r.f.'s.

Further we denote by M_{St} the subset of *stationary* r.f.'s, i. e. those which are invariant with respect to all shifts:

$$\mu \in M_{St} \quad \text{iff} \quad \mu\Theta_t^{-1} = \mu \quad \text{for every } t \in T;$$

the corresponding *shift* Θ_t is the transformation on X^T defined through: $[\Theta_t(x)]_{t'} = x_{t+t'}$ for every $x \in X^T$, $t' \in T$. (For any $V \subset T$ by X_V we mean the projection of $x \in X^T$ into the space X^V and by the same symbol we denote also the corresponding measurable cylinder.)

A stationary r.f. μ is called *ergodic* (we write $\mu \in M_E$) if its restriction to the σ-algebra $\mathcal{S}$ of invariant sets assumes only values zero or one, i.e.
$\mu \in M_E$ iff for every $F \in \mathcal{S}$ it holds: if $\mu(F) > 0$ then $\mu(F) = 1$, where $\mathcal{S} = \{F \in \mathcal{F}^T;\ \Theta_t^{-1} F = F \text{ for every } t \in T\}$.

Notation $k(T) = \{A \subset T;\ 0 < \text{card } A < \infty\}$ being used, *potential* U is a collection of maps $\{U_A : X^A \to R\}_{A \in k(T)}$.

In the sequel we shall consider only such potentials which are *stationary*: $U_A(x_A) = U_{A-t}((\Theta_t(x))_{A-t})$ for every $A \in k(T)$, $t \in T$, $x \in X^T$; and *bounded*: $||U|| = \sum_{0 \in A \in k(T)} ||U_A|| < \infty$ where $||U_A|| = \sup_{x_A \in X^A} |U_A(x_A)|$.

For fixed bounded stationary potential U a *specification* π^U is again a collection of maps

$$\pi^U = \{\pi_A^U : X^A \times X^{T\setminus A} \to R\}_{A \in k(T)}$$

every π_A^U being defined through

$$\pi_A^U(x_A | x_{T\setminus A}) = [Z_A^U(x_{T\setminus A})]^{-1} \cdot \exp\{-E_A^U(x_A | x_{T\setminus A})\},$$

where

$$E_A^U(x_A | x_{T\setminus A}) = \Sigma_{V \in k(T), V \cap A \neq \emptyset}\ U_V(x_V),\quad Z_A^U(x_{T\setminus A}) =$$

$$= \Sigma_{y_A \in X^A} \exp\{-E_A^U(y_A | x_{T\setminus A}\} .$$

The procedure of constructing specifications generates an *equivalence relation* on the Banach space of bounded stationary potentials $(\mathcal{U}, ||\cdot||)$. We shall write $U^1 \approx U^2$ (saying the potentials are equivalent) for $U^1, U^2 \in \mathcal{U}$ if the corresponding specifications π^{U^1}, π^{U^2} are equal. For $U^0 \in \mathcal{U}$ we denote by $[U^0]$ the class of potentials equivalent to U^0. As $[0]$ ($0_A \equiv 0$ for every $A \in k(T)$) is obviously a linear subspace of U, and $U^1 \approx U^2$ iff $U^1 - U^2 \in [0]$, the set of equivalence classes is a quotient space $\mathcal{U}/[0]$.

A r.f. μ is called *Gibbs* r.f. with respect to a class of potentials $[U]$, $U \in \mathcal{U}$ (we shall write $\mu \in G([U])$) if the family of its finite-dimensional conditional distributions is given by the specification π^U, i.e.

$$\mu \in G([U]) \text{ iff } \mu(x_A | \mathcal{F}^{T\setminus A})(y) = \pi_A^U(x_A | y_{T\setminus A})$$

for every $A \in k(T)$, $x_A \in X^A$, and a. e. $y \in X^T[\mu]$.

REMARK. According to Theorem 4.1 (Preston (1976)) there always exists at least one ergodic Gibbs r. f. For fixed $U \in \mathcal{U}$ all r. f.'s from $G([U])$ are, in a sense suitable for out purposes, equivalent. Thus, the non-uniqueness of definition of Gibbs r.f. will cause no substantial problem.

For $a \in T$ we denote $V(a) = \{t \in T;\ 0 \le t_i \le a_i,\ i=1,\dots,d\}$ and $|V(a)| = \text{card } V(a)$.

Writing $a \to \infty$, we mean $\min_{i=1,\dots,d} a_i \to \infty$.

PROPOSITION 1. For every $U \in \mathcal{U}$

$$p(U) = \lim_{a\to\infty} |V(a)|^{-1} \log Z_{V(a)}^U(y_{T\setminus V(a)}) \text{ exists,}$$

(where $y_{T\setminus V(a)} \in X^{T\setminus V(a)}$ may be taken arbitrarily), and $p: \mathcal{U} \to R$ is continuous convex functional.

Proof: Cf. 4.24 (Föllmer (1973)) and Lemma 8.6, Lemma 8.7 (Preston (1976)). □

For $\mu, \nu \in M$ we denote by $H(\mu|\nu) = \lim_{a\to\infty} |V(a)|^{-1} \cdot$

$$\cdot \int \log \frac{\mu(x_{V(a)})}{\nu(x_{V(a)})}\, d\mu(x) \text{ and } H(\mu) = \lim_{a\to\infty} |V(a)|^{-1} \int -\log \mu(x_{V(a)})\, d\mu(x)$$

(providing the expressions make sense and the limits exist) the corresponding *I-divergence* (information gain) and *entropy rate*, respectively.

For $U \in \mathcal{U}$ we define *function* $g_U: X^T \to R$ through the following formula:

$$g_U(x) = \Sigma_{0 \in A \in k(T)} |A|^{-1} U_A(x_A) \text{ for every } x \in X^T.$$

PROPOSITION 2. For $\mu \in M_{St}$, $\nu \in G([U])$, $U \in \mathcal{U}$, it holds

i) $H(\mu)$, $H(\mu|\nu)$ exist;

ii) $H(\mu|\nu) = p(U) + \int g_U \, d\mu - H(\mu) \geq 0$;

iii) $\mu \in G([U])$ iff $H(\mu|\nu) = 0$.

Proof: Cf. Formulas 2.4, 4.25, 4.26, 4.27. (Föllmer (1973)). □

Now, we shall restrict our considerations to some finite-dimensional subspace L of the Banach space $(\mathcal{U}, \|\cdot\|)$. Let $L = \text{Lin}\{U^1, \dots, U^N\}$ be spanned by some collection $U^1, \dots, U^N \in \mathcal{U}$ of mutually non-equivalent potentials, i.e. if $\Sigma_{i=1}^N c_i U^i \in [0]$ then $c_1 = \dots = c_N = 0$.

(Hence, obviously, if $\bar{U}, \bar{\bar{U}} \in L$, $\bar{U} \approx \bar{\bar{U}}$ then $\bar{U} = \bar{\bar{U}}$.)
Functional p restricted to L is *strictly convex* (Proposition 8.5 in Preston (1976)).

PROPOSITION 3. Let $\bar{U}, \bar{\bar{U}} \in L$, $\bar{\mu} \in G([\bar{U}]) \cap M_{St}$, $\bar{\bar{\mu}} \in G([\bar{\bar{U}}]) \cap M_{St}$. If $\int g_{U^i} \, d\bar{\mu} = \int g_{U^i} \, d\bar{\bar{\mu}}$ for every $i=1,\dots,N$ then $\bar{U} = \bar{\bar{U}}$.

Proof: Let us denote $\bar{U} = \Sigma_{i=1}^N \bar{\alpha}_i U^i$, $\bar{\bar{U}} = \Sigma_{i=1}^N \bar{\bar{\alpha}}_i U^i$.
Then $0 \leq H(\bar{\mu}|\bar{\bar{\mu}}) = p(\bar{\bar{U}}) - p(\bar{U}) + \Sigma_{i=1}^N (\bar{\bar{\alpha}}_i - \bar{\alpha}_i) \int g_{U^i} \, d\bar{\mu} =$

$$= p(\bar{\bar{U}}) - p(\bar{U}) + \Sigma_{i=1}^N (\bar{\bar{\alpha}}_i - \bar{\alpha}_i) \int g_{U^i} \, d\bar{\bar{\mu}} = -H(\bar{\bar{\mu}}|\bar{\mu}) \leq 0 \text{ holds.}$$

Therefore $\bar{\mu} \in G([\bar{\bar{U}}])$ which, according to assertion (iii) of PROPOSITION 2, implies $\bar{U} \approx \bar{\bar{U}}$, hence $\bar{U} = \bar{\bar{U}}$. □

3. LARGE DEVIATIONS THEOREM

It is obvious that the "Sanov-like" large deviations theorem describing asymptotic behaviour of the probability that the empirical distribution of some i.i.d. sample belongs to a given convex set, should be re-formulated in our case when the independence is not supposed. Thus, instead of the empirical distribution we should consider "empirical r.f.". This "empirical r.f." should be of the same type as the underlying r.f. (which "measures the probability") but it would be uneasy to construct directly an appropriate r.f. from every sample. Therefore, we shall follow a slightly different approach.

Let us suppose the underlying r.f. μ to be Gibbs r.f. with respect to a potential U^o from some fixed finite-dimensional subspace $L \subset\subset \mathcal{U}$ spanned by a N-tuple $(U^1, \ldots, U^N)$ of mutually non-equivalent potentials. According to PROPOSITION 3 of the previous section the N-tuple of expectations $(\int g_{U^1} d\mu, \ldots, \int g_{U^N} d\mu)$ is sufficient to determine the potential $U^o \in L$. Therefore, it seems natural to consider some empirical version $\hat{g}_{U^1}^{V(a)}, \ldots, \ldots, \hat{g}_{U^N}^{V(a)}$ of these expectations, and to study their asymptotic behaviour.

If we define $\hat{g}_{U^i}^{V(a)} = |V(a)|^{-1} \Sigma_{t \in V(a)} g_{U^i} \cdot \Theta_t$ for every $i=1,\ldots,N$, then assuming now $\mu \in M_E \cap G([U^o])$, according to the d-dimensional version of ergodic theorem (cf. Dunford-Schwartz (1958), Theorem VIII, 6.9) it holds

$$\hat{g}_{U^i}^{V(a)} \to \int g_{U^i} d\mu \quad \text{a. s.}[\mu] \text{ for every } i=1,\ldots,N.$$

As, according to 4.16, 4.17, 4.19 in Föllmer (1973), for every $U \in \mathcal{U}$ it holds

$$|\hat{g}_U^{V(a)}(x_{V(a)}, x_{T\setminus V(a)}) - \hat{g}_U^{V(a)}(x_{V(a)}, y_{T\setminus V(a)})| \le$$

$$\le |V(a)|^{-1} K_{V(a)}(U)$$

for every $x,y \in X^T$, where $|V(a)|^{-1} K_{V(a)}(U) \xrightarrow[a\to\infty]{} 0$,

we may substitute some fixed $y_{T\setminus V(a)} \in X^{T\setminus V(a)}$ "outside" of the sample region $V(a)$. Hence, we may consider the empirical expectations to be defined on the proper sample space $X^{V(a)}$. Anyhow, we denote $\hat{g}^{V(a)} = (\hat{g}_{U^1}^{V(a)}, \ldots, \hat{g}_{U^N}^{V(a)})$.

For fixed $U^o \in L$ and $\lambda \in R^N$ we define function $F_\lambda : R^N \to R$ through the formula

$$F_\lambda(\alpha) = p(U^o - \Sigma_{i=1}^N \alpha_i U^i) - p(U^o) - \Sigma_{i=1}^N \lambda_i\alpha_i \text{ for every } \alpha\in R^N$$

Further, for every $\lambda \in R^N$ we denote

$$B_\lambda = \{\alpha \in R^N;\ \alpha_i \geq \lambda_i \text{ for every } i=1,\dots,N\},$$

and

$$D_\lambda = \inf\{H(\mu|\mu_{U^o});\mu \in M_{St}, (\int g_{U^1}d\mu,\dots, \int g_{U^N}d\mu) \in B_\lambda\}$$

THEOREM

Let $\mu_{U^o} \in G([U^o])$, then

i) $\limsup_{a\to\infty} |V(a)|^{-1} \log\mu_{U^o}(\hat{g}^{V(a)} \in B_\lambda) \leq \inf_{\alpha \in B_0} F_\lambda(\alpha)$

holds for every $\lambda\in R^N$. If the infimum is reached then $\min_{\alpha \in B_0} F_\lambda(\alpha) = D_\lambda$.

ii) If the function F_λ assumes its minimum in $\bar{\alpha} \in \text{Int } B_0$ (i.e. $\bar{\alpha}$ is an inner point of B_0) then

$$D_\lambda=-\inf\{H(\mu|\mu_{U^o})+c.\max(0,\lambda_1-\int g_{U^1}d\mu,\dots,\lambda_N-\int g_{U^N}d\mu);\mu\in M_E\}\in(-\infty,0)$$

for every $c \geq \Sigma_{i=1}^N \bar{\alpha}_i$,

and

$$\lim_{a\to\infty} |V(a)|^{-1} \log\mu_{U^o}(\hat{g}^{V(a)} \in B_\lambda) = D_\lambda .$$

Proof: For every $\alpha \in B_0$ it holds (by Chebyshev's inequality)

$$\mu_{U^o}(\hat{g}^{V(a)} \in B_\lambda) \leq \exp\{-|V(a)|\ \Sigma_{i=1}^N \lambda_i\alpha_i\} \cdot$$

$$\cdot \int\exp\{\Sigma_{t \in V(a)}(\Sigma_{i=1}^N \alpha_i\, g_{U^i}) \circ \Theta_t\}d\mu_{U^o}$$

hence, according to Proposition 4.4 in Ruelle (1978), the inequality in the first statement is proved.

Furthermore, for every $\mu \in M_{St}$ with $(\int g_{U^1}d\mu,\dots,\int g_{U^N}d\mu) \in B_\lambda$ it holds

$$-H(\mu|\mu_{U^o}) - \inf_{\alpha\in B_0} F_\lambda(\alpha) =$$

$$= \sup_{\alpha\in B_0} [-p(U^o)-\int g_{U^o}d\mu+H(\mu)+p(U^o)-p(U^o-\Sigma_{i=1}^N\alpha_i U^i)+\Sigma_{i=1}^N\alpha_i\lambda_i] =$$

$$= \sup_{\alpha\in B_0} [-H(\mu|\mu_\alpha) + \Sigma_{i=1}^N \alpha_i\ (\lambda_i - \int g_{U^i}\ d\mu)] \le 0$$

due to the assumptions and the non-negativity of the I-divergence. If the infimum of F_λ over the set B_0 is reached in $\bar{\alpha} \in B_0$, there exists $\mu_{\bar{\alpha}} \in G([U^o - \Sigma_{i=1}^N \bar{\alpha}_i U^i]) \cap M_{St}$ satisfying

$$\int g_{U^i} d\mu_{\bar{\alpha}} = \lambda_i \text{ for every } i \in \{1,\dots,N\} \text{ such that } \bar{\alpha}_i>0.$$

(Cf. Proposition 8.4 in Preston (1976) and Formula 5.5 in Ekeland-Temam (1976)).

Therefore $D_\lambda = -H(\mu_{\bar{\alpha}}|\mu_{U^o})$ and the first statement is proved.

Let $\bar{\alpha} \in \text{Int } B_0$. Then, similarly, for fixed $c \ge \Sigma_{i=1}^N\bar{\alpha}_i$ and $\mu \in M_{St}$ (denoting $M_\mu = \max(0,\lambda_1-\int g_{U^1}d\mu,\dots,\lambda_N-\int g_{U^N}d\mu)$) we obtain

$$-H(\mu|\mu_{U^o})-cM_\mu-D_\lambda = -H(\mu|\mu_{\bar{\alpha}})+\Sigma_{i=1}^N\bar{\alpha}_i(\lambda_i- g_{U^i}d_\mu-M_\mu)-(c-\Sigma_{i=1}^N\bar{\alpha}_i)M_\mu\le 0$$

because $M_\mu \ge 0$ by definition.

And again $-H(\mu_{\bar{\alpha}}|\mu_{U^o}) = D_\lambda$ for some $\mu_{\bar{\alpha}} \in G([U^o-\Sigma_{i=1}^N\bar{\alpha}_i U^i])\cap M_{St}$.

If $\mu_{\bar{\alpha}} \in M_E$ the equality is proved. If not we may follow the construction of Perez (1974) to obtain a sequence $\{\mu_n\}_{n=1}^\infty$ of r.f.'s a satisfying: $\mu_n \in M_E$ for every n and $H(\mu_n|\mu_U)\to H(\mu_{\bar{\alpha}}|\mu_U)$ for every $U \in \mathcal{U}$.

As $\bar{\alpha} \neq 0$ and $F_\lambda(0) = 0$, it follows from the strict convexity of p that $D_\lambda < 0$.

For the sake of simplicity the rest of the proof will be carried out for the case N = 1.
Hence, assuming N = 1,

$$C_{V(a)} = \{x_{V(a)};\ \hat{g}^{V(a)}(x_{V(a)}) \in B_\lambda\} = \{x_{V(a)};\hat{g}^{V(a)}_{U^1}(x_{V(a)}) \geq \lambda_1\} =$$

$$= \{x_{V(a)};\ \exp\{c \cdot \sum_{t \in V(a)} g_{U^1} \bullet \Theta_t(x_{V(a)},\ y_{T \setminus V(a)})\} \geq \exp\{|V(a)| \cdot c \cdot \lambda_1\}\} =$$

$$= \{x_{V(a)}; \mu_{U^o}(x_{V(a)}) \leq \mu_{U^o}(x_{V(a)}) \cdot \exp\{c \cdot \sum_{t \in V(a)} g_{U^1} \bullet \Theta_t(x_{V(a)}, y_{T \setminus V(a)})\} \cdot$$

$$\cdot \exp\{-|V(a)| \cdot c \cdot \lambda_1\}\}$$

holds for every $c>0$. Let us take $c>\bar{\alpha}$.
By Formula 4.22 in Föllmer (1973) we have

$$\mu_{U^o}(x_{V(a)}) \leq \pi^{U^o}_{V(a)}(x_{V(a)} | y_{T \setminus V(a)}) \cdot \exp\{2K_{V(a)}(U^o)\} .$$

By Formula 4.16 in Föllmer (1973) we have

$$\sum_{t \in V(a)} g_{U^1} \bullet \Theta_t(x_{V(a)}, y_{T \setminus V(a)}) \leq E^{U^1}_{V(a)}(x_{V(a)} | y_{T \setminus V(a)}) + 2K_{V(a)}(U^1).$$

Putting the terms together, we obtain

$$\mu_{U^o}(x_{V(a)}) \cdot \exp\{c \cdot \sum_{t \in V(a)} g_{U^1} \bullet \Theta_t(x_{V(a)},\ y_{T \setminus V(a)})\} \leq$$

$$\leq \exp\{2(K_{V(a)}(U^o) + cK_{V(a)}(U^1))\} \cdot [Z^{U^o}_{V(a)}(y_{T \setminus V(a)})]^{-1} \cdot$$

$$\cdot \exp\{-E^{U^o}_{V(a)}(x_{V(a)} | y_{T \setminus V(a)}) + cE^{U^1}_{V(a)}(x_{V(a)} | y_{T \setminus V(a)})\} =$$

$$= \exp\{2(K_{V(a)}(U^o) + cK_{V(a)}(U^1))\} \cdot \pi^{U^o - cU^1}_{V(a)}(x_{V(a)} | y_{T \setminus V(a)}) \cdot \frac{Z^{U^o - cU^1}_{V(a)}(y_{T \setminus V(a)})}{Z^{U^o}_{V(a)}(y_{T \setminus V(a)})}$$

and, again by Formula 4.22 in Föllmer (1973), it holds

$$\pi^{U^o - cU^1}_{V(a)}(x_{V(a)} | y_{T \setminus V(a)}) \leq \exp\{2K_{V(a)}(U^o - cU^1)\} \cdot \mu_{U^o - cU^1}(x_{V(a)})$$

for every $\mu_{U^o - cU^1} \in G([U^o - cU^1])$.

Therefore $C_{V(a)} \subset \{x_{V(a)};\ \mu_{U^o}(x_{V(a)}) \leq \mu_{U^o-cU^1}(x_{V(a)}) \cdot$

$$\cdot \exp\{-|V(a)| \cdot \gamma^+_{V(a)}\}\} = C^+_{V(a)},$$

where

$$\gamma^+_{V(a)} = c\lambda_1 + |V(a)|^{-1}[\log Z^{U^o}_{V(a)}(y_{T\setminus V(a)}) - \log Z^{U^o-cU^1}_{V(a)}(y_{T\setminus V(a)}) -$$

$$- |V(a)|^{-1}(2K_{V(a)}(U^o) + 2cK_{V(a)}(U^1) + 2K_{V(a)}(U^o-cU^1)),$$

and $\gamma^+_{V(a)} \to c\lambda_1 + p(U^o) - p(U^o-cU^1)$ for $a\to\infty$ due to Formulas 4.17 and 4.24 in Föllmer (1973).
Similarly, using the same formulas, we obtain

$$C_{V(a)} \supset \{x_{V(a)};\mu_{U^o}(x_{V(a)}) \leq \mu_{U^o-cU^1}(x_{V(a)}) \cdot \exp\{-|V(a)| \cdot \gamma^-_{V(a)}\}\} = C^-_{V(a)}$$

where

$$\gamma^-_{V(a)} = \gamma^+_{V(a)} + |V(a)|^{-1}(4K_{V(a)}(U^o) + 4cK_{V(a)}(U^1) + 4K_{V(a)}(U^o-cU^1))$$

and again $\gamma^-_{V(a)} \to c\lambda_1 + p(U^o) - p(U^o-cU^1)$ for $a\to\infty$.

Thus, to cover both the $C^+_{V(a)}$ and $C^-_{V(a)}$ cases, we shall study the asymptotic behaviour of $\mu_{U^o}(\Gamma_{V(a)})$, where

$$\Gamma_{V(a)} = \{x_{V(a)};\ \mu_{U^o}(x_{V(a)}) \leq \mu_{U^o-cU^1}(x_{V(a)}) \cdot \exp\{-|V(a)| \cdot \gamma_{V(a)}\}\}$$

and $\{\gamma_{V(a)}\}$ is arbitrary fixed generalized sequence tending to $c\lambda_1 + p(U^o) - p(U^o-cU^1)$ as $a\to\infty$.
From now, similarly as in Janžura (1984), we shall follow the approach developed in Perez (1974).
We shall write

$$J^1_{V(a)} = \mu_{U^o}(\Gamma_{V(a)}) \text{ and } J^2_{V(a)} = \mu_{U^o-cU^1}(\Gamma^c_{V(a)}) \cdot \exp\{-|V(a)| \cdot \gamma_{V(a)}\}$$

(here $\Gamma^c_{V(a)}$ denotes the complement).

Then it holds

$$J^1_{V(a)}+J^2_{V(a)} = \sum_{x_{V(a)}\in X^{V(a)}} \min[\mu_{U^o}(x_{V(a)}), \exp\{-|V(a)|\gamma_{V(a)}\}$$

$$\cdot \mu_{U^o-cU^1}(x_{V(a)})] .$$

Hence, using the same properties of Gibbs r.f.'s as above and the obvious inequality $\min(u,v)\le u^{\alpha}v^{1-\alpha}$ for $u,v>0$ and $\alpha\in[0,1]$, we obtain

$$\limsup_{a\to\infty}|V(a)|^{-1}\log[J^1_{V(a)}+J^2_{V(a)}]\le-(1-\alpha)[c\lambda_1+p(U^o)-p(U^o-cU^1)] +$$

$$+ p(\alpha(U^o) + (1-\alpha)(U^o-cU^1)) - \alpha p(U^o) - (1-\alpha)p(U^o-cU^1) =$$

$$= -(1-\alpha)c\lambda_1 + p(U^o-(1-\alpha)cU^1) - p(U^o) =$$

$$= F_{\lambda}((1-\alpha)c) \text{ for every } \alpha\in[0,1].$$

Then it holds also for $\alpha = 1 - \frac{\bar{\alpha}}{c}$ and $\limsup_{a\to\infty}|V(a)|^{-1}\log(J^1_{V(a)}+ + J^2_{V(a)}) \le D_{\lambda}$ follows.

For every $\nu\in M_E$ it holds $\liminf_{a\to\infty}|V(a)|^{-1}\log(J^1_{V(a)}+J^2_{V(a)}) \ge$

$$\ge \lim_{a\to\infty}\int\min\left(-|V(a)|^{-1}\log\frac{\nu(x_{V(a)})}{\mu_{U^o}(x_{V(a)})}, -\gamma_{V(a)}-|V(a)|^{-1}\log\frac{\nu(x_{V(a)})}{\mu_{U^o-cU^1}(x_{V(a)})}\right)\cdot$$

$$\cdot d\nu(x) = \min(-H(\nu|\mu_{U^o}), -c\lambda_1-p(U^o) + p(U^o-cU^1) - H(\nu|\mu_{U^o-cU^1})) =$$

$$= -H(\nu|\mu_{U^o}) + \min(0, -c(\lambda_1 - \int g_{U^1}d\nu)) .$$

The limit exists due to generalized version of McMillan's theorem the proof of which is contained in proofs of Theorem 3.2, Theorem 4.11 and Remark 4.28 in Föllmer (1973).
The inequality remains valid if we take the supremum over all $\nu\in M_E$ which is equal to D_{λ}.

Thus, we have proved that $\lim_{a\to\infty}|V(a)|^{-1}\log(J^1_{V(a)}+J^2_{V(a)}) = D_\lambda$.

As an immediate consequence we have

$$\lim_{a\to\infty} |V(a)|^{-1} \log\max(J^1_{V(a)}, J^2_{V(a)}) = D_\lambda .$$

For $\delta>0$ it holds

$$\max[J^1_{V(a)},J^2_{V(a)}] + \exp\{c.\delta.|V(a)|\} \cdot \min[J^1_{V(a)},J^2_{V(a)}] \geq$$

$$\min[(J^1_{V(a)} + \exp\{c.\delta.|V(a)|\} \cdot J^2_{V(a)}),(\exp\{c.\delta.|V(a)|\}\cdot J^1_{V(a)}+J^2_{V(a)})].$$

Further

$$J^1_{V(a)}+\exp\{c.\delta.|V(a)|\}\cdot J^2_{V(a)}\geq J^1_{V(a)}(\gamma_{V(a)}-c\delta)+J^2_{V(a)}(\gamma_{V(a)}-c.\delta)$$

and

$$\exp\{c.\delta.|V(a)|\} \cdot J^1_{V(a)} + J^2_{V(a)} \geq$$

$$\geq \exp\{c.\delta.|V(a)|\}[J^1_{V(a)}(\gamma_{V(a)} + c.\delta) + J^2_{V(a)}(\gamma_{V(a)}+ c\delta)]$$

(here the right hand side terms are defined in the same way as the left hand side ones, only with another constants as indicated by the added indeces).

Using what was proved above, we obtain that for sufficiently small $\delta>0$ it holds

$$\liminf_{a\to\infty}|V(a)|^{-1}\log[\max(J^1_{V(a)},J^2_{V(a)})+\exp\{c.\delta.|V(a)|\}\cdot\min(J^1_{V(a)},J^2_{V(a)})]\geq$$

$$\geq \min(D_{\lambda-\delta}, c.\delta + D_{\lambda+\delta}) > D_\lambda .$$

The existence of $D_{\lambda-\delta}$ and $D_{\lambda+\delta}$ and the last inequality follow from properties of convex functions and the assumption $c>\bar{\alpha}>0$.

Therefore it must hold

$$c.\delta+\liminf_{a\to\infty}|V(a)|^{-1}\log\min(J^1_{V(a)}, J^2_{V(a)}) \geq \min(D_{\lambda-\delta}, c.\delta + D_{\lambda+\delta})$$

and taking limit for $\delta \to 0$, we obtain

$$\liminf_{a\to\infty} |V(a)|^{-1} \log \min(J^1_{V(a)}, J^2_{V(a)}) \geq D_\lambda .$$

Hence $\lim_{a\to\infty} |V(a)|^{-1} \log J^1_{V(a)} = D_\lambda$ holds which completes the proof.

The proof for N>1 is the same, only with much more complicated expressions and notation. □

REMARK 1. The assumption $U^o \in L$, made to ensure the same type of distribution of both the underlying and the empirical r.f.'s plays no role within the proof. Hence, we may consider arbitrary $U^o \in \mathcal{U}$, and the preceding theorem remains true.

REMARK 2. According to original Cramér's formulation of the large deviations theorem the following limit exists

$$\lim_{n\to\infty} n^{-1} \log \Pr(n^{-1} S_n > \lambda) = s(\lambda),$$

where $S_n = Y_1 + \ldots + Y_n$ is the n'th partial sum of i.i.d. random variables $Y_1, Y_2, \ldots$.

Now, suppose $S_{V(a)} = \Sigma_{t\in V(a)} Y_t$, where $Y_t = f \circ \Theta_t : X^T \to R$ for every $t \in T$, and the probability Pr is given by a r.f. $\mu_{U^o} \in G([U^o])$, $U^o \in \mathcal{U}$.

Furthermore, let us suppose f to be continuous function on X^T (X being considered with discrete topology, X^T with the product topology is metrizable compact space).

According to PROPOSITION 2 we may write

$$p(U) = \max_{\mu \in M_{St}} (H(\mu) - \int g_U \, d\mu) \text{ for every } U \in \mathcal{U}.$$

This definition may be extended to all continuous functions (not only of the type g_U).

Thus, let us define

$$p(g) = \max_{\mu \in M_{St}} (H(\mu) - \int g \, d\mu) \text{ for every } g \in C(X^T).$$

We denote $G_{St}(g)$ the set of r.f.'s for which the maximum is reached ($G_{St}(g) \neq \emptyset$ due to Theorem 3.12 in Ruelle (1978)). Since

$$\lim_{a\to\infty} |V(a)|^{-1} \log \int \exp\{\Sigma_{t\in V(a)} g\circ\Theta_t\} d\mu_{U^o} = p(g_{U^o}-g) - p(g_{U^o})$$

for every $g \in C(X^T)$ (Proposition 4.4 in Ruelle (1978)), it remains to guarantee strict convexity of the function $F_\lambda(\alpha) = p(g_{U^o}-\alpha f) - p(g_{U^o}) - \lambda\alpha$ and existence of $\bar{\alpha}>0$ minimizing F_λ.

We may sum up one considerations in the following THEOREM:

Let $f \in C(X^T)$, $\max_{\mu\in G_{St}(g_{U^o})} \int f d\mu < \lambda < \min_{\mu\in G_{St}(g_{U^o}-cf)} \int f d\mu$

for some $c>0$.

Then

$$\lim_{a\to\infty} |V(a)|^{-1} \log \mu_{U^o}(|V(a)|^{-1} S_{V(a)} > \lambda) = s(\lambda) = D_\lambda = \min_{\alpha\in R} F_\lambda(\alpha) < 0. \square$$

Our assumption says that the function F_λ is decreasing in zero point and increasing in some positive point c. And, simultaneously, this condition can not be satisfied for $f = f_1 + \text{const.}$, where $f_1 \in \overline{\text{Lin}}\,\{g-g\Theta_t;\ g \in C(X^T),\ t \in T\}$

(i.e. f_1 from closed subspace of $C(X^T)$ generated by functions of type $g-g\circ\Theta_t$). Thus, according to Proposition 4.7 in Ruelle (1978), the strict convexity of F_λ is implicitely ensured.

REMARK 3. On the base of the result contained in the preceding remark we may conclude that for every

$$\varepsilon \in (\max(\max_{\mu\in G_{St}(g_{U^o})} \int f d\mu - \int f d\mu_{U^o}, \int f d\mu_{U^o} - \min_{\mu\in G_{St}(g_{U^o})} \int f d\mu),$$

$$\min(\sup\{\int f d\mu;\ \mu \in G_{St}(g_{U^o}-cf),\ c>0\} - \int f d\mu_{U^o},$$

$$\int f d\mu_{U^o} - \inf\{\int f d\mu;\ \mu \in G_{St}(g_{U^o}+cf),\ c>0\})$$

it holds

$$\lim_{a\to\infty} |V(a)|^{-1} \log \mu_{U^o}(\,|\,|V(a)|^{-1} S_{V(a)} - \int f d\mu_{U^o}| > \varepsilon) =$$

$$= \max(\min_{\alpha>0}\{p(g_{U^o} - \alpha f) - p(g_{U^o}) - \alpha(\int f d\mu_{U^o} + \varepsilon)\},$$

$$\min_{\alpha>0}\{p(g_{U^o} + \alpha f) - p(g_{U^o}) - \alpha(\varepsilon - \int f d\mu_{U^o})\}) = D \in (-\infty, 0) .$$

Here, the upper bound for ε occurs just to ensure existence of the limit. The more important lower bound is equal to zero if $G([U^o]) = \{\mu_{U^o}\}$, i.e. if there is exactly one Gibbs r.f. with respect to the potential U^o. This condition is satisfied e. g. for potentials from certain open subset of $\mathcal{U}$ containing the zero potential (cf. e. g. Gross (1981)).

REFERENCES

Averintsev, M. B. (1972). The description of Markov random fields by Gibbs conditional distributions, Teor. Ver. i ee Prim. 17, 21-25 (in Russian).

Bahadur, R. R. and Zabell, S. L. (1979). Large deviations of the sample mean in general vector spaces. Ann. Prob. 7, 587-621.

Chernoff, H. (1952). A measure of Asymptotic Efficiency for Tests of a Hypothesis Based on the Sum of Observations. Ann. Math. Statist. 23, 493-507.

Cramér, H. (1938). Sur un nouveau théorème-limite de la théorie des probabilités. Actualités Sci. Indust. 736, 5-23.

Csiszár, I. (1984). Sanov property, generalized I-projection and conditional limit theorem. Ann. Probab. 12, 768-793.

Dunford, N. and Schwartz, J. T. (1958). Linear Operators I. New York: Interscience.

Ekeland, I. and Temam R. (1976). Convex Analysis and Variational Problems. Amsterdam, North-Holland.

Föllmer, H. (1973). On entropy and Information Gain in Random Fields. Z. Wahrs. verw. Geb. 26, 207-217.

Gross, L. (1981). Absence of second-order phase transition in the Dobrushin Uniqueness Region. J. Stat. Phys. 25, 57-72.

Janžura, M. (1984). Discrimination rate of Gibbs random fields for unfitted decision procedure. Proc. 3rd Prague Symp. on Asympt. Stat. (eds. Mandl, P., Hušková, M.) SNTL, Prague, 303-308.

Perez, A. (1974). Generalization of Chernoff's Result on the Asymptotic Discernibility of Two Random Processes. Proc. European Meeting of Statisticians 1972, Budapest, North Holland, vol. 2, 619-632.

Preston, C. (1976). Random Fields. Springer-Verlag, Lecture Notes in Math. 534.

Ruelle, D. (1978). Thermodynamic Formalism. Reading, Mass. Addison-Wesley.

Sanov, I. N. (1957). On the probability of large deviations of random variables. Math. Sb. 42, 11-44 (in Russian).

Martin Janžura
Institute of Information Theory and Automation
Czechoslovak Academy of Sciences
Pod vodárenskou věží 4
182 08 Praha 8 - Libeň
Czechoslovakia

Proc. of the 5th Pannonian Symp. on Math. Stat., Visegrád, Hungary 1985
W. Grossmann, J. Mogyoródi, I. Vincze, W. Wertz, eds

ON SOME KINDS OF EVENT ALGEBRAS IN NON-KOLMOGOROVIAN PROBABILITY THEORY

I.G. KALMÁR

Debrecen

The proposition that different kinds of event-algebras can be investigated is due to G. Birkhoff and J. von Neumann. We accepted their point of view that the events of a (physical) system form a complemented lattice with some other properties. Our aim was to give such models of event-algebras in which the events are not necessarily observable at the same experiment, but some natural and acceptable conditions are satisfied. With the help of the precise definition of outcome, experiment, observable events and logical implication we succeeded in constructing three different kinds of event-algebras.
AMS subject classification: 60A05

The Kolmogorov model of probability is based on the notion of a sample space and the events form a set algebra. The idea that it is justified to study other kinds of event-algebras was proposed by J. von Neumann and G. Birkhoff in 1936, [1]. On physical grounds they proposed a modular lattice as event algebra. Later on it came to light that the modularity of the lattice is too strong a condition, but orthomodular lattices are general enough to be adopted as event algebras. Orthomodular lattices (or orthomodular partially ordered sets) with some special properties are generally accepted in modern theoretical physics.

Probability theories can be distinguished from each other on the ground of their fundamental notions. In the "logical approaches" events are taken as fundamental (see the Boolean probability theory, for example), and in the "manual approaches" outcomes are taken as fundamental and the events are derived from them (e.g. see the classical model of probability).

We can remark that in case of the logical approach it is not customary to define precisely such notions as outcome, experiment. But in Boolean probability theory this was done by Loš [5].

The following table gives a dictionary of Kolmogorov and Boolean probability theory to clarify their characteristic features.

Notions	Prob. theory of Kolmogorov (manual approach)	Boolean probability th. (logical approach)
1. experiment	sample space $((\Omega, \mathfrak{F}))$	Boolean algebra ($\mathbb{B}$)
2. outcomes of an experiment	elements of the basic space $(\omega \in \Omega)$	maximal filters of $\mathbb{B}$ $(\mathcal{N} \subset \mathbb{B})$
3. observable events with resp. to an experiment	measurable sets $(A \in \mathfrak{F})$	elements of $\mathbb{B}$ $(a \in \mathbb{B})$
4. certain event	the full set (Ω)	the unit element of $\mathbb{B}$ $(1 \in \mathbb{B})$
5. impossible event	empty set (Φ)	the zero element of $\mathbb{B}$ $(0 \in \mathbb{B})$
6. opposite event of an event	complemented set $(A' = \Omega \setminus A)$	the complement of the element $(a^{\perp} \in \mathbb{B})$
7. logical implication of events	set theoretical inclusion $(A \subseteq B)$	the natural ordering of $\mathbb{B}$ $(a \leq b)$

In both approaches one can define the compound of the experiments by their product (see, Rényi [6]). So in case of a given system and two or more experiments we may consider the carrying out of the experiments as a simple experiment. In addition, the observable events of the former experiments will be observable with respect to this compound experiment as well.

Basically we can interpret the Kolmogorov and Boolean probability theories as probability models for which:

B1. Given a "system" there exists an experiment in which all events connected with the system are observable.

B2. The natural ordering of the events ($\subseteq$ and $\leq$ respectively) and the logical implication of the events are identical.

(Under logical implication we mean that if the event x and y are observable with respect to an experiment then the occurence of x implies that of y.)

B3. If x is not the impossible event then there exists such an outcome which implies the occurence of x. (Mathematically this means that $x \neq 0$ and every non-zero element of $\mathbb{B}$ is included in a maximal filter, respectively.)

Practically B3 means that each non-impossible event may occur, so B3 is fairly acceptable. So is B2.

As for B1, it is not sufficiently justified, so we wish to restrict our consideration to such event-algebras in which only B2 and B3 are satisfied. The mathematical formalism of Boolean probability theory gives a good tool for us to generalize these notions.

Before going on we should mention that we will use some lattice theoretical notions which can be found in [4] and [7]. Moreover we need some new comcepts.

Let $\mathcal{L} = \mathcal{L}(\vee, \wedge, \ , 0, 1)$ be a complemented lattice. A nonempty subset $\mathcal{N}$ of $\mathcal{L}$ will be called a *proper *-filter*, if

(i) $0 \notin \mathcal{N}$

(ii) if $a \in \mathcal{N}$ and $a \leq b$ then $b \in \mathcal{N}$

(iii) if $a, b \in \mathcal{N}$ then $a * b \in \mathcal{N}$

If $\mathcal{N}$ satisfies also

(iv) $a \in \mathcal{N}$ or $a^{\perp} \in \mathcal{N}$ for every $a \in \mathcal{L}$,

then $\mathcal{N}$ is called a **-ultrafilter*.

We have to recall that the *star-product* $*$ is defined by the so called Sasaki projection: $a*b = (a\vee b^{\perp})\wedge b$; $a, b\in\mathcal{L}$.

It is clear that a $*$-ultrafilter $\mathcal{N}$ is maximal in the sense that it is not a proper subset of another proper $*$-filter.

A *block* of a complemented lattice $\mathcal{L}$ is a maximal Boolean subalgebra of $\mathcal{L}$. If two elements a, b of $\mathcal{L}$ are in a block then a and b will be called *compatible*.

Let $\hat{\mathbb{B}}_j$, $j\in J$ be the class of the blocks of a complemented lattice . Let $J'\subseteq J$ and let us suppose that for each $j\in J'$ there is a maximal filter $\mathcal{N}_j$ of $\hat{\mathbb{B}}_j$ such that

(i) $\bigcup_{j\in J'}\mathcal{N}_j$ does not contain orthogonal elements

(ii) $\bigcup_{j\in J'}\mathcal{N}_j$ is maximal i.e. for every $J''\supset J'$ and for each maximal filter $\mathcal{N}_i$ of $\hat{\mathbb{B}}_i$, $i\in J''\setminus J'$, $(\bigcup_{j\in J'}\mathcal{N}_j)\cup\mathcal{N}_i$ has orthogonal elements. Then $\mathcal{R}=\bigcup_{j\in J'}\mathcal{N}_j$ is called a *maximal set of filters*.

Let $J'=J$ then we say that $\mathcal{R}=\bigcup_{j\in J'}\mathcal{N}_j$ is a *full set of filters*. On an orthocomplemented lattice $\mathcal{L}$ we can define also a *probability measure* as a mapping m of $\mathcal{L}$ into $[0,1]$ with the following two properties: (i) $m(1)=1$, (ii) $m(a\vee b)=m(a)+m(b)$ for all $a, b\in\mathcal{L}$ and $a\perp b$ ($a^{\perp}\geq b$). We call a probability measure a *two valued measure* if it has only two values (0 and 1).

Let $\mathcal{G}$ denote a class of subsets of a complemented lattice $\mathcal{L}$. We say that $\mathcal{G}$ is *order determining* if for every $A\in\mathcal{G}$ which contains a A contains also b then $a\leq b$.

We shall use the following valuable connection between $*$-ultafilters and two valued measures.

<u>Lemma 1</u>. Let $\mathcal{L}$ be an orthocomplemented lattice. Then

a) if m is a two valued measure on $\mathcal{L}$ then $\mathcal{R} = \{a\in\mathcal{L}\mid m(a)=1\}$ is a $*$-ultrafilter

b if $\mathcal{R}$ is a $*$-ultrafilter then the characteristic function m of $\mathcal{R}$ ($m(a)=0$ if $a\notin\mathcal{R}$ and $m(a)=1$ if $a\in\mathcal{R}$) is a two valued measure.

Proof. It can be omitted because of its simplicity.

Our next table contains a terminology of notions used in the sequel:

1. event algebra	complemented lattice ($\mathcal{L}$)
2. certain event	the unit element of $\mathcal{L}$ ($1 \in \mathcal{L}$)
3. impossible event	the zero element of $\mathcal{L}$ ($0 \in \mathcal{L}$)
4. opposite event	the complement of the element ($a^{\perp} \in \mathcal{L}$)
5. experiments	blocks of $\mathcal{L}$ ($\hat{\mathbb{B}} \in \mathcal{L}$)
6. outcomes of an experiment	maximal filters of $\hat{\mathbb{B}}$ ($\mathcal{N}_j \subset \hat{\mathbb{B}}$, $j \in J$)
7. observable events with respect to an experiment	elements of $\hat{\mathbb{B}}$
8. realization	a maximal set of filters ($\mathcal{R} = \bigcup_{j \in J' \subseteq J} \mathcal{N}_j$)
9. complete realization	a full set of filters ($\mathcal{R} = \bigcup_{j \in J} \mathcal{N}_j$)

Now we are able to define some special kinds of event--algebras keeping in mind that we would like to save the validity of B2 and B3. Of course there are a lot of possible ways to do this which depend partly on the notion of logical implication of events.

Let $\mathcal{L}$ be an event algebra. $\mathcal{L}$ is an *event algebra of the first type* if

F1. For each $a \in \mathcal{L}$, $a \neq 0$ there exists an experiment $\hat{\mathbb{B}}$ of $\mathcal{L}$ and an outcome $\mathcal{N}$ of $\hat{\mathbb{B}}$ containing a.

F2. If $a \in \mathcal{L}$, $a \neq 0$ then $a \leq b$ holds if and only if

a) there exists such an experiment $\hat{\mathbb{B}}$ of $\mathcal{L}$ that $a, b \in \hat{\mathbb{B}}$ and

b) if $\hat{\mathbb{B}}$ is an experiment of $\mathcal{L}$ and $a, b \in \hat{\mathbb{B}}$ then $a \in \mathcal{N}$ implies $b \in \mathcal{N}$ for all outcomes of $\hat{\mathbb{B}}$. This shows that under the logical implication of a and b we mean that a and b are observable with respect to an experiment and for every such experiment if a is in an outcome then so is b.

We have to remark that F1 and F2 are not independent from each other because F2 implies F1. In addition, if $\mathcal{L}$ is distributive then it should be called *Boolean event algebra* because in this case $\mathcal{L}$ is a Boolean algebra.

Theorem 2. $\mathcal{L}$ is an event algebra of the first type iff $\mathcal{L}$ is orthomodular.

Proof. Let $\mathcal{L}$ be a complemented lattice. Then for all $a \leq b$, $a, b \in \mathcal{L}$ there exists a block $\hat{\mathbb{B}}$ of $\mathcal{L}$ containing a and b. So $b = a \vee (b \wedge a\ )$ in $\hat{\mathbb{B}}$ which implies the weak modularity of $\mathcal{L}$. Moreover the complementation in $\hat{\mathbb{B}}$ is an orthocomplementation which means that $\mathcal{L}$ is an orthocomplemented lattice. Thus $\mathcal{L}$ is orthomodular. Conversely, if $\mathcal{L}$ is orthomodular and $a \leq b$, $a, b \in \mathcal{L}$ then it is easy to see that $a, b \in \hat{\mathbb{B}}$ for a block $\hat{\mathbb{B}}$ of $\mathcal{L}$. If $\mathcal{N} \subset \hat{\mathbb{B}}$ is a maximal filter of $\hat{\mathbb{B}}$ and $a \in \mathcal{N}$ then $b \in \mathcal{N}$ by definition. Furthermore, the converse is true in the following sense. Let $\hat{\mathbb{B}}$ be a block of $\mathcal{L}$ and suppose that $a, b \in \hat{\mathbb{B}}$, $a \neq 0$. If for each maximal filter $\mathcal{N}$ which contains a $\mathcal{N}$ contains also b then $a \leq b$. Indeed, if $a \leq b$ does not hold then we can construct a maximal filter $\mathcal{N}$ of $\hat{\mathbb{B}}$ containing $a \wedge b^{\perp}$. So $a, b^{\perp} \in \mathcal{N}$ which contradicts $b \in \mathcal{N}$. This means that F2 holds, which implies also F1 by our previous remark.

We remark that if $\mathcal{L}$ is an event algebra of the first type then by the Zorn lemma every event (exept $a=0$) is included in a realization $\mathcal{R}$ of $\mathcal{L}$. But it is not necessary that $\mathcal{R}$ should be a complete realization.

This remark leads us to define the following type of event--algebras.

Let $\mathcal{L}$ be a complemented lattice. $\mathcal{L}$ will be called an *event algebra of the second type* if

S1. For each $a \in \mathcal{L}$, $a \neq 0$ there exists a complete realization containing a.

S2. This condition is the same as F2.

From the preceding theorem we can easily get the following.

Theorem 3. If $\mathcal{L}$ is an event algebra then the following conditions are equivalent:

a) $\mathcal{L}$ is of the second type

b) $\mathcal{L}$ is orthomodular and S1 holds in $\mathcal{L}$.

We have seen that orthomodularity holds for both types of event algebras, so the following statement will be interesting for us.

Theorem 4. If $\mathcal{L}$ is an orthomodular lattice then the following two conditions for a subset $\mathcal{R}$ of $\mathcal{L}$ are equivalent:

a) $\mathcal{R}$ is a complete realization

b) $\mathcal{R}$ is a $*$-ultrafilter.

Before beginning the proof we note one fact which can be easily proved in an orthocomplemented lattice. Namely, if $\mathcal{L}$ is an orthomodular lattice then $\mathcal{N} \subset \mathcal{L}$ is a proper $*$-filter iff (i) $0 \in \mathcal{N}$, (ii) if $a \in \mathcal{N}$ and $a \leq b$ then $b \in \mathcal{N}$, (iii) if $a, b \in \mathcal{N}$, a and b are compatible, then $a \wedge b \in \mathcal{N}$.

This technical remark will be used in the following Proof of Theorem 4.

Let $\mathcal{R} = \bigcup_{j \in J} \mathcal{N}_j$ be a complete realization. Then $0 \notin \mathcal{R}$. Furthermore, if $a \in \mathcal{R}$ and $a \leq b$ then there exists a block $\hat{\mathbb{B}}_j$ containing a and b. Then $a \in \mathcal{N}_j \subset \hat{\mathbb{B}}_j$ because $a \in \mathcal{N}_j$ implies $a^{\perp} \in \mathcal{N}_j \subset \mathcal{R}$ which contradicts $a \in \mathcal{R}$. Then $a \in \mathcal{N}_j$ and by $a \leq b$ $b \in \mathcal{N}_j \subset \mathcal{R}$.

Moreover, if $a, b \in \mathcal{R}$ then $a \vee b^{\perp} \in \mathcal{R}$. Since $a \vee b^{\perp}$ and b are compatible, there exists a block $\hat{\mathbb{B}}_k$, $k \in J$ such that $a \vee b^{\perp}$, $b \in \hat{\mathbb{B}}_k$. Then $a \vee b^{\perp}$, $b \in \mathcal{R}$ and $a \vee b^{\perp}$, $b \in \mathcal{N}_k \subset \hat{\mathbb{B}}_k$ which implies $(a \vee b^{\perp}) \wedge b = a * b \in \mathcal{N}_k$. Finally, if $c \in \mathcal{L}$ then $c \in \mathcal{R}$ or $c^{\perp} \in \mathcal{R}$. These remarks indicate that is a complete realization. Conversery, let us suppose that $\mathcal{R}$ is a $*$-ultrafilter of $\mathcal{L}$. Denote by $\hat{\mathbb{B}}_j$, $j \in J$ the family of the blocks of $\mathcal{L}$ and let $\mathcal{N}_j = \mathcal{R} \cap \hat{\mathbb{B}}_j$. Then $\mathcal{N}_j$ is a maximal filter of $\hat{\mathbb{B}}_j$ and $\mathcal{R} = \bigcup_{j \in J} \mathcal{N}_j$ is a complete realization in $\mathcal{L}$.

By our last two theorems and Lemma 1. we can state now

<u>Theorem 5</u>. A necessary and sufficient condition that an event algebra $\mathcal{L}$ be of the second type is that the following two conditions are satisfied.

a) $\mathcal{L}$ is an orthomodular lattice

b) For all $a \in \mathcal{L}$, $a \neq 0$ there exists a two valued measure m on $\mathcal{L}$ such that $m(a)=1$.

The two conditions of this theorem are independent from each other because of the following well-known result.

Let $\mathcal{H}_n$ $(n \geq 3)$ be an n-dimensional Hilbert space and let $\mathcal{L}(\mathcal{H}_n)$ be the orthocomplemented lattice of its closed subspaces taking the set-theoretical inclusion as ordering and the orthogonal subspace as an orthocomplement. Then $\mathcal{L}(\mathcal{H}_n)$ is a modular lattice, but there is no two valued measure on $\mathcal{L}$ at all.

Thus we can state that there exists an event algebra of the first type which is not of the second type. But the class of the event algebras of the second type is not empty as the following orthomodular lattice satisfying S1 shows:

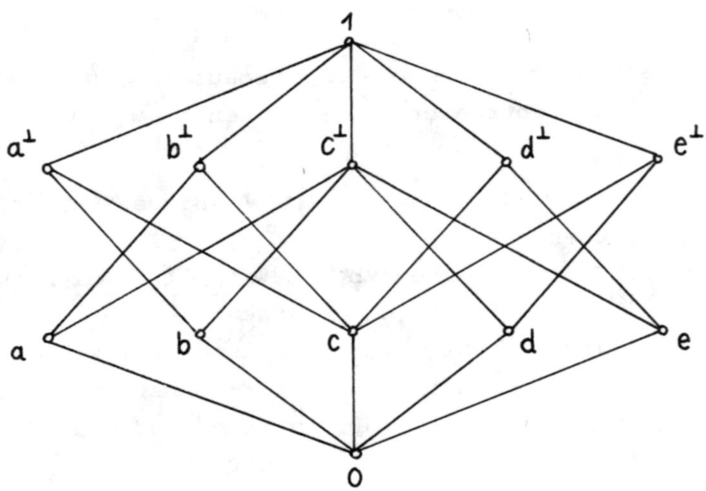

Finally we give a third type of event algebras. An orthocomplemented lattice $\mathcal{L}$ will be called an *event algebra of the third type* iff

T1. If $a \in \mathcal{L}$, $a \neq 0$ then there exists a complete realization of $\mathcal{L}$ containing a.

T2. If a, $b \in \mathcal{L}$ and $a \leq b$ then there exists an experiment of which contains a and b.

T3. The set of the complete realizations of $\mathcal{L}$ is order determining.

One can find some characterizations of third type event algebras in [3].

Let us remark that an event algebra is of the third type if and only if it is isomorphic to an n-class. (A family S of subsets of a nonempty set Ω is called an *n-class* if (i) $\Omega \in S$ (ii) if $A \in S$ then $A' = \Omega \setminus A \in S$, (iii) if A_1, $A_2 \in S$ and $A_2 \cap A_2 = 0$ then $A_1 \cup A_2 \in S$.)

REFERENCES

1. G. Birkhoff - J. von Neumann, The logic of quantum mechanics, Ann. of Math. 37 (1936), 823-843.
2. I. G. Kalmár, *-structures and orthomodular lattices, Publ. Math. 32 (1985), 1-5.
3. I. G. Kalmár, Lattice theoretical characterizations of quantum probability space I., Publ. Math. 32 (1985), 179-185.
4. G. Kalmbach, Orthomodular lattices, Academic Press, 1983.
5. J. Loš, On the axiomatic treatment of probability, Coll. Math. 3 (1955), 125-137.
6. A. Rényi, Foundation of probability, Holden-Day, San Francisco, 1970.
7. R. Sikorski, Boolean Algebras, Springer, Berlin, 1964.

I. G. Kalmár
Kossuth Lajos University
Department of Mathematics
4010. Debrecen, Pf. 12.
Hungary

Proc. of the 5th Pannonian Symp. on Math. Stat., Visegrád, Hungary 1985
W. Grossmann, J. Mogyoródi, I. Vincze, W. Wertz, eds

ON BIVARIATE GEOMETRIC COMPOUNDING

A. KOVÁTS

Budapest

Let (X_j, Y_j), $j = 1,2,\dots$ be nonnegative i.i.d random vectors and let (N_1, N_2) be independent of (X_j, Y_j), $j = 1,2,\dots$ with bivariate geometric distribution (see (3)). The random vector $\left(Z_1 = \sum_{j=1}^{N_1} X_j,\ \ Z_2 = \sum_{j=1}^{N_2} Y_j\right)$ is called the bivariate geometric compounding of (X_j, Y_j)-s. In this note we characterize some two-dimensional exponential distributions by bivariate geometric compounding.

AMS subject classification: 62E10, 60F05

1. First we consider a well-known characterization of univariate exponential distribution.

Let $X_1, X_2, \dots$ be nonnegative i.i.d. rv's, $P(X_j > x) = \overline{F}(x)$, $E(X_j) = 1$, $j = 1,2,\dots$ and let N be independent of X_j, $j = 1,2,\dots$ with geometric distribution, i.e.

$$P(N = k) = p(1-p)^{k-1}, \quad k = 1,2,\dots$$

The random variable $Z = \sum_{j=1}^{N} X_j$ is called the geometric compounding of X_j-s. Denote

$$\overline{G}_p(x) = P(pZ > x).$$

Rényi [1] characterized the exponential distribution proving the following two assertions:

$$\lim_{p\to 0} \bar{G}_p (x) = e^{-x} \tag{1}$$

$$\bar{G}_p (x) = \bar{F} (x) \iff \bar{F} (x) = e^{-x} \tag{2}$$

This theorem was generalized by several authors. A survey on this topic can be found in [6].

2. Now we want to formulate two-dimensional versions of (1) and (2). First we have to define the bivariate geometric distribution. Let A_1, A_2 be arbitrary events and $\underline{p} = (p_1, p_2, p_{12})$ means the following probabilities

$$P(A_1\bar{A}_2) = p_1, \quad P(\bar{A}_1 A_2) = p_2, \quad P(A_1 A_2) = p_{12},$$

and $q = 1 - p_1 - p_2 - p_{12}$.

Trying independent experiments let N_1, N_2 be the serial numbers of trials when the events A_1, resp. A_2 occur at first. We say the random vector (N_1,N_2) to have bivariate geometric distribution. One can compute that

$$P(N_1 = k_1, N_2 = k_2) = \begin{cases} q^{k_2-1} p_2 (1-p_1-p_{12})^{k_1-k_2-1} (p_1+p_{12}), \text{ if } k_1 > k_2 \\ q^{k_1-1} p_{12}, \text{ if } k_1 = k_2 \\ q^{k_1-1} p_1 (1-p_2-p_{12})^{k_2-k_1-1} (p_2+p_{12}), \text{ if } k_1 < k_2. \end{cases} \tag{3}$$

Let (X_j,Y_j), $j = 1,2,\ldots$ be nonnegative i.i.d. random vectors, $P(X_j > x, Y_j > y) = \bar{F}(x,y)$, $E(e^{i(t_1X_j+t_2Y_j)}) = \varphi(t_1,t_2)$, $E(X_j) = E(Y_j) = 1$ and let (N_1,N_2) be independent of (X_j,Y_j), $j = 1,2,\ldots$ with bivariate geometric distribution.

The random vector $\left(Z_1 = \sum_{j=1}^{N_1} X_j, \quad Z_2 = \sum_{j=1}^{N_2} Y_j\right)$ is called the bivariate geometric compounding of (X_j, Y_j)-s.

Denote $\overline{G}_{\underline{p}}(x,y) = P\left((p_1+p_{12}) Z_1 > x, (p_2+p_{12}) Z_2 > y\right)$.

Some of its properties were investigated by Arnold [2], Paulson [3] and Block [4]. Generalizing (1) we get the following result.

Theorem 1.

$$\lim_{\underline{p}\to\underline{0}} \overline{G}_{\underline{p}}(x,y) = e^{-\lambda_1 x-\lambda_2 y-\max(\lambda_{11}x,\lambda_{22}y)} \tag{4}$$

where

$$\frac{p_1}{p_1+p_{12}} \to \lambda_1, \qquad \frac{p_{12}}{p_1+p_{12}} \to \lambda_{11}$$

$$\frac{p_2}{p_2+p_{12}} \to \lambda_2, \qquad \frac{p_{12}}{p_2+p_{12}} \to \lambda_{22}$$

Remark. The limit law is a linear transform of the well-known Marshall-Olkin distribution.

After this theorem one can believe that this limit distribution is invariant under bivariate geometric compounding. The following theorem however shows it is not so.

Theorem 2.

$\overline{G}_{\underline{p}}(x,y) = \overline{F}(x,y)$ if and only if

$$\varphi(t_1,t_2) = \left[1 - it_1 - it_2 + \sum_{n=1}^{\infty}\sum_{k=1}^{\infty} (-i)^{n+k} a_{n,k} t_1^n t_2^k\right]^{-1}$$

where

$$a_{1,1} = \frac{p_1 + p_2}{p_1 + p_2 + p_{12} - (p_1 + p_{12})(p_2 + p_{12})} \quad ,$$

$$a_{1,2} = \frac{p_2 - a_{11}(p_2 + p_{12})(1 - p_1 - p_{12})}{p_1 + p_2 + p_{12} - (p_1 + p_{12})(p_2 + p_{12})^2} \quad ,$$

$$a_{2,1} = \frac{p_1 - a_{11}(p_1 + p_{12})(1 - p_2 - p_{12})}{p_1 + p_2 + p_{12} - (p_1 + p_{12})^2(p_2 + p_{12})}$$

and for $n \geqq 2$, $k \geqq 2$

$$a_{n,k} = \left[p_1 + p_2 + p_{12} - (p_1 + p_{12})^n (p_2 + p_{12})^k\right]^{-1} \cdot$$

$$\left\{a_{n-1,k-1}\left[(p_1 + p_{12})^{n-1}(p_2 + p_{12})^{k-1} - p_{12}\right] + \right.$$

$$+ a_{n,k-1}\left[(p_1 + p_{12})^n (p_2 + p_{12})^{k-1} - p_2 - p_{12}\right] +$$

$$\left. + a_{n-1,k}\left[(p_1 + p_{12})^{n-1}(p_2 + p_{12})^k - p_1 - p_{12}\right]\right\}$$

<u>Remark:</u> We can't give explicite formulas for the distribution or density function.

<u>Proof of Theorem 1.</u>

Let us denote the characteristic function of (Z_1, Z_2) by $\psi(t_1, t_2)$. Between $\psi(t_1, t_2)$ and $\varphi(t_1, t_2)$ the following relation holds:

$$\psi(t_1, t_2) = \varphi(t_1, t_2)(p_{12} + p_1 \psi(0, t_2) + p_2 \psi(t_1, 0) + q\, \psi(t_1, t_2)) \quad (5)$$

From (1) we know that

$$\lim_{\underline{p}\to\underline{0}} \psi\,(0,(p_2+p_{12})t) = \lim_{\underline{p}\to\underline{0}} \psi\,((p_1+p_{12})t,0) = \frac{1}{1-it}$$

therefore from (5) we get

$$\lim_{\underline{p}\to\underline{0}} \psi\,((p_1+p_{12})t)\ t_1,(p_2+p_{12})t_2) =$$

$$= \lim_{\underline{p}\to\underline{0}} \frac{p_{12}+\dfrac{p_1}{1-it_2}+\dfrac{p_2}{1-it_2}}{1-q\,\varphi\,((p_1+p_{12})t_1,(p_2+p_{12})t_2)}$$

Using the Taylor-expansion we can approximate

$\varphi((p_1+p_{12})t_1,(p_2+p_{12})t_2)$ by $1+i(p_1+p_{12})t_1+i(p_2+p_{12})t_2$

and straightforward calculations give that

$$\lim_{\underline{p}\to\underline{0}} \psi\,((p_1+p_{12})t_1,(p_2+p_{12})t_2) = \tag{6}$$

$$= \frac{\lambda_{11}+\lambda_{22}-\lambda_{11}\lambda_{22}-i\lambda_{22}t_1-i\lambda_{11}t_2-\lambda_{11}\lambda_{22}t_1t_2}{(1+t_1)(1+t_2)(\lambda_{11}+\lambda_{22}-\lambda_{11}\lambda_{22}-i\lambda_{22}t_1-i\lambda_{11}t_2)}$$

The characteristic function of limit distribution stated in (4) is the following

$$\frac{\lambda_{22}(\lambda_1-it_1)+\lambda_{11}(\lambda_2-it_2)+\lambda_{11}\lambda_{22}(1-t_1t_2)}{(\lambda_1+\lambda_{11}-t_1)(\lambda_2+\lambda_{22}-it_2)(\lambda_{11}\lambda_{22}+\lambda_{11}\lambda_2+\lambda_{22}\lambda_1-i\lambda_{22}t_1-i\lambda_{11}t_2)} \tag{7}$$

As $\lambda_1 = 1-\lambda_{11}$ and $\lambda_2 = 1-\lambda_{22}$ so (6) equals (7) consequently the theorem is proved.

Proof of Theorem 2.

The characteristic function of the invariant distribution under bivariate geometric compounding must satisfy the following functional equation (see (5)):

$$\varphi(t_1,t_2)=\varphi((p_1+p_{12})t_1,(p_2+p_{12})t_2)(p_{12}+p_1\varphi(0,t_2)+ \qquad (8)$$

$$+p_2\varphi(t_1,0)+q\varphi(t_1t_2))$$

It follows from (2) that

$$\varphi(0,t_2)=\frac{1}{1-it_2} \quad \text{and} \quad \varphi(t_1,0)=\frac{1}{1-it_1} \qquad (9)$$

From (8) using (9) and putting

$$\varphi(t_1,t_1)=\left(1-it_1-it_2+\sum_{n=1}^{\infty}\sum_{k=1}^{\infty}(-i)^{n+k}a_{n,k}t_1^n\,t_2^k\right)^{-1}$$

we get recurrence formulas for the coefficients $a_{n,k}$ stated in Theorem 2.

REFERENCES

1. Rényi A. (1976): A characterization of the Poisson process. Magyar Tud. Akad. Mat. Kut. Int. Köz., 1956.1. 519-527 (in Hungarian). (Translated into English in Selected Papers of Alfred Rényi, Vol.1., Akadémiai Kiadó, Budapest).
2. Arnold, B. C. (1975): A characterization of the exponential distribution by multivariate geometric compounding, Sankhyā, Ser.A, 37, 164-173.
3. Paulson, A. S. (1973): A characterization of the exponential distribution and a bivariate exponential distribution, Sankhyā, Ser. A, 35, 69-78.
4. Block, H. W. (1977): A family of bivariate exponential

distribution. Theory and Applications of Reliability, Vol. I. 349-371. (C.P.Tsokos and I.M.Shimi, eds.) Academic Press, New York.

5. Galambos, J. - Kotz, S. (1978): Characterizations of Probability Distribution. Lecture Notes in Mathematics, 675. Springer Verlag.

6. Azlarov, T. - Atakuziev, D. - Dzamirzaev, A. (1977): A scheme on sum of rv's with geometrically distributed index. Limit theorems for stochastic processes. Tashkent, (in Russian).

A. Kováts
Department of Probability Theory
L. Eötvös University
Budapest, Muzeum krt. 6-8.
H-1000, Hungary

Proc. of the 5th Pannonian Symp. on Math. Stat., Visegrád, Hungary 1985
W. Grossmann, J. Mogyoródi, I. Vincze, W. Wertz, eds

ON INHOMOGENEOUS MARKOV CHAIN WITH BOUNDED POSITIVE JUMPS AND HOMOGENEOUS SECOND COMPONENT

L. LAKATOS

Budapest

In this note we consider an inhomogeneous Markov chain on the integers 0,1,2,... with arbitrary negative and bounded positive jumps. Our aim is to determine the transition probabilities of this chain for a given time interval [t,s]. We associate with the Markov chain a homogeneous second component introduced by Ezhov and Skorohod. The generating function of transition probabilities /for the second component we take its Fourier transform/ satisfies a linear differential equation. The determination of an arbitrary $k \to l$ transition probability we reduce to determining the first $c+1$, $i \to l$ $/0 \le i \le c/$ transition probabilities. The latters as it shown satisfy a system of Volterra integral equations.
AMS subject classification: 60J27

1. Let $\xi(t)$, $t \ge 0$ be an inhomogeneous Markov chain with state space $\{0,1,\dots\}$ and transition probabilities for $[t,t+\Delta]$ $(\Delta \to 0)$:

$$P\left\{0 \xrightarrow{[t,t+\Delta]} r\right\} = \delta_{or} + b_r(t)\Delta + o(\Delta),$$

$$r = 0,1,\dots,c;$$

$$P\left\{k \xrightarrow{[t,t+\Delta]} k+c-r\right\} = \delta_{cr} + a_r(t)\Delta + o(\Delta), \tag{1}$$

$$r = 0,1,\dots,k+c-1;$$

$$P\left\{k \xrightarrow{[t,t+\Delta]} 0\right\} = a_{k+c}(t)\Delta + o(\Delta),$$

$$\hat{a}_{k+c}(t) = \sum_{j=k+c}^{\infty} a_j(t),$$

where $b_o(t)<0$, $a_c(t)<0$, $b_r(t)$ $(r \neq 0)$ and $a_r(t)$ $(r \neq c)$ nonnegative continuous functions for which the relations

$$\sum_{r=0}^{c} b_r(t) = 0, \quad \sum_{r=0}^{\infty} a_r(t) = 0$$

hold.

We will consider a more general process. We associate with $\xi(t)$ a homogeneous second component $\eta(t) \in R^d$ and create the process $\zeta(t) = \{\xi(t), \eta(t)\}$. The second component will be additive. Such Markov processes with homogeneous second component i.e. for which

$$P\{\xi(s) = r, \eta(s) \in A | \xi(t) = k, \eta(t) = x\} =$$

$$= P\{\xi(s) = r, \eta(s) \in A_{-x} | \xi(t) = k, \eta(t) = 0\},$$

where

$$A_{-x} = \{y \in R^d : x + y \in A\},$$

were introduced and investigated by Ezhov and Skorohod in [1] and [3]. We determine $\eta(t)$ depending on the value of $\xi(t)$ as follows:

1. if $\xi(t-0) = k$, $\xi(t+0) = k$, then $\eta(t) = \varepsilon(t)$;
2. if $\xi(t-0) = 0$, $\xi(t+0) = k$, then $\eta(t) = \varepsilon_{ok}(t)$;
3. if $\xi(t-0) = k$, $\xi(t+0) = r$, then $\eta(t) = \varepsilon_{r-k}(t)$, $k \neq r$,

where $\varepsilon(t)$, $\varepsilon_{ok}(t)$ and $\varepsilon_j(t)$ $(j \neq 0)$ are homogeneous processes, $\eta(t)$ coincides with one of them depending on the actual jump of the Markov chain.

Moreover, let

$$M\left\{e^{i(\lambda, \varepsilon(s) - \varepsilon(t))} | \xi(x) = \text{const}, x \in [t,s]\right\} = \exp\left\{\int_t^s f(x,\lambda)\, dx\right\};$$

$$M\left[e^{i(\lambda,\varepsilon_{ok}(t+0)-\varepsilon_{ok}(t-0))}\,|\,\xi(t-0)=0,\xi(t+0)=k\right] = \varphi_{ok}(t,\lambda),\quad 1\leq k\leq c;$$

$$M\left[e^{i(\lambda,\varepsilon_{c-r}(t+0)-\varepsilon_{c-r}(t-0))}\,|\,\xi(t-0)=k,\xi(t+0)=k+c-r\right] = \varphi_{c-r}(t,\lambda),$$

$$k\geq 1,\ r=0,1,\ldots,c-1,c+1,\ldots,k+c,$$

where $f(t,\lambda)$ is cumulant of a process with independent increments in R^d, $\varphi_{ok}(t,\lambda)$ and $\varphi_j(t,\lambda)$ are characteristic functions for certain distributions in R^d.

2. The described Markov chain with homogeneous second component is regular. We derive the inverse system of Kolmogorov differential equations for it. Let

$$P_{kr}(t,s,A) = P\{\xi(s)=r,\eta(s)-\eta(t)\in A\,|\,\xi(t)=k\},$$

where A is a set of possible values of the second component and

$$P_{kr}(t,s,\lambda) = \int_{R^d} e^{i(\lambda,z)}P\{\xi(s)=r,\ \eta(s)-\eta(t)\in dz\,|\,\xi(t)=k\}.$$

Determine $P_{kr}(t,t+\Delta,\lambda)$:

I. $0 \longrightarrow r$

1. $r\neq 0$

$$P_{or}(t,t+\Delta,\lambda) = \left[b_r(t)\Delta + o(\Delta)\right]Me^{i(\lambda,\varepsilon_{or}(t+\Delta)-\varepsilon_{or}(t))} =$$
$$=\left[b_r(t)\Delta + o(\Delta)\right]\ {}_{or}(t,\lambda) = b_r(t,\lambda)\Delta + o(\Delta);$$

2. $r=0$

$$P_{oo}(t,t+\Delta,\lambda) = \left[1+b_0(t)\Delta + o(\Delta)\right]Me^{i(\lambda,\varepsilon(t+\Delta)-\varepsilon(t))} =$$
$$=\left[1+b_0(t)\Delta + o(\Delta)\right]\exp\left\{\int_t^{t+\Delta} f(x,\lambda)dx\right\} =$$
$$= 1+\left[b_0(t)+f(t,\lambda)\right]\Delta + o(\Delta) = 1 + b_0(t,\lambda)\Delta + o(\Delta);$$

II. $k \longrightarrow k+c-r$

1. $r \neq c$

$$P_{kk}(t,t+\Delta,\lambda) = \left[1 + a_c(t)\Delta + o(\Delta)\right] Me^{i(\lambda,\varepsilon(t+\Delta)-\varepsilon(t))} =$$

$$= \left[1 + a_c(t)\Delta + o(\Delta)\right] \exp\left\{\int_t^{t+\Delta} f(x,\lambda)dx\right\} =$$

$$= 1 + \left[a_c(t) + f(t,\lambda)\right]\Delta + o(\Delta) = 1 + a_c(t,\lambda)\Delta + o(\Delta);$$

III. $k \longrightarrow 0$

$$P_{ko}(t,t+\Delta,\lambda) = \left[\hat{a}_{k+c}(t)\Delta + o(\Delta)\right] Me^{i(\lambda,\varepsilon_{-k}(t+\Delta)-\varepsilon_{-k}(t))} =$$

$$= \left[\hat{a}_{k+c}(t)\Delta + o(\Delta)\right] \varphi_{-k}(t,\lambda) = \hat{a}_{k+c}(t,\lambda)\Delta + o(\Delta).$$

We have

$$P_{o\ell}(t,s,\lambda) = \sum_{r=0}^{c} \left[\delta_{or} + b_r(t,\lambda)\Delta + o(\Delta)\right] P_{r\ell}(t+\Delta,s,\lambda),$$

$$P_{k\ell}(t,s,\lambda) = \sum_{r=0}^{k+c-1} \left[\delta_{cr} + a_r(t,\lambda)\Delta + o(\Delta)\right] P_{k-r+c,\ell}(t+\Delta,s,\lambda) +$$

$$+ \left[\hat{a}_{k+c}(t,\lambda)\Delta + o(\Delta)\right] P_{o\ell}(t+\Delta,s,\lambda), \quad k \geq 1. \tag{3}$$

To establish e.g. the first equality from (3) it is necessary to represent it in the form

$$P_{o\ell}(t,s,A) =$$

$$= \int_{R^d} \left[1 + b_0(t)\Delta + o(\Delta)\right] P\left\{\varepsilon(t+\Delta)-\varepsilon(t) + x \; A\right\} P_{o\ell}(t+\Delta,s,dx) +$$

$$+ \sum_{r=1}^{c} \int_{R^d} \left[b_r(t)\Delta + o(\Delta)\right] P\left\{\varepsilon_{or}(t+\Delta)-\varepsilon_{or}(t) + x \in A\right\} P_{r\ell}(t+\Delta,s,dx)$$

and to apply the Fourier transform. From (3) we get

$$-\frac{\partial P_{o\ell}(t,s,\lambda)}{\partial t} = \sum_{r=0}^{c} b_r(t,\lambda) P_{r\ell}(t,s,\lambda),$$

$$-\frac{\partial P_{k\ell}(t,s,\lambda)}{\partial t} = \sum_{r=0}^{k+c-1} a_r(t,\lambda) P_{k-r+c,\ell}(t,s,\lambda) + \tag{4}$$

$$+ \hat{a}_{k+c}(t,\lambda) P_{o\ell}(t,s,\lambda), \qquad k > 0.$$

Introduce the notations

$$P_\ell(t,s,\lambda,\Theta) = \sum_{k=1}^{\infty} P_{k\ell}(t,s,\lambda)\Theta^k,$$

$$a(t,\lambda,\Theta) = \sum_{k=0}^{\infty} a_k(t,\lambda)\Theta^k.$$

Now we have

Theorem 1. The generating function of transition probabilities $P_\ell(t,s,\lambda,\Theta)$ satisfies the equation

$$\frac{\partial P_\ell(t,s,\lambda,\Theta)}{\partial t} = -\frac{1}{\Theta^c} a(t,\lambda,\Theta) P_\ell(t,s,\lambda,\Theta) -$$

$$- P_{o\ell}(t,s,\lambda)\hat{a}_{k+c}(t,\lambda,\Theta) + \sum_{i=1}^{c-1} P_{i,\ell}(t,s,\lambda)\left[\frac{a(t,\lambda,\Theta)}{\Theta^c}\Theta^{c-i} -\right.$$

$$\left. - \frac{a(t,\lambda,\Theta)}{\Theta^c}\Theta^i + \sum_{j=0}^{c-i} a_j(t,\lambda)\Theta^{i+j-c}\right] + a_0(t,\lambda) P_{c,\ell}(t,s,\lambda), \tag{5}$$

where

$$\hat{a}_{k+c}(t,\lambda,\Theta) = \sum_{k=1}^{\infty} \Theta^k \hat{a}_{k+c}(t,\lambda).$$

3. (5) is a linear differential equation for $P_\ell(t,s,\lambda,\Theta)$, so its solution according to [7] is

$$P_\ell(t,s,\lambda,\Theta) = \exp\left\{\int_t^s \frac{a(x,\lambda,\Theta)}{\Theta^c}\,dx\right\}\cdot$$

$$\cdot\Big[c + \int_t^s \exp\left\{-\int_y^s \frac{a(x,\lambda,\Theta)}{\Theta^c}\,dx\right\}\Big\{P_{o\ell}(y,s,\lambda)a_{k+c}(y,\lambda,\Theta) +$$

$$+\sum_{i=1}^{c-1} P_{i,\ell}(y,s,\lambda)\left[\frac{a(y,\lambda,\Theta)}{\Theta^c}\Theta^i - \frac{a(y,\lambda,\Theta)}{\Theta^c}\Theta^{c-i} - \sum_{j=0}^{c-i} a_j(y,\lambda)\Theta^{i+j-c}\right] - \tag{6}$$

$$- a_0(y,\lambda)P_{c,\ell}(y,s,\lambda)\Big\}dy\Big].$$

The initial condition we get at $t = s : C = (1-\delta_{o\ell})\Theta^\ell$.

Introduce the auxiliary process with independent increments $\xi^*(t)$ with state space $\{0,\pm1,\pm2,\dots\}$ and transition probabilities for $[t,t+\Delta]$

$$P\left\{k \xrightarrow{[t,t+\Delta]} k + c - r\right\} = \delta_{cr} + a_r(t)\Delta + o(\Delta).$$

Then for the process $\zeta^*(t) = \{\xi^*(t),\eta(t)\}$

$$M\Theta^{\xi^*(s)-\xi^*(t)}e^{i(\lambda,\eta(s)-\eta(t))} = \exp\left\{\int_t^s \frac{a(x,\lambda,\Theta)}{\Theta^c}\,dx\right\} =$$

$$= A(t,s,\lambda,\Theta),$$

and let

$$\rho_k(t,s,\lambda) = \frac{1}{2\pi i} \oint_{|\Theta|=1} \exp\left\{\frac{a(x,\lambda,\Theta)}{\Theta^c}\, dx\right\} \frac{d\Theta}{\Theta^{k+1}} \quad . \tag{7}$$

Taking into account that

$$\frac{a(s,\lambda,\Theta)}{\Theta^c} \exp\left\{\int_t^s \frac{a(x,\lambda,\Theta)}{\Theta^c}\, dx\right\} = \frac{\partial A(t,s,\lambda,\Theta)}{\partial s} =$$

$$= \sum_{j=-\infty}^{\infty} \frac{\partial \rho_j(t,s,\lambda)}{\partial s} \Theta^j ,$$

and comparing the coefficients in (6) at Θ^k we obtain

<u>Theorem 2.</u> The transition probabilities $P_{k\ell}(t,s,\lambda)$ are equal to

$$P_{k\ell}(t,s,\lambda) = (1 - \delta_{o\ell})\, \rho_{k-\ell}(t,s,\lambda) +$$

$$+ \int_t^s \left\{ \sum_{i=-\infty}^{k-1} \rho_i(t,y,\lambda) \hat{a}_{k-i+c}(y,\lambda) P_{o\ell}(y,s,\lambda) + \right.$$

$$+ \sum_{i=1}^{c-1} P_{i,\ell}(y,s,\lambda) \left[\frac{\partial \rho_{k-i}(t,y,\lambda)}{\partial y} - \frac{\partial \rho_{k+i-c}(t,y,\lambda)}{\partial y} - \right. \tag{8}$$

$$\left. \left. - \sum_{j=0}^{c-i} a_j(y,\lambda) \rho_{k+c-i-j}(t,y,\lambda) \right] - \rho_k(t,y,\lambda) a_0(y,\lambda) P_{c,\ell}(y,s,\lambda) \right\} dy,$$

where

$$\hat{a}_{k-i+c}(y,\lambda) = \hat{a}_{k-i+c}(y)\, \varphi_{i-k}(y,\lambda).$$

4. (8) shows that the transition probabilities $P_{k\ell}(t,s,\lambda)$

$(k \geq 1)$ may be expressed by means of the first $c+1$ probabilities $P_{o\ell}(t,s,\lambda), P_{1\ell}(t,s,\lambda), \ldots, P_{c,\ell}(t,s,\lambda)$ from the desired distribution. From the first equation of (4)

$$P_{o\ell}(t,s,\lambda) = \delta_{o\ell} \exp\left[\int_t^s b_0(x,\lambda)\,dx\right] + \int_t^s \left\{\sum_{r=1}^{c} b_r(y,\lambda) P_{r,\ell}(y,s,\lambda) \exp\left[\int_t^y b_0(x,\lambda)\,dx\right]\right\} dy. \tag{9}$$

Introduce the notations

$$\vec{P}_\ell(t,s,\lambda) = \left\{P_{o\ell}(t,s,\lambda),\ P_{1\ell}(t,s,\lambda), \ldots, P_{c,\ell}(t,s,\lambda)\right\} ;$$

$$\vec{\rho}_\ell(t,s,\lambda) = \left\{\delta_{o\ell} \exp\left\{\int_t^s b_0(x,\lambda)\,dx\right\},\right.$$

$$\left.(1-\delta_{o\ell})\rho_{1-\ell}(t,s,\lambda), \ldots, (1-\delta_{o\ell})\rho_{c-\ell}(t,s,\lambda)\right\} ;$$

$$\pi_{00}(t,s,\lambda) = 0;$$

$$\pi_{m0}(t,s,\lambda) = \exp\left\{\int_t^s b_0(x,\lambda)\ dx\right\} b_m(s,\lambda), \qquad 1 \leq m \leq c;$$

$$\pi_{on}(t,s,\lambda) = \sum_{i=-\infty}^{n-1} \rho_i(t,s,\lambda)\ a_{n-i+c}(s,\lambda), \qquad n \geq 1;$$

$$\pi_{cn}(t,s,\lambda) = -a_0(s,\lambda)\rho_n(t,s,\lambda), \qquad n \geq 1;$$

$$\pi_{mn}(t,s,\lambda) = \frac{\partial \rho_{n-m}(t,s,\lambda)}{\partial s} - \frac{\partial \rho_{n+m-c}(t,s,\lambda)}{\partial s} - \sum_{j=0}^{c-m} a_j(s,\lambda)\rho_{n+c-m-j}(t,s,\lambda), \tag{10}$$

$$1 \leq m \leq c-1, \quad n \geq 1;$$

$$F(t,s,\lambda) = \| \pi_{mn}(t,s,\lambda) \| \qquad 0 \leq m,n \leq c.$$

Now from (8) and (9) we obtain the system of equations

$$\vec{P}_\ell(t,s,\lambda) = \vec{\rho}_\ell(t,s,\lambda) + \int_t^s \vec{P}_\ell(y,s,\lambda) F(t,y,\lambda)\, dy.$$

We fix certain T , then according to the theory of Volterra integral equations [7] in $0 \leq t \leq s \leq T$

$$\vec{P}_\ell(t,s,\lambda) = \vec{\rho}_\ell(t,s,\lambda) + \int_t^s \vec{\rho}_\ell(y,s,\lambda) G(t,y,\lambda)\, dy,$$

where $G(t,s,\lambda)$ is the resolvent of the matrix $F(t,s,\lambda)$. Let

$$\begin{bmatrix} \pi_{ok}(t,s,\lambda) \\ \pi_{1k}(t,s,\lambda) \\ \vdots \\ \pi_{ck}(t,s,\lambda) \end{bmatrix} = \vec{g}_k(t,s,\lambda), \tag{11}$$

then

$$P_{k\ell}(t,s,\lambda) =$$

$$= (1-\delta_{o\ell})\rho_{k-\ell}(t,s,\lambda) + \int_t^s \vec{P}_\ell(y,s,\lambda)\vec{g}_k(t,y,\lambda)\, dy =$$

$$= (1-\delta_{o\ell})\rho_{k-\ell}(t,s,\lambda) + \int_t^s \vec{\rho}_\ell(y,s,\lambda)\vec{g}_k(t,y,\lambda)\, dy + \tag{12}$$

$$+ \iint_{t \leq x \leq y \leq s} \vec{\rho}_\ell(y,s,\lambda) G(x,y,\lambda) \vec{g}_k(t,x,\lambda)\, dx dy,$$

$$k \geq c+1.$$

We have proved

Theorem 3. The transition probabilities of $\zeta(t)$ are determined by (12), where $\rho_k(t,s,\lambda)$ are given by (7), $\vec{g}_k(t,s,\lambda)$ by (10) and (11), $G(t,s,\lambda)$ is the resolvent of matrix $F(t,s,\lambda)$.

5. Similar model in case of bounded negative and arbitrary positive jumps was considered by Tairov in [6] for description of queueing systems of $M_t/M_t/c$ type. Earlier Olynik (5) investigated this problem for semi-continuous case /i.e. the negative jumps are equal to one/, but he associated with the Markov chain a homogeneous second component, too. In [2] we gave the solution of this problem for the case of bounded positive jumps. At last, in [8] this problem is solved for the case when the Markov chain is determined on the integers 0,1,...,N and a homogeneous second component is used, too. The present paper gives a natural generalization of models for description of queueing systems of certain types, the second component may characterize the costs connected with the functioning of the system.

REFERENCES

1. Ежов И. И. (1967). Исследования по теории случайных процессов с дискретной компонентой, докторская диссертация, Киев.
2. Ежов И. И., Лакатош Л. (1980). Про ймовірності станів системи масового обслуговування, що описується неоднорідним ланцюгом Маркова, Доповіді АН УРСР, сер. А, 6, 13-16.
3. Ежов И.И., Скороход А.В. (1969). Марковские процессы, однородные по второй компоненте, ч. I.,П., Теория вероятностей и ее применения ,№ 1., 3-14, № 4., 679-692.
4. Камке Э. (1971). Справочник по обыкновенным дифференциальным уравнениям, изд. Наука, Москва.
5. Олийнык И.Д. (1975). Исследование одного класса стохастических систем с доходами, кандидатская диссертация, КГУ, Киев.

6. Таиров М.Ф. (1978). Исследование систем массового обслуживания, локальные характеристики которых зависят от времени, кандидатская диссертация, КГУ, Киев.

7. Трикоми Ф.Дж. (1960). Интегральные уравнения. Изд. иностр. лит., Москва.

8. Lakatos L. (1984). On inhomogeneous Markov chain describing random walk in a band with homogeneous second component, Statistics and Probability, Proc. 3rd Pannonian Symp., Visegrád, Hung., 1982.

László Lakatos
Eötvös Loránd University
Computer Center
Budapest, XI.
Bogdánfy u. 10/b.
H-1117.

Proc. of the 5th Pannonian Symp. on Math. Stat., Visegrád, Hungary 1985
W. Grossmann, J. Mogyoródi, I. Vincze, W. Wertz, eds

THE EXTREMAL LIMIT PROBLEM - EXTENSIONS

E.G. MARTINS, D.D. PESTANA

Lisboa

Extensions of the classical extremal limit problem are considired, by relaxing the identical distribution hypothesis. In particular, refinements of the self-decomposable Mejzler's M class, similar to Urbanik's L_r classes in the additive scheme, are characterized. Extensions of the max-stable class S, in the spirit of Gnedenko's conjecture, are discussed.
AMS subject classification: 60F05, 60E07.

1. INTRODUCTION AND PRELIMINARIES

Let $\{X_i\}_{i\in N}$ be a sequence of independent random variables (r.v.'s), $\{F_i\}_{i\in N}$ the corresponding sequence of distribution functions (d.f.'s), $G_n(x) = \prod_{i=1}^{n} F_i(x)$ the d.f. of $M_n = \max(X_1,\ldots,X_n)$. Let us assume that there exists a proper d.f. $G(x)$ and attraction coefficients $\{(a_n,b_n)\}_{n\in N}$ with $a_n>0$, $b_n\in R$, such that

$$G_n(a_n x+b_n) \to G(x) \tag{1.1}$$

for all continuity point of G.

In the classical extreme value scheme the X_i's are assumed to be independent and identically distributed (i.i.d.). Hence $G_n(x) = F^n(x)$, where $F=F_i$, $i=1,2,\ldots$ Therefore the possible limit d.f.'s $G(x)$ are in the max-stable class S, i.e. belong to the set of solutions of the functional equation

$$G^n(\alpha_n x+\beta_n) = G(x) \tag{1.2}$$

for some $\alpha_n>0$, $\beta_n \in R$ (Fréchet's stability postulate). In the above situation we say that the d.f. $F(x)$ is in the domain of attraction of the max-stable d.f. $G(x)$, and denote this by $F\in \mathcal{D}(G)$. Max-stable d.f.'s are of the type of the Generalized Extreme Value (GEV) or von Mises-Jenkinson d.f. $G_\gamma(x)$, with support $\{x: 1+\gamma x\geq 0\}$,

$$G_\gamma(x) = \exp\{-(1+\gamma x)^{-1/\gamma}\} \tag{1.3}$$

for some $\gamma\in R$. Observe that for $\gamma>0$, $\gamma=0$, $\gamma<0$, the GEV d.f. $G(x)$ is in the Fréchet type family Φ_α with $\alpha=\gamma^{-1}$, resp. in the Gumbel type $\Lambda(x)$, resp. in the Weibull type family Ψ_α with $\alpha=-\gamma^{-1}$, where

$$\begin{aligned}
\Phi_\alpha(x) &= \begin{cases} 0 & x<0 \\ \exp\{-x^{-\alpha}\} & x\geq 0 \end{cases} & (\alpha>0) \\
\Lambda(x) &= \exp\{-\exp(-x)\} & x\in R \qquad\qquad (1.4) \\
\Psi_\alpha(x) &= \begin{cases} \exp\{-(-x)^{\alpha}\}, & x<0 \\ 1 & x\geq 0 \end{cases} & (\alpha>0)
\end{aligned}$$

See Gnedenko (1943) and Galambos (1978) for further discussion and references.

On the other hand, any univariate d.f. may arise as a limit d.f. in the sense (1.1), in case we assume no more than independence of the X_i's (Mejzler and Weissman, 1969; Green, 1976). This is related to the fact that any univariate d.f. $G(x)$ is max-infinitely divisible (max-I.D.), since $G^{1/n}(x)$, $n=1,2,\ldots$, is again a d.f. Observe however that max-infinite divisibility in the multivariate case is closely related to extremal processes (cf. Balkema and Resnick, 1977, Deheuvels, 1980).

The present note is concerned with the study of Mejzler's $\mathcal{M}$ class (see e.g. Galambos, 1978, (3.10.) which arises as limits in the sense of (1.1) with additional uniformity conditions. We shall define a sequence of subclasses $M_1 \supset M_2 \supset \ldots$ which appear as analogues of Urbanik's classes $L_0 \supset L_1 \supset L_2 \supset \ldots$ (Urbanik, 1973, Kumar and Schreiber, 1978) in the additive scheme. Likewise, we consider extensions of the maxstable class $S_1 \subset S_2 \subset S_3 \subset \ldots$, where S_r is the set of all G's limits in (1.1) where the F_i's come from at most r different types. This is in turn an analogue of Gnedenko's problem (Zolotarev and Koroljuk, 1961, Zinger, 1965, Mason, 1970 Jurek, 1981) in the max-scheme.

2. REFINEMENTS OF MEJZLER's M class

2.1 Definition

Let F(x) be a distribution function, $R = \sup \{x: F(x)<1\}$. The θ- function of F is

$$\theta_F(x) = \begin{cases} F(x) & \text{if } R=\infty \\ F(R-e^{-x}) & \text{if } R<\infty \end{cases} \tag{2.1}$$

2.2 Definition

The function K(x) bounded away from the origin is monotone of order r in $[0, \infty)$ iff $K^{(r-1)}(x)$ exists a.e., and $(-1)^{r-1} K^{(r-1)}(x)$ is non-negative and non-increasing.

From definition 2.2 it follows that if K(x) is monotone of order r, then $(-1)^n K^{(n)}(x)$ is non-negative, non-increasing and convex, n=0,...,r-2. If K(x) is monotone of order r for every r=1,2,..., we say that K(x) is completely monotone. It is well-known that completely monotone functions may be characterized as Laplace transforms of bounded non-decreasing functions. A similar Choquet type integral representation holds for monotone functions of order r:

$$K(x) = \int_0^{x^{-1}} (1-xt)^{r-1} \, dG(t) \tag{2.2}$$

where G(t) is bounded and non-decreasing (Vincze, 1980, Pestana, 1981).

Observe that $(-1)^{r-2} K^{(r-2)}(x)$ non-negative, non-increasing and convex is a necessary and sufficient condition for K(x) monotone of order r, $r \geq 2$.

2.3 Definition

Let $\{X_{nk}\}$, k=1,2,...,n, n=1,2,... be a triangular array such that for each n the X_{nk}, k=1,2,...,n are independent. The array is generated by a sequence $\{X_n\}$ if for each n and k, X_{nk} is distributed as X_k, and two triangular arrays are equivalent if they have the same limiting distribution.

It follows that $\{X_{nk}\}$ is equivalent to an array generated by $\{X_n\}$ if and only if

$$\frac{1}{n} \max_{1 \leq k \leq n} \{X_{nk} - b_n\} \quad \text{and} \quad \frac{1}{n} \max_{1 \leq k \leq n} \{X_n - b_n^*\} \tag{2.3}$$

have the same limiting distribution, where b_n and b_n^* are suitable centering constants.

2.4 Definition

A_1 is the class of sequences $\{X_n\}$ of independent random variables generating convergent triangular arrays. A_r is the class of sequences $\{X_n\} \in A_1$ such that for every c>0, the triangular array $\{X_{nk} = X_{[cn]+k}\}$ is equivalent to an array generated by a sequence from S_{r-1}. $\mathcal{M}_r$, r=1,2,... is the set of all possible limit distributions of suitably normed maxima of sequences $\{X_n\} \in A_r$, and $\mathcal{M}_\infty = \bigcap_{r=1}^{\infty} \mathcal{M}_r$. We shall denote $\mathcal{M}_0$ the set of all (max-infinitely divisible) univariate distributions.

Mejzler (1956) identified A_1 as the set of sequences of random variables whose distribution functions satisfy the uniformity condition for the maximum

i) $\lim \max_{1\le i\le n} \{1-F_i(a_n x+b_n)\} = 0$

ii) $\lim_{n\to\infty} \sum_{i=1}^{[nt]} [1-F_i(a_n x+b_n)] = w(t;x)$ exists and is

finite for any fixed $t\in(0,1]$ whenever it is finite for $t=1$ ($[y]$ the greatest integer less than or equal to y).

The same author characterized M_1 as the set of distribution functions F whose θ-function satisfies the functional equation

$$\theta_F(x) = \theta_F(x+a)\ \theta_a(x) \quad \text{for all } a>0$$

where $\theta_a(x)$ is non-decreasing. This may be extended as follows:

Theorem 2.1

A distribution function F belongs to M_r if and only if for every a>0 the corresponging θ-function satisfies the functional equation

$$\theta_F(x) = \theta_F(x+a)\ \theta_a(x) \tag{2.4}$$

where θ_a is the θ-function of some function proportional to $F_a \in M_{r-1}$.

Proof:

The result follows from Mejzler's (1956) characterization of M_1, and from Proposition 1 in Urbanik (1973), with the necessary adaptations from the additive to the max-scheme.

Theorem 2.2

The distribution function F belongs to M_r if and only if $F(x) = \exp[-K(x)]$, with K(x) monotone of order r+1.

Proof:

Assume that the result is true for r=n. For $F\in M_{n+1}$, we have then

$$\theta_F(x) = \theta_F(x+a)\exp[-K_a(x)] \quad \text{for all } a>0$$

where $K_a(x)$ is monotone of order n+1. Therefore, with $K(x) = -\log\theta_F(x)$,

$$(-1)^n K_a^{(n)}(x) = (-1)^n [K^{(n)}(x) - K^{(n)}(x+a)]$$

is non-increasing for every a>0. Therefore, for $x_1<x_2$

$$(-1)^n [K^{(n)}(x_1) - k^{(n)}(x_1+a)] \geq (-1)^n [K^{(n)}(x_2) - K^{(n)}(x_2+a)].$$

Putting $x=x_2=x_1+a$

$$(-1)^n K^{(n)}(x) \leq \frac{(-1)^n K^{(n)}(x-a) + (-1)^n K^{(n)}(x+a)}{2}$$

and hence $(-1)^n K^{(n)}(x) = (-1)^{(n+2)-2} K^{((n+2)-2)}(x)$ is convex. From the observations on definition 2.2, it follows that K(x) is monotone of order n+2.

For r=0, observe that Mejzler's result (1956) may be rephrased saying that θ_F is log-concave, x>0.

From the integral representation of monotone functions of order r, the proof that the θ-function of $\exp[-K(x)]$ with K(x) monotone of order r+1 satisfies the functional equation (2.4) is straighforward.

Corollary 2.1

$M_\infty = \{F: \theta_F(x) = \exp[-K(x)],\ K(x) \text{ completely monotone}\}$ is the smallest class containing all max-stable distributions which is closed under products and passages to the limit.

Therefore, the class M_∞ is a non-trivial extension of the class of max-stable distributions. This may have some practical relevance, since there is a long-standing claim, mainly from geophysicists, that some non-extreme value d.f.'s such as normal, lognormal, log-gamma, provide a much better fit to experimental data than the classical extreme value d.f.'s (1.4) (cf. Galambos, 1978). Mejzler's M class and its refinements M_r, $r=1,2,\ldots$, and specially M_∞ provide a theoretical framework for developments in this area - in fact, the reason for the above mentioned claims may indeed lie in the failure of the identical distribution hypothesis.

In particular the case $G(x) = \exp\left[-K(x)\right]$ with $K(x)$ completely monotone generalizes nicely the widely used Gumbel model, where $K(x) = e^{-x}$. Here, an ultimate adjustement of $K(.)$ to real data could use Gumbel model in a first step. Furthermore observe that any d.f. $F(x)$ with c.m. probability density function (known to be I.D., in the additive scheme) is in M_∞: $K'(x) = F'(x)/F(x)$, and $1/F(x)$ is c.m., being the composition of the c.m. function $1/y$ with the positive function, having c.m. derivative, $F(x)$.

Since we have been stressing similarities of approach and results in the additive and max schemes, it seems worthwile to record that Kumar and Schreiber's (1978) results on Urbanik's L_r classes may be rephrased in terms of $(r+1)$-monotonicity of $e^x\,\theta(e^x)$, where $\theta(x)$ is Lévy's spectral measure (Pestana, 1984). It is well known that $e^x\theta(e^x)$ is convex whenever $\phi(t)\in L$.

3. GNEDENKO'S PROBLEM

Consider independent r.v.'s $\{X_i\}_{i\in N}$, let $M_n = \max(X_1,\ldots,X_n)$ and assume that n_1 of the n X_i's have d.f. $F_1(X)$, $n_2=n-n_1$ of the n X_i's have d.f. $F_2(x)$. Assume further that $F_1\in D(G_{\gamma_1})$, $F_2\in D(G_{\gamma_2})$, with $|\gamma_1| \geq |\gamma_2|$, and that there exist coefficients $\alpha_n>0$, $\beta_n\in R$ such that, with $n_i\to\infty$, $i=1,2$, $(F_n^{n_1}\,F_2^{n_2})\,((x-\beta_n)/\alpha_n)$ converges to some proper d.f. $F(x)$.

As Resnick (1972) points out, in a slightly different context (domains of attraction for products of d.f.'s), the above scheme is uninteresting but for the case where F_1 and F_2 have the same right-endpoint - and therefore either $\gamma_1, \gamma_2 \geq 0$, or $\gamma_1, \gamma_2 \leq 0$.

Tucker's (1968) results are readily rephrased in terms of the max scheme, and from this it follows that if $n_1 \approx n_2$ then $G \equiv G_{\gamma_1}$.

On the other hand, if $n_2 >> n_1$, and in particular if $n_2/n \to 1$, the situation becomes completely different as shown in the following example:

$F_1(x)=1-x^{-1}$, $x \geq 1$, $F_2(x)=1-x^{-2}$, $x \geq 1$ (and hence $F_i \in D(\Phi_i)$, $i=1,2$), $n_1=[\sqrt{n}\,]$, $\alpha_n=\sqrt{n}$, $\beta_n=0$. We thus have

$$G(x) = \lim_{n\to\infty} \left[(1-x^{-1}/\sqrt{n})^{\sqrt{n}}\right]^{[\sqrt{n}\,]/\sqrt{n}} \cdot \left[(1-x^{-2}/n)^{n}\right]^{(n-[\sqrt{n}\,])/n} =$$

$$\exp(-x^{-1}) \exp(-x^{-2}) = (\Phi_1\Phi_2)(x)$$

(the choice from n_1 follows from a suggestion in Zinger, 1965).

This leads us to Gnedenko's conjecture that any d.f. in S_r (i.e. the class of limit distributions of suitably normed maxima of independent r.v.'s whose d.f.'s come from at most r different types) is the product of at most r max-stable d.f.'s.

Gnedenko's conjecture has been originaly put forward in the additive scheme, where it has been shown to hold true for r=2 (Zolotarev and Koroljuk, 1961, for real-valued r.v.'s; Jurek, 1981, for Banach spece valued r.v.'s), but to be false for $r \geq 3$ (Zinger, 1965).Mason (1970) investigated thoroughly additional hypotheses for the validity of Gnedenko's conjecture and related problems.

Results in the max-scheme are quite similar to the ones in the additive scheme, if we restrict ourselves to generalized Pareto tails with either $\gamma>0$ or $\gamma<0$. The reason for that similarity is clear: tail regular variation (both tails, in the additive scheme; upper tail, in the max scheme) underlies all interesting properties. The case $\gamma=0$ brings in some oddities, easily understood in the light of Resnick's (1972) concept of A-equivalence.

As in the additive scheme, F_1F_2 may be attracted by some $G(x)$ without either F_1 or F_2 being in any domain of attraction (Resnick, 1972, prop. 1). On the other hand, an example due to de Haan (Resnick, 1972, prop. 2.)

$$F_1(x) = 1 - \exp(-x), \quad x \geq 0$$

$$F_2(x) = 1 - \exp\{-x[1 - \tfrac{1}{4}\sin(\log x) - \tfrac{1}{4}\cos(\log x)]\}, \quad x \geq 1$$

exhibits one of the oddities of the case $\gamma=0$: both $F_1, F_2 \in \mathcal{D}(\Lambda)$, but F_1F_2 isn't attracted by $\Lambda(x)$; hence, even the case $r=2$ requires, in the max scheme, additional restrictions on tail-behaviour.

4. FINAL COMMENTS

In the chains

$$\mathcal{S} \equiv \mathcal{S}_1 \subset \mathcal{S}_2 \subset \ldots \subset \mathcal{S}_r \subset \mathcal{S}_{r+1} \subset \ldots; \quad \mathcal{M}_\infty \subset \ldots \subset \mathcal{M}_{r+1} \subset \mathcal{M}_r \subset \ldots \subset \mathcal{M}_1 \subset \mathcal{M}_0 \equiv \mathcal{M} \subset \mathcal{M}_{-1}$$

the classes $\mathcal{M}_r$ (and in particular $\mathcal{M}_\infty$) seem worthwile to exploit from a modelling viewpoint; on the other hand, Gnedenko's problem is a nice mathematical problem, but the "unnatural" fact that $\mathcal{S}_r$ is in general broader that the class of products of products of r max-stable d.f.'s (and hence in particular $\mathcal{M}_\infty \subset \mathcal{S}_\infty = \bigcup_{r=1}^{\infty} \mathcal{S}_r$) is discouraging, for practical purposes.

An interesting open problem is to characterize the smallest class under pointwise products and limits generated by $G_\gamma(x)$, $\gamma \in \mathcal{C} \subset \mathbb{R}$; this seems worth to be investigated, along with tests for shifts on the regular variation exponent of tails (for instance score tests, cf. Pestana, 1984a).

Delphic semigroups are perhaps the correct frame to investigate sums and maxima; Zempléni (1985) has recently prosented some interesting results on the Delphic approach to the extreme value scheme. The self-decompasibility property (2.5) seems worth to be looked into in the light of Székely's (1985) concepts of autophagy and allelophagy.

AKNOWLEDGEMENTS

The authors are grateful to Calouste Gulbenkian Foundation (Lisbon) for financial support to present this paper at the V. Pannonian Symposium on Mathematical Statistics, Visegrád, May 1985, and to the anonymous referee for helpful and stimulating comments.

REFERENCES

Balkema, A. A. and Resnick, S.I. (1977) Max-infinite divisibility, J. Appl. Probab. 14, 309-319.

Deheuvels, P. (1980) The decomposition of infinite order and extreme multivariate distributions, Asymptotic Theory of Statistical Tests and Estimation, 259-286 (I.M. Chakravarti, ed.), Academic Press, New York.

Galambos, J. (1978) The Asymptotic Theory of Extreme Order Statistics, Wiley, New York.

Gnedenko, B. V. (1943) Sur la distribution limite du terme maximum d'une série aléatoire, Ann. Math. 44. 423-453.

Green, R. F. (1976) Partial attraction of maxima, J. Appl. Probab. 14, 159-163.

Jurek, Z. J. (1981) The problem of B. V. Gnedenko on Banach spaces. Bull. Acad. Polonaise Sc., Ser. Sc. Math., 29, 399-407.

Kumar, A. and Schreiber, B. M. (1978) Characterization of subclasses of class L probability distributions, Ann. Probab. 6, 279-293.

Mason, J. D. (1970) Convolutions of stable laws as limit distributions of partial sums. Ann. Math. Statist. 41, 101-114.

Mejzler, D. (1956) On the problem of the limit distribution for the maximal term of a variational series, L'vov Politechn. Inst. Naucn Zp. (Fiz.-Mat.) 38, 90-109 (in Russian).

Mejzler, D. and Weissman, I. (1969) On some results of N. V. Smirnov concerning limit distributions for variational series, Ann. Math. Statist. 40, 480-491.

Pestana, D.D. (1981) Extreme point methods and functions with monotone derivatives in probability theory. Abs. 14th European Meeting Statist., Wroclaw, 249-250.

Pestana, D.D. (1984) Urbanik's L_r classes, Mejzler's M_r classes, and higher order monotone functions. Actas III Coloquio Estatistica Investigacao Operacional, Lagos.

Pestana, D.D. (1984a) Time series fitting of almost stable stochastic processes, to appear in Proc. J. Bolyai Math Soc. Coll. Goodness of Fit, Debrecen.

Resnick, S. I. (1972) Products of distribution functions attracted to extreme value laws, J. Appl. Probab. 8, 781-793.

Székely, G. J. (1985) Autophage and allelophage distributions. Abs. V Pannonian Symp. Math. Statist., Visegrad, 76.

Tucker, H. G. (1968) Convolutions of distributions attracted to stable laws, Ann. Math. Statist. 39, 1381-1390.

Urbanik, K. (1973) Limit laws for sequences of normed sums satisfying some stability conditions, Multivariate Analysis, III (P.R. Krishnaiah, ed.) 225-237, Academic Press, New York.

Vincze, I. (1980) On a certain class of multiply monotone and completely monotone functions, Coll. Math. Soc. J. Bolyai. 35 - Functions, Series, Operators, 1259-1266.

Zempléni, A. (1985) On the arithmetical properties of the multiplicative structure cs probability distribution functions, Abs. V. Pannonian Symp. Math. Statist., Visegrad, 98.

Zinger, A.A. (1965) On a class of limit distributions for normed sums of independent random variables, Theor. Probab. Appl. 10, 607-620.

Zolotarev, V. M. and Koroljuk, V. S. (1961) On a hypothesis proposed by B. V. Gnedenko. Theor. Probab. Appl. 6, 431-435.

E. G. Martins and D. D. Pestana
Department of Statistics and Computation (DEIOC)
Centro de Estatistica e Aplicacoes (INIC)
Faculdade de Ciencias da Universidade de Lisboa
58 Rua da Escola Politecnica
1294 LISBOA CODEX
Portugal

Proc. of the 5th Pannonian Symp. on Math. Stat., Visegrád, Hungary 1985
W. Grossmann, J. Mogyoródi, I. Vincze, W. Wertz, eds

HITTING A SMALL GROUP OF MIDDLE RANKED CANDIDATES IN THE SECRETARY PROBLEM

T.F. MÓRI

Budapest

In (1982) Rose dealt with the optimal stopping of a random permutation of size n in order to select the median object or an object whose rank lies in a small neighbourhood of $\frac{n}{2}$, only basing on relative ranks. He found that

$$P(\text{selecting the median}) \gtrsim \frac{1}{n},$$

$$P(\text{hitting the set } \tfrac{n}{2} \pm \tfrac{1}{2} n^{1/2+\varepsilon}) \to 1,$$

as $n\to\infty$. Improving his results we show that for the optimal strategy

$$P(\text{selecting the median}) \sim 0.57 n^{-1/3}$$

$$\lim_{c\to\infty} \lim_{n\to\infty} P(\text{hitting the set } \tfrac{n}{2} \pm cn^{1/3}) = 1.$$

Sharp asymptotic results are proved in more general models.

AMS subject classification: 60G40, 62L15

1. INTRODUCTION

In the classical secretary problem n rankable objects (candidates) appear in a random order. The decision-maker is to select the rank 1 object with maximal probability, under the following constraints:

- at each arrival it must be decided immediately whether the object just appeared is to be accepted or rejected, later recall is not permitted;
- the only information that can be utilized in the selecting procedure is the sequence of relative ranks with respect to the preceding arrivals.

Several versions and generalizations of the original problem have been investigated since the classical paper by Gardner (1960). Secretary problems are commonly thought the most popular part of the theory of optimal stopping.

There exist generalizations imputing the decision-maker a general preference function of the ranks according to which his choice should be optimized. In other papers the number of objects is supposed to be a random variable. There are investigations allowing uncertainty in selection and/or recall of rejected objects. These examples constitute but a minor part of the huge literature on secretary type problems (for further references see the survey paper by Freeman(1983)).

In his paper (1982) Rose introduced the problem of selecting the median object of a random sequence, on the suggestion of DeGroot. The novelty of his investigations was the appearance of a non-monotone preference function. In most of the earlier generalizations of the classical best choice problem the preference ordering on the set of ranks was monotone: the object of rank r was at least as desirable as that of rank r+1. An exception of the monotony appears in Szajowski (1982), where the object is to select the rank r candidate, r>1 fixed. The main difficulty of the median problem is to recognize the median object when it appears in the sequence of observations. If the rank r candidate appears then clearly he will rank not greater than r among the objects observed so far, but the relative rank of the median object when it appears is not necessarily even close to the relative median; in fact, the size of the set of possible relative ranks is proportional to n. It is no surprise that the probability of selecting the median object tends to 0 as n gets large, while the probability of selecting the object of rank r is kapt bounded from 0.

Rose found that the probability of selecting the median was bounded below by $n^{-1/2}$. He also showed that the probability of selecting an object whose rank lies within the $\pm n^{1/2+\varepsilon}$ neighbourhood of $\frac{n}{2}$ was asymptotically unity, for any positive ε. Since all his estimations were one-sided, it remained open if they were asymptotically sharp. Rose conjectured but could not proved it. The aim of the present paper is to give a negative answer to this question. In fact, we can prove that if the decision-maker is guided by the optimal strategy, the probability of success is proportional to $n^{-1/3}$ and the asymptotic probability of hitting the $\pm c\, n^{1/3}$ neighbourhood of the median tends to 1 as $c\to\infty$.

2. GENERALIZATION OF THE MEDIAN PROBLEM

Let X_s denote the rank of the object observed at stage s (i.e. $X_1,\dots,X_n$ is a random permutation of $1,2,\dots,n$). Let Y_s denote its relative rank with respect to the preceding arrivals:

$$Y_s = 1 + |\{i:\ 1\leqq i<s,\ X_i< X_s\}|.$$

Then Y_s is uniformly distributed on $\{1,\dots,s\}$ and $Y_1,\dots Y_s$, $X_{s+1},\dots,X_n$ are independent for $s=1,2,\dots,n$. Let

$$u(a,r,s)= P(X_s=a\,|\,Y_s=r),\quad r\leqq s\leqq n,\qquad r\leqq a\leqq n,$$

then one can easily verify that

$$u(a,r,s) = \frac{\binom{a-1}{r-1}\binom{n-a}{s-r}}{\binom{n}{s}} .$$

<u>Lemma 1.</u> Let $D=r-a\frac{s}{n}$, $B=\left[s(1-\frac{s}{n})\frac{a}{n}(1-\frac{a}{n})\right]^{-1}$ and suppose that B, BD and B^2D^3 tend to 0. Then

$$u(a,r,s)\sim\frac{s}{n}\sqrt{B}\,\varphi(D\sqrt{B}),$$

where $\varphi(x)=\frac{1}{\sqrt{2\pi}}\exp\left(-\frac{1}{2}x^2\right)$, the standard normal density.

<u>Proof.</u> Since $B=\left(\frac{1}{s}+\frac{1}{n-s}\right)\left(\frac{1}{\frac{a}{n}}+\frac{1}{1-\frac{a}{n}}\right)$ and $BD\to 0$, we obtain that

$$\frac{D}{s\frac{a}{n}},\quad \frac{D}{s(1-\frac{a}{n})},\quad \frac{D}{(n-s)\frac{a}{n}},\quad \frac{D}{(n-s)(1-\frac{a}{n})}$$

converge to 0. These terms also appear in the following identities:

$$\frac{r}{s} = \frac{a}{n}\left(1 + \frac{D}{s\frac{a}{n}}\right), \qquad 1 - \frac{r}{s} = \left(1 - \frac{a}{n}\right)\left(1 - \frac{D}{s(1-\frac{a}{n})}\right),$$

$$\frac{a-r}{n-s} = \frac{a}{n}\left(1 - \frac{D}{(n-s)\frac{a}{n}}\right), \qquad 1 - \frac{a-r}{n-s} = \left(1 - \frac{a}{n}\right)\left(1 + \frac{D}{(n-s)(1-\frac{a}{n})}\right), \tag{1}$$

which imply

$$\frac{1}{r} + \frac{1}{s-r} + \frac{1}{a-r} + \frac{1}{n-s-a+r} = B(1 + O(DB)). \tag{2}$$

Remembering the Stirling formula we can write

$$\binom{m}{k} = \sqrt{\frac{m}{2\pi k(m-k)}} \left(\frac{k}{m}\right)^{-k}\left(1 - \frac{k}{m}\right)^{-(m-k)} \exp\left\{O\left(\frac{1}{k} + \frac{1}{m-k}\right)\right\},$$

and by this

$$u(a,r,s) = \frac{r}{a}\,\frac{\binom{s}{r}\binom{n-s}{a-r}}{\binom{n}{a}} = \left[\frac{s(n-s)r(n-a)}{2\pi a(s-r)(a-r)(n-s-a+r)n}\right]^{1/2}.$$

$$\cdot\left(\frac{a}{n}\right)^{a}\left(1-\frac{a}{n}\right)^{n-a}\left(\frac{r}{s}\right)^{-r}\left(1-\frac{r}{s}\right)^{-(s-r)}\left(\frac{a-r}{n-s}\right)^{-(a-r)}\left(1-\frac{a-r}{n-s}\right)^{-(n-s-a+r)}.$$

$$\cdot \exp\{O(B)\}.$$

Applying (1) to the first row on the right-hand side of this equality we get

$$\frac{s}{n}\left(\frac{\frac{r}{s}\left(1-\frac{a}{n}\right)\frac{n}{n-s}\,\frac{1}{s}}{2\pi\frac{a}{n}\left(1-\frac{r}{s}\right)\frac{a-r}{n-s}\left(1-\frac{a-r}{n-s}\right)}\right)^{1/2} = \frac{s}{n}\sqrt{\frac{B}{2\pi}}\,(1 + O(DB)).$$

The formula in the second row can be transformed again by the help of (1) into

$$\left(1+\frac{D}{s\frac{a}{n}}\right)^{-D-s\frac{a}{n}}\left(1-\frac{D}{s(1-\frac{a}{n})}\right)^{D-s(1-\frac{a}{n})}\cdot$$

$$\cdot\left(1-\frac{D}{(n-s)\frac{a}{n}}\right)^{D-(n-s)\frac{a}{n}}\left(1+\frac{D}{(n-s)(1-\frac{a}{n})}\right)^{-D-(n-s)(1-\frac{a}{n})}.$$

Let us take the logarithm of this expression and expand it into Taylor series. Using (2) we obtain

$$\left(1+\frac{D}{s\frac{a}{n}}\right)^{-D}\left(1-\frac{D}{s(1-\frac{a}{n})}\right)^{D}\left(1-\frac{D}{(n-s)\frac{a}{n}}\right)^{D}\left(1+\frac{D}{(n-s)(1-\frac{a}{n})}\right)^{-D}=$$

$$=\exp\{-D(DB+O(D^2B^2))\},$$

further

$$\left(1+\frac{D}{s\frac{a}{n}}\right)^{-s\frac{a}{n}}\left(1-\frac{D}{s(1-\frac{a}{n})}\right)^{-s(1-\frac{a}{n})}=$$

$$=\exp\left\{-D+\frac{1}{2}\frac{D^2}{s\frac{a}{n}}+O\left(\frac{D^3}{(s-\frac{a}{n})^2}\right)+D+\frac{1}{2}\frac{D^2}{s(1-\frac{a}{n})}+O\left(\frac{D^3}{s^2(1-\frac{a}{n})^2}\right)\right\},$$

and similarly

$$\left(1-\frac{D}{(n-s)\frac{a}{n}}\right)^{-(n-s)\frac{a}{n}}\left(1+\frac{D}{(n-s)(1-\frac{a}{n})}\right)^{-(n-s)(1-\frac{a}{n})}=$$

$$= \exp\left\{-D + \frac{1}{2}\frac{D^2}{(n-s)\frac{a}{n}} + O\left(\frac{D^3}{(n-s)^2(\frac{a}{n})^2}\right) + D + \frac{1}{2}\frac{D^2}{(n-s)(1-\frac{a}{n})} + \right.$$

$$\left. + O\left(\frac{D^3}{(n-s)^2(1-\frac{a}{n})^2}\right)\right\} .$$

Finally, all these together give

$$\exp\{-\tfrac{1}{2}D^2B + O(D^3B^2)\},$$

thus

$$u(a,r,s) = \frac{s}{n}\sqrt{B}\,\varphi(D\sqrt{B})(1 + O(B + BD + B^2D^3)),$$

which was to be proved.

The decision-maker is to select the rank a object with maximal probability. He has to count on some uncertainty at the selection; that is, any chosen candidate can refuse to be selected with probability 1-p.

Let $v(s)$ denote the maximum probability of success when no attempt can be made to stop before stage s. The standard backward induction argumentation leads to the following recurrence relations:

$$v(n) = \frac{1}{n},$$

$$v(s) = \sum_{r=1}^{s} \frac{1}{s} \max\{pu(a,r,s) + (1-p)v(a+1),\ v(s+1)\}, \quad s<n.$$

or equivalently,

$$v(s) - v(s+1) = \frac{p}{s} \sum_{r=1}^{s} [u(a,r,s) - v(s+1)]^+ , \tag{3}$$

where $x^+ = \max\{x,0\}$.

We are interested in the probability of success $v(1)$ and in the optimal strategy, which can also be read from (3): observing relative rank r at stage s, try to stop if $u(a,r,s) > v(s+1)$.

With the notation

$$\Gamma = \{(s,r): \text{ either } s=n \text{ and } 1\leq r\leq n \text{ or } 1\leq r\leq s<n \text{ such that } u(a,r,s) > v(s+1)\}$$

the optimal strategy can be rewrited as "try to stop whenever the process visits Γ". It is easy to see that $u(a,r,s)$, as a function of r, is unimodal for fixed a and s, with mode $\approx a\frac{s}{n}$, thus one can define the boundaries $h_-(s)$ and $h_+(s)$ such that

$$\Gamma = \{(s,r): h_-(s) \leq r - \frac{a}{n}s \leq h_+(s), \quad s\geq s_0\} .$$

Suppose a and n-a tend to infinity. Then the same holds for

$$m = \left(\frac{1}{a} + \frac{1}{n-a}\right)^{-1/3} ,$$

but at a lower rate. In the median problem $a=[n/2]$ and $m\sim(\frac{n}{4})^{1/3}$. In order to describe the asymptotic form of Γ and the success probability let us define the function $f(y)$, $y\geq 0$, as the solution of the differential equation

$$f'(y) = p \int_{+\infty}^{-\infty} \left(\frac{1}{\sqrt{y}}\varphi\left(\frac{x}{\sqrt{y}}\right) - f(y)\right)^+ dx, \qquad f(0) = 0. \tag{4}$$

Since f increases, there exists a unique value y_0 at which $f(y_0) = \frac{1}{\sqrt{2\pi y_0}}$ and f is constant for $y\geq y_0$. Let φ^{-1} denote the inverse of the density function restricted to the positive reals. Then the integration in (4) can be taken between $\pm\sqrt{y}\,\varphi^{-1}(\sqrt{y}\, f(y))$, provided $y<y_0$.

<u>Theorem 1.</u> Suppose $\frac{m}{n}(n-s)\to y$. Then

$$\frac{1}{m}h_\pm(s) \to \pm\sqrt{y}\,\varphi^{-1}(\sqrt{y}f(y)), \text{ if } y<y_0, \tag{5}$$

and

$$mv(s) \to f(y), \tag{6}$$

particulary,

$$m\,P\,(\text{success}) \to \frac{1}{\sqrt{2\pi y_0}} \,.$$

Before proving these assertions we outline a heuristic method how to find the proper norming factors for s, v(s) and $h_{\pm}(s)$. If we arrive at stage s (s is supposed comparable with n) then in Lemma 1 $B=O\left(\frac{n}{(n-s)m^3}\right)$ and it follows the existence of an $O\left(\frac{1}{\sqrt{B}}\right)$ sized subset of relative ranks, where the probability of success is of $O(\sqrt{B})$. It is no worth stopping at relative rank <u>outside</u> this set. Then the probability that sooner or later after stage s we hit the subset "worth stopping" is

$$1-\left(1-O\left(\frac{1}{n\sqrt{B}}\right)\right)^{O(n-s)} = O\left(\min\left\{\frac{n-s}{n\sqrt{B}}\,,\,1\right\}\right)\,,$$

thus the probability of success is $O\left(\min\left\{\frac{n-s}{n}\,,\,B\right\}\right)$. This is of maximal order if $n-s=O\left(\frac{n}{m}\right)$, when the success probability is $O\left(\frac{1}{m}\right)$ and the size of "worth stopping" subset is $O(m)$.

<u>Proof.</u> We are going to apply the differential equation method developed by Mucci(1973). Set

$$n-s = \frac{n}{m}y\,, \qquad \text{that is,} \qquad y = m\frac{n-s}{n}\,,$$

$$r-\frac{a}{n}s = mx, \qquad \text{that is,} \qquad x = \frac{1}{m}(r-\frac{a}{n}s)\,,$$

and for the above s denote m v(s) by $f_n(y)$. Then the increments of y and x are $\Delta y = \frac{m}{n}$ and $\Delta x = \frac{1}{m}$, resp. The increment of f_n is derived from the recurrence relation (3) and Lemma 1:

$$B = \frac{1}{m^2 y}(1 + O(\tfrac{1}{m}))\,, \qquad D = mx,$$

hence

$$\frac{\Delta f_n(y)}{\Delta y} = p\left(1 + O(\tfrac{1}{m})\right) \sum_x \left[\frac{1}{\sqrt{y}} \varphi(\frac{x}{\sqrt{y}} + O(\tfrac{1}{m}) - f_n(y-\Delta y)\right]^+ \Delta x.$$

Now the standard Picard method for ordinary differential equations gives the existence of lim $f_n = f$ and the validity of (4). Then (5) immediately follows.

Table 1 contains the limiting value $\frac{1}{\sqrt{2\pi y_0}}$ as a function of p. In Fig.1 the asymptotic shape of the stopping area Γ is plotted for different values of p.

3. SELECTION FROM A SMALL INTERVAL

In Section 2 we have seen that the probability of finding the rank a object approaches 0 at rate $\frac{1}{m}$. In the sequel the following question will be answered. What is the minimum size of intervals containing a in the middle which can be hit with fixed probability? It is no surprise that the length of such an interval should be at least of O(m). On the one hand, it is easy to see that the optimal for a policy selects any given object within the $\pm$ cm neighbourhood of a with probability $O(\frac{1}{m})$. On the other hand, for arbitrary strategy it is impossible to hit the interval a $\pm$ cm with probability asymptotically greater than $\frac{2c}{\sqrt{2\pi y_0}}$

Let A = [a-cm, a+cm], the winning set of ranks. Denote by U(A,r,s) the conditional probability that the rank of the object observed at stage s belongs to A supposed it has relative rank r:

$$U(A,r,s) = \sum_{a\in A} u(a,r,s).$$

Next, let V(s) denote the probability of hitting the set A according to an optimal stopping rule, given that the process has not been stopped during the first s-1 stages. Analogously to (3) we have

$$V(n) = \frac{2cm}{n}$$

$$V(s)-V(s+1) = \frac{p}{s}\sum_{r=1}^{s}\left[U(A,r,s)-V(s+1)\right]^{+}. \qquad (7)$$

Since U is unimodal in r, the support set of the optimal strategy is of similar type as it has been in Section 2,

$$\Gamma = \{(s,r):\ H_{-}(s) \leq r - \frac{a}{n}s \leq H_{+}(s),\quad s \geq s_0\},$$

where s_0 is the first stage at which we should not automatically pass the observation.

Let us parametrize our model with $\nu = c^3 p$. For fixed $\nu > 0$ define f(y) as the solution of the differential equation

$$f'(y) = \nu \int_{-\infty}^{+\infty} \left[\Phi\left(\frac{x+1}{\sqrt{y}}\right) - \Phi\left(\frac{x-1}{\sqrt{y}}\right) - f(y)\right]^{+} dx, \quad f(0) = 0, \qquad (8)$$

where $\Phi(x) = \int_{-\infty}^{x} \varphi(t)dt$, the standard normal distribution function.

Let y_0 denote the only positive solution of the equation $\Phi\left(\frac{1}{\sqrt{y}}\right) - \Phi\left(\frac{-1}{\sqrt{y}}\right) = f(y)$. Then the integrand in (8) is identically equal to 0 for $y \geq y_0$, thus f is constant there. For $y \leq y_0$ there exists a unique positive $x = H(y)$ for which

$$\Phi\left(\frac{x+1}{\sqrt{y}}\right) - \Phi\left(\frac{x-1}{\sqrt{y}}\right) = f(y).$$

<u>Theorem 2.</u> Suppose $\frac{m}{nc^2}(n-s) \to y$. Then $V(s) \to f(y)$ and for $y \leq y_0$, $\frac{1}{mc} H_{\pm}(s) \to \pm H(y)$.

Particularly,

$P(\text{success}) \to f(y_0)$.

<u>Proof.</u> The proof is similar to that of Theorem 1. Define x and y by

$$n-s = \frac{nc^2}{m}y, \quad r - \frac{a}{n}s = mcx$$

and let $f_n(y) = V(s)$ with the above s. Then from Lemma 1 one can derive

$$U(A,r,s) = \Phi\left(\frac{x+1}{\sqrt{y}}\right) - \Phi\left(\frac{x-1}{\sqrt{y}}\right) + O\left(\frac{1}{n}\right).$$

With $x = \frac{1}{mc}$ and $y = \frac{m}{nc^2}$, (7) can be rewritten as

$$\frac{\Delta f_n(y)}{\Delta y} = \nu\left(1+O\left(\frac{1}{m}\right)\right) \sum_x \left[\Phi\left(\frac{x+1}{\sqrt{y}}\right) - \Phi\left(\frac{x-1}{\sqrt{y}}\right) + C\left(\frac{1}{m}\right) - f_n(y-\Delta y)\right]^+ \Delta x,$$

from which (8) follows as $n\to\infty$. Now the proof can easily be completed.

Finally, we show that the limit of the success probability can be arbitrarily close (but never equal) to 1 for sufficiently large ν. Let us denote $f(y_0)$ by $\psi(\nu)$ so as to indicate the dependence on ν.

<u>Theorem 3.</u>

$$\left(1- e^{-\sqrt{\frac{\nu}{8}}}\right)^2 \leq \psi(\nu) \leq 1 - e^{-\sqrt{\nu}}.$$

<u>Proof.</u> Suppose p=1, then the probability V(s) of selecting an object from A after stage s is clearly less than the probability of observing such an object after stage s. That is,

$$V(s) \leq 1 - \prod_{i=s}^{n}\left(1 - \frac{2cm}{i}\right) \leq 1 - \left(1 - \frac{2cm}{s}\right)^{n-s} \to 1 - e^{-2c^3y}$$

if $\frac{n-s}{n} m \to c^2 y$. Applying the estimation $1 - \Phi(x) \leq \frac{1}{2}e^{-x^2/2}$ (x>0) we obtain

$$1 - e^{-\frac{1}{2y_0}} \leq \Phi\left(\frac{1}{\sqrt{y_0}}\right) - \Phi\left(\frac{-1}{\sqrt{y_0}}\right) = f(y_0) \leq 1 - e^{-2\nu y_0},$$

hence $y_0 \geq \frac{1}{2\sqrt{\nu}}$ and

$$\psi(\nu) = f(y_0) \leq 1 - e^{-2\nu y_0} = 1 - e^{-\sqrt{\nu}} .$$

On the other hand, consider the following strategy: observing relative rank r at stage s stop if

$$n - s \leq \frac{n}{m}\sqrt{\frac{c}{8}} \quad \text{and} \quad |r - \frac{a}{n}s| \leq \frac{1}{2}mc . \tag{9}$$

The probability of stopping is estimated by

$$1 - \prod_{i=s}^{n} \left(1 - \frac{mc}{i}\right) \geq 1 - \left(1 - \frac{mc}{n}\right)^{n-s} \to 1 - \exp\{-\sqrt{\frac{\nu}{8}}\}.$$

Once stopped, the (conditional) probability of hitting A is clearly not less than $U(A, r_0, s_0)$, where r_0 and s_0 fulfil inequality (9) with equality. From Lemma 1 it follows that

$$\lim U(A, r_0, s_0) = \Phi\left(3c\sqrt[4]{\frac{c}{2}}\right) - \Phi\left(-c\sqrt[4]{\frac{c}{2}}\right) \geq$$

$$\geq 1 - 2\left(1 - \Phi\left(\sqrt[4]{\frac{\nu}{2}}\right)\right) \geq 1 - \exp\{-\sqrt{\frac{\nu}{8}}\},$$

which immediately provides the lower estimate of the theorem.

<u>Remark.</u> One can readily verify in a similar way that in the problem of selecting the rank <u>a</u> object the probability of success is asymptotically less than $\frac{1}{m}\left(\frac{p}{2\pi}\right)^{1/3}$, from which another upper estimate is obtained for $\psi(\nu)$:

$$\psi(\nu) \leq cm \frac{1}{m}\left(\frac{p}{2\pi}\right)^{1/3} = \left(\frac{\nu}{2\pi}\right)^{1/3},$$

but it is uniformly worse than that of Theorem 3.

For different ν's the values of y_0 and $f(y_0) = \psi(\nu)$ are provided (Table 2). In Fig.2 the asymptotic shape of the stopping area is shown. In order to demonstrate the true dependence on the length of the target interval, $p = 1$ thus $\nu = c^3$ is supposed and $cH(c^{-2}y)$ is plotted instead of $H(y)$.

Table 1. $\psi(p) = \frac{1}{\sqrt{2\pi y_0}}$

p	ψ(p)	p	ψ(p)	p	ψ(p)	p	ψ(p)
0.10	0.167	0.35	0.254	0.60	0.304	0.85	0.342
0.15	0.191	0.40	0.266	0.65	0.313	0.90	0.349
0.20	0.211	0.45	0.276	0.70	0.321	0.95	0.355
0.25	0.227	0.50	0.286	0.75	0.328	1.00	0.361
0.30	0.241	0.55	0.296	0.80	0.335		

Table 2.

ν	y_0	ψ(ν)	ν	y_0	ψ(ν)	ν	y_0	ψ(ν)
0.2	*[1]	*	2.0	0.85	0.721	5	0.48	0.848
0.4	*	*	2.2	0.80	0.735	6	0.43	0.869
0.6	*	*	2.4	0.76	0.748	7	0.39	0.886
0.8	*	*	2.6	0.72	0.759	8	0.36	0.900
1.0	1.32	0.617	2.8	0.69	0.770	9	0.34	0.912
1.2	1.17	0.644	3.0	0.66	0.780	10	0.32	0.921
1.4	1.06	0.668	3.5	0.60	0.801	15	0.25	0.952
1.6	0.98	0.688	4.0	0.55	0.819	20	0.21	0.968
1.8	0.91	0.705	4.5	0.52	0.834	25	0.18	0.979

[1] The computation was carried out on C64 personal computer. Items replaced by asterisk were unreliable due to round-off error and therefore omitted.

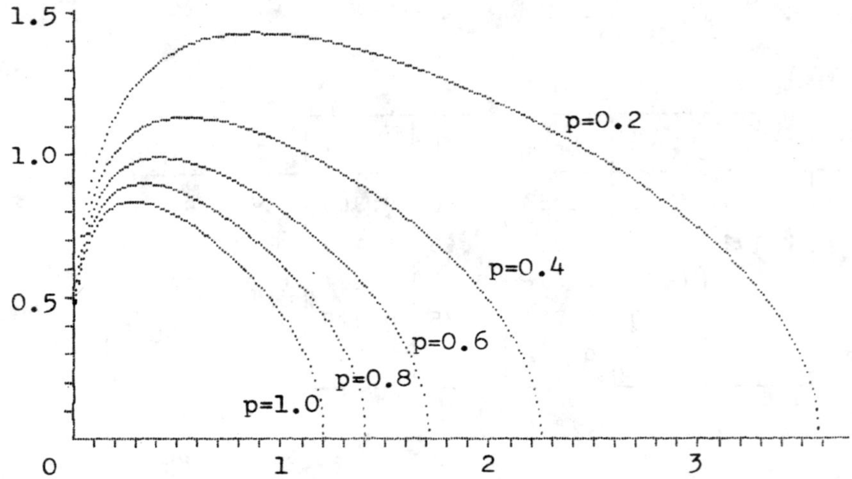

Fig. 1. Asymptotic shape of the stopping area.

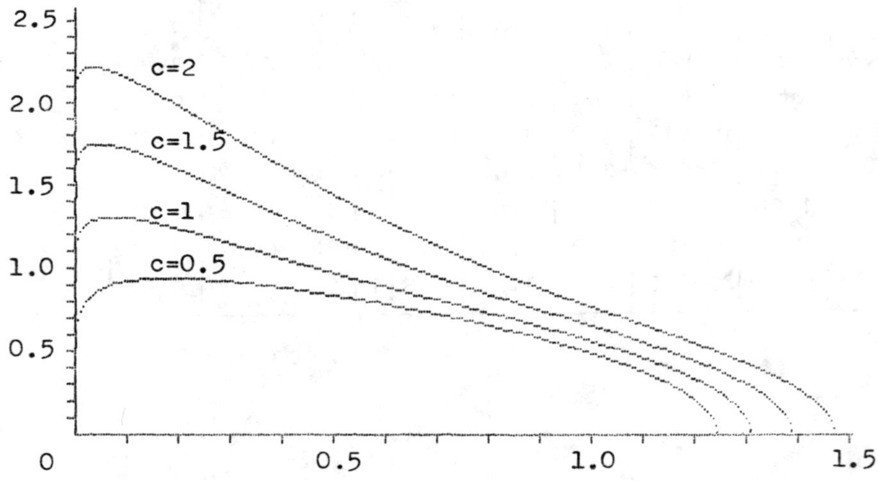

Fig. 2. Asymptotic shape of the stopping area.

REFERENCES

Freeman, P.R. (1982) The secretary problem and its extensions: A review, Internat. Statist. Rev., 51, 189-206.

Gardner, M. (1960) Mathematical games, Sci. Amer., 202, 150-156, 173-182.

Mucci, A.G. (1973) Differential equations and optimal choice problems, Ann. Statist., 1, 104-113.

Rose, J.S. (1982) Selection of nonextremal candidates from a random sequence, J. Optim. Theory Appl., 38, 207-219.

Szajowski, K. (1982) Optymalny wybór obiektu o a-tej randze, Mat. Stos. 3, XIX, 51-65.

Tamás F. Móri
Dept. of Probability Theory
L. Eötvös University
Budapest, Muzeum krt. 6-8.
H-1088, Hungary

Proc. of the 5th Pannonian Symp. on Math. Stat., Visegrád, Hungary 1985
W. Grossmann, J. Mogyoródi, I. Vincze, W. Wertz, eds

A RANDOM LINEAR SECRET-KEY ENCRYPTION

T. NEMETZ J. URECZKY

Budapest Kaposvár

A new linear encryption algorithm is proposed and discussed. The crypto-scheme utilizes artificial, unpredictable redundancy to prevent even "known plaintext attacks"
AMS subject classification: 94B05

1. INTRODUCTION

Linear codes are easy to implement and use. That is why they were popular over the centuries. The first secret code of the history was linear (it is attributed to Caesar, the emperor). The only theoretically unbreakable code is linear, as well ("One time pad", due to Vernam, see Kahn (1969)). Nevertheless there is a general abstaining from applying linear cryptosystems. Some reasons for this are as follows:

- A perfect random encryption, like the Vernam-cipher, applies a key-stream matching in length the clear text, and this secret key must be sent trough a safe channel and stored.
- A repeated application of the same key-stream in the Vernam-cipher results in an easy unauthorized deciphering, as a consequence of the preservation of the statistical and structural dependence of the original texts.
- If pseudo-random keys are used, then a cryptanalist needs to solve linear equations over a finite field in order to break the system. Systems of linear equations are, however, easy to solve. Especially, linear shift register based cryptosystems are breakable, see Siegenthaler (1984).

Fighting with the drawbacks we propose a safe new procedure, where the length of the key stream is just a fraction of that of the cleartext, at the expense of increasing the length of the cryptotext. Our algorithm can be summarized as follows:

The cleartext (viewed as a sequence over a finite field) is divided into blocks. These blocks are supplemented by random symbols, and the new blocks are scrambled by linear operations. Describing and discussing this algorithm we mention two open problems, as well.

2. THE ENCRYPTION ALGORITHM

Let p be a prime number larger than or equal to the alphabet size, and let the cleartext symbols (letters) be assigned to different mod-p integers 0,1,... . This way we can identify the cleartexts as strings over the mod-p integers, what is supposed from now on.

The algorithm uses 3 key-elements:

- N: integer, the length of the cleartext-blocks
- K: integer, the number of random characters assigned to the blocks
- A: scrambler, an $n \times n$ mod-p matrix with $A^2 = E$, where E is the unit matrix. n is an integer, uniquelly given by N,K,p, see later.

STEP 1: The cleartext is divided into blocks of equal length N.
Remark: If necessary, the last block can be filled up arbitrarily. The blocks are processed consecutively.

STEP 2: A "perfect" random combination of K places from $N+K$ consecutive places is chosen. These K places are filled in with "perfect" random characters, while the remaining ones with the next cleartext block (in the given order).
Remark: "perfect" random means uniformly distributed independent random variables. Both the random combinations and the characters are unknown to the receiver.

STEP 3: The index-number of the applied combination is determined, where the index varies from 0 to $\binom{N+K}{K} - 1$ in alphabetic order. The index is expressed in the numbersystem of base p and this p-adic number is augmented on the left by zeros to make a block of fixed length $\lceil \log_p \binom{N+K}{K} \rceil$. This will be added to the block of length N+K as a prefix.

Remark: The information block of length N is transformed to a block of length

$$n := \left\lceil \log_p \binom{N+K}{K} \right\rceil + N + K.$$

It is easy to see that

$$n \approx (N+K)\left(1+h\left(\frac{K}{N+K}\right) \log_p 2\right),$$

where $h(t) := t \log_2 t + (1-t)\log_2(1-t)$ is the entropy function.

STEP 4: The block formed in STEP 3 is multiplied (in modulo arithmetic) by the key matrix A to get the next block of the cryptotext.

Remark: Information characters, random ones and those of the index are scrambled. The scrambler matrix A is chosen randomly with uniform distribution over all n x n, mod-p self inverse matrices, and is known to the receiver, as well.

The *deciphering* algorithm is straightforward. The receiver applies the scrabler matrix A to the blocks of length n of the cryptotext. The first $n - N - K$ characters obtained, show the place of the random numbers and thus, also the places of the information symbols.

It belongs to the technical side of the transmission that the rate increase

$$\left(1 + \frac{K}{N}\right)\left(1 + h\left(\frac{K}{K+N}\right) \cdot \log_p 2\right)$$

must be tolerable.

Mathematical problems, raised by the proposal, are the following:

A) Fast algorithm is needed for determining the index of a random combination and vice versa. Also, there is a need for uniform random combination generators.
B) The number of n x n mod-p self inverse matrices must be large enough to prevent "brute force" methods. A generator of uniformly distributed such matrices has to be found.
C) The unicity point of the proposed cryptosystem is to be determined
D) It should be shown that statistical methods and "known plaintext attacks" are hardly feasible.

3. DISCUSSION

3.1. Indexing combinations; generating random combinations

We define the index of a combination essentially the same way as Schalkwijk (1972) and Cover (1973) had done.

The choices of K elements from M given elements are characterized 1-to-1 by zero-one vectors of length M with K ones and M-K zeros, where a "one" means that the corresponding element has been chosen. Therefore we will index just these vectors.

Let M_j be defined as the number of bits following the from-right-counted j'th "1", j=1,2,...,K. Obviously

$$M_1 \in \{0,1,\ldots,M-K\},$$
$$M_j \in \{M_{j-1},\ldots,M-K+j-1\}, \quad 2 \le j \le K. \qquad (*)$$

It is clear, that $M_1,\ldots,M_K$ uniquelly define a combination and vice versa.

The index of a combination $(M_1,\ldots,M_K)$ is given by

$$I(M_1,\ldots,M_K) = \sum_{j=1}^{K} \binom{M_j}{j}. \qquad (**)$$

Example: $M=6,\ K=3.$

$$\begin{array}{cc|cc|cccc} 0 & 1 & & 1 & & 0 & 0 & 1 \,|\, M_1=0 \\ & & & & M_2=3 & & & \\ & & M_3=4 & & & & & \end{array}$$

$$I(0,3,4) = \binom{0}{1} + \binom{3}{2} + \binom{4}{3} = 7.$$

It easy to see that for all $0 \leq I \leq \binom{M}{K} - 1$ there is one and only one vector $(M_1,\ldots,M_K)$ satisfying both (*) and (**).

The determination of the index of a combination is very fast. For recovering the combination from its index we can even store the binomial coefficients, since the values N and K are moderate. A dictionary look-up method of these coefficients provides an effective way for finding the combinations.

For the generation of a random combination, it suffices to generate its index randomly. One can make use, however, the vector representation $(M_1,\ldots,M_K)$, as well. If the combination is uniformly distributed, then M_1 follows the distribution

$$P(M_1=J) = \binom{M-J-1}{K-1} \Big/ \binom{M}{K}, \quad J = 0,1,\ldots,M-K.$$

Observing that $P(M_1=0) = \frac{K}{M}$, and

$$P(M_1=J) = P(M_1=J-1)\cdot\frac{N-K+1-J}{N-J},$$

it is easy to get a random value for M_1 from a common random number, that is distributed uniformly over $[0,1]$.

Setting $M_{new} := M_{old} - M_{i-1} - 1,$

$K_{new} := K_{old} - 1,$

one can consecutively find all random values M_i.

3.2 The number of n x n mod-p self-inverse matrices; generating them randomly

This question is intensively dealt with in the Ph.D. thesis of the second author, Ureczky (1985). Here we restrict ourselves to a short summary thereof.

Theorem: Let S(p) denote the number of n x n mod-p self-inverse matrices. Let furthermore $C_1(p) = 1$, $S_0(p) = 1$, and

$$S_1(p) = \begin{cases} 1, & \text{if } p=2 \\ 2, & \text{if } p>2. \end{cases}$$

Then

$$S_n(p) = p^{n-1}(p^{n-1} - 1)S_{n-2}(p) + S_1(p)\, C_n(p),$$

where

$$C_n(p) = (p^{n-1} - 1)C_{n-1}(p) + S_{n-1}(p).$$

This recursion yields the asymptotics:

$$S_n \approx p^{\frac{n^2}{2}}$$

This asymptotics can be interpreted as follows: the number of the n x n mod-p self-inverse matrices equals roughly the square root of the number of all n x n mod-p matrices. The total number of the self-inverse matrices is large enough, e.g. it is $\approx 10^{750}$, if $n = 32$ and $p = 31$.

Ureczky gives also an algorithm, that can generate a random n x n self-inverse matrix in $\frac{1}{2}n^3$ steps, in average, independently of p.

3.3. The unicity point of the system

As was pointed out by Hellmann (1977), the unicity point is a very important characteristics of a cryptosystem. Roughly speaking it provides a critical length. If the cleartext is shorter than that, then the cryptotest is even theoretically unbreakable, but there is no such assurance, if it is longer one. By its definition:

$$\text{Unicity point} = \frac{\text{entropy of the key-variable}}{\log p - \text{entropy of the text to be coded}} .$$

In our case:

$$\text{Entropy of the text to be coded} \approx$$

$$\approx \text{Entropy of the supplemented blocks} =$$

$$= \frac{1}{n}\,(\text{Entropy of combinations} + K \log p + N.\text{entropy of cleartext}).$$

From here, making use of

$$n \approx (N+K)(1+h\left(\frac{K}{N+K}\right)\log_p 2):$$

$$\text{Entropy of Key} \approx \frac{n^2}{2} \log p,$$

We get, after some manipulation, that

$$\text{Unicity point} \approx \left(\frac{n^2}{2} \log p\right) \cdot \frac{(1+r)(1+h(\frac{r}{1+r})\cdot\frac{1}{\log p})}{\log p - \text{entropy of cleartext}}$$

where $r = \frac{K}{N}$.

Let us define the key-rate as the average number of key variables necessary to encipher safely one cleartext letter, i.e. $\text{Key-rate} = R_K := \frac{\text{number of p-adic key-variables}}{\text{unicity point}}$.

In our case:

$$R_K = \frac{\log p - \text{entropy of cleartext}}{(1+r)(1+h(\frac{r}{1+r})\frac{1}{\log p})} = \frac{\text{Redundancy}}{R_{crypto}} ,$$

where

$$R_{crypto} = R_C := \frac{\text{length of the cryptotext}}{\text{length of the cleartext}} = (1+r)(1+h\,(\frac{r}{1+r})\,\frac{1}{\log p})$$

is the cryptotext rate.

In the case of the "one time pad" $R_K = R_C = 1$. In our case R_K can be made as small as whished at the expense of increasing R_C. The possible trade-offs are characterized by the following set, which is worth of investigation:

Definition: For a given source the region

$$\{(R_K, R_C): \text{there is an encryption with key-rate} = R_K, \text{ and cryptotext-rate} = R_C\}$$

is called *unicity-region*.

Remark: Our results shows, that the set

$$\left\{\left(\frac{\text{Redundancy of the source}}{x}, x\right); \ x \geq 1\right\}$$

is a subset of the unicity region. We have no further results in this relation. Let us formulate

OPEN PROBLEM 1: Find a characterization of the unicity region for stationary (ergodic) sources!

3.4 Statistical considerations

The essence of introducing artificial uncertainty is that this way one can make the cryptotext independent of the clear-text. This is motivated by an earlier results of us, see Nemetz (1981):

Let A be an n x n mod-p invertable matrix, x and e be mod-p vectors of length n. If A is uniformly random and $e \neq 0$, then Ae is uniformly distributed over the mod-p non-zero vectors, and x and $A^{-1}(Ax + e)$ are independent. This indicates that the statistical dependence within the text must be destroyed. This seems to be backed by our first empirical results, which show that the cryptotext fullfils the usual tests for randomness. This results are, however, not final.

It is also desirable to investigate, how can one solve linear equations, if the right hand sides are loaded with random errors. In this direction our efforts failed. As a first step we have tried to solve the following "easy" problem:

Let $x = (x_1, x_2, \ldots, x_n)$ be a vector of mod-p integers, that is unknown to us. We can observe an independent sequence of vectors $a_j = (a_{j1}, a_{j2}, \ldots, a_{jn})$ $j = 1, 2, \ldots, N$, $N >> n$, consisting of independent, over $\{0, 1, \ldots, p-1\}$ uniformly distributed random variables.

Let $b_j = \langle a_j, x \rangle$ be the mod-p inner product of a_j and x. Let furthermore, e_j be a sequence of independent random variables with common distribution

$$P(e_j=0) = \text{large},\ P(e_j=1) = P(e_j=2) = \ldots = P(e_j=p-1) = \text{small}.$$

Suppose, that instead of b_j we know just the noisy versions

$$Y_j = b_j + e_j.$$

This "easy" problem turned out for us, not so easy at all, thus we formulate it as

OPEN PROBLEM 2. Give an estimation $\tilde{x}$ of x from the data (a_j, Y_j), $j = 1,2,\ldots,N$, such that

a) $\tilde{x}$ and x coincides with high probability

b) or: The Hamming distance $d_H(\tilde{x}, x) \leq A.n$ with small A and with high probability. How large N do we need?

REFERENCES

Cover, T.M. (1973): Enumerative Source Coding, IEEE Trans. Inf. Th. IT-19, 73-77.

Hellman, M.E. (1977): An Extension of the Shannon Theory Approach to Cryptography, IEEE Trans. Inf. Th. IT-23, 289-294.

Kahn, D. (1969): The codebreakers, 5-th edition, MacMillan Comp. Toronto

Nemetz, T. (1981): Run-test discrimination between written Hungarian and random sequences, Lecture Notes in Statistics: The first Pannonian Symposium on Math. Stat. 182-194 (P. Révész et al. eds.), Springer

Schalkwijk, J.P.M. (1972): An Algorithm for Source Coding, IEEE Trans. Inf. Th. IT-18, 395-399.

Siegenthaler, T. (1984): Decrypting a Class of Stream Ciphers Using Ciphertext Only, IEEE Trans. Comp. C-34, 81-85.

Ureczky, J. (1985): Mod-prime self-inverse matrices and their application to cryptography, Ph.D. Thesis, Eötvös Lorand University, Budapest (in Hungarian)

Tibor Nemetz
Mathematical Institute of the
Hungarian Academy of Sciences
H-1053 Budapest
Reáltanoda u. 13-15.

Jozsef Ureczky
Agricultural College
Department of Mathematics
H-7400 Kaposvár
Dénesmajor 2.

Proc. of the 5th Pannonian Symp. on Math. Stat., Visegrád, Hungary 1985
W. Grossmann, J. Mogyoródi, I. Vincze, W. Wertz, eds

ON A FUNCTIONAL RAREFACTION PROCEDURE

P.M. PERUNIČIĆ

Beograd

In this paper we consider a non-constant rarefaction procedure of renewal processes, using the conditional expectation technique. The concept of the rarefaction as a "measurable procedure" may be of some interest.

AMS subject classification: 60K05

1. BASIC CONCEPTS

1.1. Let $\{X_n(\omega),\ n\geq 1\}$ be a sequence of i.i.d.r.v. with the common distribution function F such that $F(0+)=0$, and put $S_n = \sum_{k=1}^{n} X_k$, $n\geq 1$. In view of the general rarefaction procedure, the following formalism seems to be natural.

We consider a trajectory

$$\omega_1 = t(\omega) = (S_1(\omega),\ S_2(\omega),\ \dots)$$

as elementary event. Let $\mathfrak{M} = \{\omega_1\}$ be the set of all elementary events, and let τ be the minimal σ-field generated by the cylinders. On the measurable space $\langle\mathfrak{M},\tau\rangle$ we introduce the product probability measure F^{∞}. The probability space $\langle\mathfrak{M},\tau,F^{\infty}\rangle$ is a "sample space".

Let $p\colon \mathbb{R}^+ \to [0,1]$ be a measurable function /in the general sense, i.e. mapping the measurable space $\langle \mathbb{R}^+,\mathcal{B}_{\mathbb{R}^+}\rangle$ into $\langle [0,1],\mathcal{B}_{[0,1]}\rangle$ /, provided that $E_F(p(X))=E(p(X_1)) = p\in(0,1)$.

The function p is called a rarefying function and denote $q(x) = 1-p(x)$. We abbreviate $E(p(X_1))$ by p and $E(q(X_1))$ by q.

1.2. Let $<\{0,1\}^{\mathbb{N}}, \tau'>$ be a measurable space, where τ' is the minimal σ-field generated by the cylinders from $\{0,1\}^{\mathbb{N}}$. This is called the "decision space". Let us emphasize here the fact that there is no an a priori probability measure defined on that space.

We consider the cylinder $O_n = \{0,1\}^{n-1} \times \{0\} \times \{0,1\}^{\mathbb{N}}$ as the decision "$t_n(\omega_1) = S_n(\omega)$ is erased" and $1_n = \{0,1\}^{n-1} \times \{1\} \times \{0,1\}^{\mathbb{N}}$ as "$S_n(\omega)$ is not erased".

Let us consider the function $Q: \mathfrak{M} \times \tau' \to [0,1]$ defined by

$$Q(\omega_1,B) = \int_B d \prod_{n=1}^{\infty} \mathcal{B}(1,q(t_n(\omega_1) - t_{n-1}(\omega_1))),$$

where $\mathcal{B}(1,\alpha)$ is a binomial measure. It is obvious that a/ keeping $\omega_1 \in \mathfrak{M}$ fixed, $Q,(\omega_1,.)$ is a probability measure on $<\{0,1\}^{\mathbb{N}}, \tau'>$; b/ keeping $B \in \tau'$ fixed, $Q(., B)$ is $<\tilde{\mathfrak{M}},\tau>$ -measurable. Hence, Q is a transition probability measure with respect to the spaces $<\tilde{\mathfrak{M}},\tau>$ and $<\{0,1\}^{\mathbb{N}}, \tau'>$.

Let $A \in \tau$, $B \in \tau'$. Consider $<\tilde{\mathfrak{M}} \times \{0,1\}^{\mathbb{N}}, \tau \otimes \tau'>$ where $\tau \otimes \tau'$ is the minimal σ-field which contains the cylinders $A \times B$.

It is known /Neveu, [1]/ that there is a unique probability measure P on this measurable space, with the property

$$P(A \times B) = \int_A Q(\omega_1,B)\, F^{\infty}(d\omega_1),$$

where $A \times B$ is an arbitrary cylinder. Also,

$$\int_{\tilde{\mathfrak{M}}} Q(\omega_1,B)\, F^{\infty}(d\omega_1) = P(\tilde{\mathfrak{M}} \times B) = \int_B d \prod_1^{\infty} \mathcal{B}(1,q)$$

is a probability measure on $<\{0,1\}^{\mathbb{N}}, \tau'>$ induced by the rarefaction with the function $p(x)$, $x \in \mathbb{R}^+$.

1.3. Let $out(n)_{(\omega_1,\omega_2)} = out(n)_{(\omega_2)} = I(O_n)$ be the indicator of erasing S_n, $n \geq 1$. The above construction enables us to consider the sequences $\{S_n, n \geq 1\}$ and $\{ont(n), n \geq 1\}$ on the same space. The conditional measure $Q(\omega_1 .)$ describes the behaviour of the sequence $\{out(n), n \geq 1\}$ given $\omega_1 = (S_1(\omega), S_2(\omega), \ldots)$.

1.4. The random variables

$$T_1 = \sum_{m=1}^{\infty} S_m \{ \prod_{k=1}^{m-1} out(k) \} (1-out(m)) = S_{\nu_1},$$

$$T_{n+1} = \sum_{m=\nu_n+1}^{\infty} S_m \{ \prod_{\nu_n+1}^{m-1} out(k) \} (1-out(m)) = S_{\nu_{n+1}} \qquad (n \geq 1)$$

are points of the rarefied sequence, where $\nu_1 = \inf\{m | out(m) = 0\}$ and for $n \geq 1$ $\nu_{n+1} = \inf \{m | m > \nu_n, out(m) = 0\}$.

It is easy to show that

$$P(T_1 < x) = P(S_{\nu_1} < x) = \sum_{k=1}^{\infty} Q_k(x) \qquad (x>0),$$

where

$$Q_1(x) = \int_0^x q(t) dF(t), \quad Q_{k+1}(x) = \int_0^x Q_k(x-t) \; p \; (t) dF(t) \quad (k \geq 1)$$

1.5 Let $\mathcal{H} = \mathcal{F}(X_n, n \geq 1)$. Then

$$P(T_1 = S_m | \mathcal{H}) = \{ \prod_{k=1}^{m-1} p(X_k) \} q(X_m),$$

$$P(T_1 = S_m, T_2 = S_n | \mathcal{H}) = \{ \prod_{1}^{m-1} p(X_k) \} q(X_m) \{ \prod_{m+1}^{n-1} p(X_k) \} q(X_n)$$

for $n \geq m+1$, et caetera.

1.6.
$$P(T_1 < +\infty | \mathcal{H}) = 1.$$

Proof. Put

$$G_n = \sum_{m=1}^{n} P(T_1 = S_m | \mathcal{H}).$$

It is obvious that

$$G_{n+1} \geq G_n, \qquad (\forall n) \; G_n \leq 1$$

and

$$\lim_{n\to+\infty} G_n = G = \sum_{m=1}^{\infty} \{\prod_{1}^{m-1} p(X_k)\}\, q(X_m) \text{ a.s.}$$

Then by the Monotone Convegence Theorem we have

$$E(G) = \lim_{n\to+\infty} E(G_n) = \lim_{n\to\infty} \sum_{m=1}^{n} p^{m-1} q = 1$$

Finally,

$$P(G\leqq 1) = 1 \text{ and } E(G) = 1) \Rightarrow P(G=1) = 1.$$

2. THE RAREFIED RENEWAL PROCESS

In this section we prove that $\{T_n, n \geq 1\}$ is also a renewal process.

2.1. Let us determine $P(T_1<x|\mathcal{H})$ $(x\in\mathbb{R}^+)$.

$$P(T_1<x, T_1 = S_m|\mathcal{H}) = E(I\{T_1<x, T_1 = S_m\}|\mathcal{H}) =$$

$$= E(I\{S_m<x\}I\{T_1=S_m\}|\mathcal{H}).$$

As $I\{S_m<x\}$ is $\mathcal{H}$-measurable, we have

$$P(T_1<x, T_1= S_m|\mathcal{H}) = I\{S_m<x\}\, P(T_1 = S_m|\mathcal{H}) =$$

$$= I\{S_m<x\} \{\prod_{1}^{m-1} p(X_k)\}\, q(X_m)$$

and

$$P(T_1<x|\mathcal{H}) = \sum_{m=1}^{\infty} P(T_1<x, T_1 = S_m|\mathcal{H}) =$$

$$= \sum_{m=1}^{\infty} \{\prod_{k=1}^{m-1} p(X_k)\}\, q(X_m)\, I\{S_m< x\}.$$

2.2. The independence of T_1 and $T_2 - T_1$.

For $n \geq m+1$ by the conditional independece we have

$P(T_1 = S_m, T_2 - T_1 = S_n - S_m | \mathcal{H}) = \alpha(X_1, \ldots, X_m)\alpha(X_{m+1}, \ldots, X_n)$,

where

$$\alpha(X_1, \ldots, X_l) = p(X_1) \ldots p(x_{l-1})\, q\, (X_l).$$

For $x, y \in \mathbb{R}^+$

$$P(T_1 < x,\ T_2 - T_1 < y | \mathcal{H}) =$$

$$= \sum_{m=1}^{\infty} \sum_{n=m+1}^{\infty} E(I\{T_1 < x, T_1 = S_m,\ T_2 - T_1 < y,\ T_2 - T_1 = S_n - S_m\} | \mathcal{H}) =$$

$$= \sum_{m=1}^{\infty} \sum_{n=m+1}^{\infty} I\{S_m < x\}\alpha(X_1, \ldots, X_m) I\{S_n - S_m < y\}\alpha(X_{m+1}, \ldots, X_n).$$

Thus

$$P(T_1 < x,\ T_2 - T_1 < y) = E(P(T_1 < x, T_2 - T_1 < y\ | \mathcal{H})) =$$

$$= \sum_{m=1}^{\infty} \sum_{n=m+1}^{\infty} E(I\{S_m < x\}\alpha(X_1, \ldots, X_m)) E(I\{S_n - S_m < y\}\alpha(X_{m+1}, \ldots, X_n)) =$$

$$= \sum_{m=1}^{\infty} E(I\{S_m < x\}\alpha(X_1, \ldots, X_m)) \sum_{n-m=1}^{\infty} (I\{S_n - S_m < y\}\alpha(X_{m+1}, \ldots, X_n)) =$$

$$= P(T_1 < x)\ P\ (T_2 - T_1 < y),$$

which proves the independence of T_1 and $T_2 - T_1$.
It is obvious that T_1 and $T_2 - T_1$ are identically distributed.
By induction over n we get that $\langle T_n, G\rangle$ is a renewal process, with $G(x) = P(T_1 < x)$.

2.3. Let

$$N(t) = \sum_{m=1}^{\infty} I\{S_m < t\}$$

be the number of renewal points in the interval (0,t) for a renewal process $\langle S_n, F\rangle$. Then

$$P(T_1<x|\quad) = \sum_{m=1}^{N(x)} \{\prod_{k=1}^{m-1} p(X_k)\}\, q(X_m).$$

If the rarefaction is constant ($p(x) = p\in(0,1)$), we have

$$P(T_1<x|\mathcal{H}) = 1-p^{N(x)},$$

and

$$P(T_1<x) = 1-E(p^{N(x)}),\ x>0.$$

2.4. Using the same technique as in 2.1. and 2.2., we can show that

$$H(x,y) = P(S_1<x, T_1<y) = \int_0^{\min(x,y)} (q(t)+G(y-t)p(t))\,dF(t).$$

If

$$\varphi(s) = \int_0^{+\infty} e^{-st}dF(t),\ \varphi^*(s) = \int_0^{+\infty} e^{-st}p(t)\,dF(t),\ \varphi_p(s) = \int_0^{+\infty} e^{-st}dG(t)$$

are the Laplace transforms of F, $\int pdF$ and G respectively, then we have

$$1-\varphi_p(s) = (1-\varphi(s))(1-\varphi^*(s))^{-1}.$$

2.5. The distribution of T_n-S_n, $n\geq 1$.

$$E(e^{-s(T_n-S_n)}) = E(e^{-s(T_n-S_n)}\sum_{m=n}^{\infty} I\{T_n=S_m\}) =$$

$$= \sum_{m=n}^{\infty} E(E(e^{-s(S_m-S_n)}\ I\{T_n=S_m\}|\mathcal{H})) =$$

$$= \sum_{m=n}^{\infty} E(e^{-s(S_m-S_n)}P(T_n=S_m|\mathcal{H})).$$

For m=n

$$P(T_n=S_n|\mathcal{H}) = q(X_1) \ldots q(X_n),$$

so we have

$$E(e^{-s(S_n-S_n)}P(T_n=S_n|\mathcal{H})) = q^n.$$

For $m \geq n+1$

$$E(e^{-s(S_m-S_n)} P(T_n=S_m|\mathcal{H})) =$$

$$= \Sigma E(e^{-s(S_m-S_n)} \prod_{l=0}^{n-1} p(X_{i_l+1}) \ldots p(X_{i_{l+1}-1}) q(X_{i_{l+1}}) =$$

$$= \Sigma E(e^{-sX_{n+1}} \ldots e^{-sX_m} \prod_{l=0}^{n-1} p(X_{i_l+1}) \ldots p(X_{i_{l+1}-1}) q(X_{i_{l+1}}),$$

where Σ is taken over all the possible combinations $0 = i_0 < i_1 < \ldots < i_{n-1} < i_n = m$.

Let k be the number of the $q(.)$'s with index $\leq n$ $0 \leq k \leq n-1$ /there are $\binom{n}{k}\binom{m-n}{n-k}$ possible combinations/. Since the random vectors $(X_1, \ldots, X_n)$ and $(X_{n+1}, \ldots, X_m)$ are independent and the X_i's are equally distributed and independent we have

$$E(e^{-s(S_m-S_n)} P(T_n= S_m|\mathcal{H})) =$$

$$= \Sigma \binom{n}{k}\binom{m-n}{n-k} q^k p^{n-k} (\varphi(s)-\varphi^*(s))^{n-k-1} (\varphi^*(s))^{m+k-2n} (\varphi(s)-\varphi^*(s)),$$

where the summation is taken over $0 \leq k \leq n-1$, $0 \leq n-k \leq m-n$.

Also,

$$\sum_{m=n+1}^{\infty} \sum_k E(e^{-s(S_m-S_n)} P(T_n=S_m|\mathcal{H}) = \sum_{k=0}^{n-1} \sum_m (.)$$

where the summation for m is taken for $m-n \geq n-k$.

Since

$$\sum_{m-n=n-k} \binom{m-n}{n-k} (\varphi^*(s))^{m-n-(n-k)} = (1-\varphi^*(s))^{-(n-k)}$$

we have

$$\Psi_n(s) = E(e^{-s(T_n-S_n)}) =$$

$$= q^n + \sum_{k=0}^{n-1} \binom{n}{k} q^k \left(p \frac{\varphi(s) - \varphi^*(s)}{1-\varphi^*(s)}\right)^{n-k} =$$

$$= (q+p\,\varphi_p(s))^n .$$

2.6. The joint distribution of (S_n, T_n-S_n), $n\in\mathbb{N}$. We can determine

$$\nu_n(u,v) = E(e^{-uS_n-v(T_n-S_n)}) \quad (u,v\geq 0)$$

using the same technique as in 2.5. We get

$$\nu_n(u,v) = (\varphi(u)-\varphi^*(u)+\varphi^*(u)\,\varphi_p(v))^n, \quad n\in\mathbb{N}.$$

2.7. The joint distribution of (S_n, T_n), $n\in\mathbb{N}$.

It can be easily determined from $\nu_n(u,v)$, by transformation

$$u-v \mapsto u, \qquad v \mapsto v.$$

Remark. Underlying the "surface" structure of the rarefaction procedure /which is geometric, as in Rényi's first introduction of the rarefaction in 1956./ we actually have a "deep" level described by the conditional expectation technique.

REFERENCES

Neveu, J.: Bases mathématiques du clacul des probabilités, Masson et Cie, Paris, 1964.

Galambos, J., Kotz, S.: Characterizations of Probability Distributions, Springer Verlag, Berlin-Heidelberg-N.Y., 1978.

Peruničić, P.: O nekonstantnom resetanju, Mat. vesnik 37,215-220 /1985/

Peruničić, P.: Nekonstantno resetanje procesa obnavljanja /Doctoral Thesis/, Univerzitet u Beogradu, 1984.

P.M. Peruničić
Prirodno-matematički Fakultet
Odsek za MMA nauke
Studentski trg 16.
11000 Beograd, Yugoslavia

Proc. of the 5th Pannonian Symp. on Math. Stat., Visegrád, Hungary 1985
W. Grossmann, J. Mogyoródi, I. Vincze, W. Wertz, eds

COMPLETE CONVERGENCE OF THE SAMPLE MEAN FOR EXCHANGEABLE RANDOM VARIABLES

D. PLACHKY

Münster

In the Theorem of this article we give sufficient conditions such that (*) holds with $E(X_1, \mathcal{A}_0)$ instead of $E(X_1)$ for a sequence of real valued, identically distributed, and conditionally independent random variables. The proove applies a truncation technique, which yields an exponential bound for the corresponding, sequence of truncated random variables.

AMS subject classification: 60E99

We start from the underlying probability space $(\Omega, \mathcal{A}, P)$ and a sequence $(X_n)_{n=1,2,\ldots}$ of real-valued and identically distributed random variables, which are conditionally independent relative to a sub-σ-algebra $\mathcal{A}_0$ of $\mathcal{A}$. If $(X_n)_{n=1,2,\ldots}$ is a sequence of real-valued, exchangeable random variables, $\mathcal{A}_0$ can be chosen as the σ-algebra consisting of the terminal events with respect to $(X_n)_{n=1,2,\ldots}$ (cf. Bártfai (1980) for a short and simple proof). In the case $\mathcal{A}_0 = \mathcal{A}$ and $E(X_1^2) < \infty$ Hsu and Robbins (1947) have proved complete convergence of the sample mean (cf. Lukács (1968) or Stout (1974)), i.e.:

$$\sum_{n=1}^{\infty} P(\{|\frac{\sum_{i=1}^{n} X_i}{n} - E(X_1)| > \varepsilon\}) < \infty \tag{*}$$

holds for every $\varepsilon > 0$. Asmussen and Kurtz (1980) (cf. also Gut (1985)) have given necessary and sufficient conditions for (*) in the case $\mathcal{A}_0 = \mathcal{A}$ for subsequence $(X_{n_k})_{k=1,2,\dots}$ of a sequence $(X_n)_{n=1,2,\dots}$ of i.i.d. random variables satisfying $E(|X_1|) < \infty$. The aim of this article is to find sufficient conditions, such that (*) holds with $E(X_1|\mathcal{A}_0)$ instead of $E(X_1)$ for a sequence of real-valued, identically distributed, and conditionally independent random variables by means of a truncation technique, which yields an exponential bound for the corresponding sequence of truncated random variables. For this purpose we start from a real-valued random variable Y with existing moment generating function $m(t) = E(e^{tY})$ for $t \in (-t_0, t_0)$ $(t_0 > 0)$. By E_t resp. Var_t we denote the expanctiation resp. variance relative to the conjugated distributions P_t defined by $P_t(A) = \int_A e^{tY} dP / \int d^{tY} dP$ for all measurable subsets A of the underlying σ-algebra on which P is defined. Markov's inequality yields the following simple

<u>Lemma.</u>

$m(t) = E(e^{tY}) < \infty$ for $t \in (-t_0, t_0)$ $(t_0 > 0)$ implies

$$P(\{Y - E(Y) \geq \frac{m'}{m}(t)\}) \leq \exp - \frac{t^2}{2}[2\ Var_{\tau_2}(Y) - Var_{\tau_1}(Y)]$$

for some $\tau_i \in [0,t]$, $i = 1,2$, and $t \in (0, t_0)$.

<u>Proof:</u> Markov's inequality implies

$$P(\{P \geq \frac{m'}{m}(t)\}) \leq \exp\ (-t\ \frac{m'}{m}(t) + \ln m(t)) \text{ for } t > 0$$

and $E_t(Y) = \frac{m'}{m}(t)$, $Var_t(Y) = (\frac{m'}{m}(t))'$ together with $\ln m(t) = t(\ln m(t))' + \frac{t^2}{2} Var_{\tau_1}(Y)$, $(\ln m(t))' = (\ln m(0))' + t\ Var_{\tau_2}(Y)$ for some $\tau_i \in [0,t]$, $i = 1,2$, $t \in (0, t_0)$ yields the assertion of the lemma.

This lemma will be applied to the truncated random variables $X_i^{(n)} = X_i\ I_{\{|X_i| \leq k_n\}}$, $i = 1,\dots,n$, where $(k_n)_{n=1,2,\dots}$ is a suitably chosen sequence of positive numbers with $k_n \to \infty$. The starting point is the following inequality:

$$P\left(\left\{\frac{X_1 + \ldots + X_n}{n} < \varepsilon + E(X_1|\mathcal{A}_0)\right\}\right) \leq a_n + b_n + c_n, \text{ where}$$

$$a_n = P\left(\left\{\frac{\sum_{i=1}^{n} X_i^{(n)}}{n} > E(X_1^{(n)}|\mathcal{A}_0) + E_{t_n}(X_1^{(n)}|\mathcal{A}_0)\right\}\right),$$

$$b_n = P\left(\left\{\sum_{i=1}^{n}(X_i - X_i^{(n)}) > 0\right\}\right),$$

$$c_n = P(\{E(X_1^{(n)}|\mathcal{A}_0) + E_{t_n}(X_1^{(n)}|\mathcal{A}_0) > E(X_1|\mathcal{A}_0) + \varepsilon\}).$$

Because of

$$|(X_1^{(n)})^m (e^{t_n X_1^{(n)}} - 1)| \leq |X_1^{(n)}|^m \sum_{k=1}^{\infty} |t_n X_1^{(n)}|^k/k! \leq |k_n|^m (e^{t_n k_n} - 1)$$

$(m = 0,1,2)$

and $k_n^2(e^{t_n k_n} - 1) \to 0$ for $t_n = \frac{1}{k_n^4 + \kappa}$ ($\kappa > 0$ sufficiently small) and and $n \to \infty$, we get $\mathrm{Var}_{t_n}(X_1^{(n)}|\mathcal{A}_0) \in [\frac{3}{4}\,\mathrm{Var}\,(X_1|\mathcal{A}_0), \frac{5}{4}\,\mathrm{Var}(X_1|\mathcal{A}_0)]$ simultaneously on $A_0 = \{\mathrm{Var}\,(X_1|\mathcal{A}_0) > 0\}$ P-almost everywhere for sufficiently large n, if the following assumption holds:

$$E(X_1^2\, I_{\{|X_1| > a\}}|\mathcal{A}_0) \to 0 \qquad \text{P-a.e. uniformly for } a \to \infty. \qquad (A_1)$$

Choosing now k_n such that $\frac{n}{k_n^{6+2\kappa}} = n^\alpha$ for $\alpha \in (0,1)$ fixed holds, yields together with the lemma above

$$a_n \leq E\left(\sum_{n=1}^{\infty} \exp[-\frac{n^\alpha}{8} Var(X_1|\mathcal{A}_0) \cdot I_{A_0}]\right) \leq E\left\{\int_0^\infty \exp[-\frac{x^\alpha}{8} Var'(X_1|\mathcal{A}_0) I_{A_0}]\right\} dx,$$

wich implies the convergence of $\sum_{n=1}^{\infty} a_n$ if the assumption

$$E\left(\left[\frac{I_{A_0}}{Var\ (X_1|\mathcal{A}_0)}\right]^{\frac{1}{\alpha}}\right) < \infty \qquad (A1')$$

for some $\alpha \in (0,1)$ holds because of $\int_0^\infty e^{-\beta x^\alpha} dx = \frac{1}{\alpha\beta^{1/\alpha}} \Gamma(\frac{1}{\alpha})$ for $\beta > 0$. Since Markov's inequality implies $b_n \leq \frac{n}{k_n^{6+\lambda}}$ if $E(|X_1|^{6+\lambda} < \infty$ holds, we choose $\lambda > 0$ such that $\frac{n}{k_n^{6+\lambda}} = \frac{1}{n^{1+\mu}}$ for some $\mu > 0$ sufficiently small is true, which leads to the following conditions for λ : $\lambda = \frac{(1+\mu)(8+2\kappa) + 6\alpha + 2\kappa}{1-\alpha}$, which tends to $\frac{8+6}{1-\alpha}$ for $\mu \to 0$ and $\kappa \to 0$. Therefore, convergence of $\sum_{n=1}^{\infty} b_n$ will be true, if the assumption

$$E(|X_1|^{\frac{16}{1-\alpha}+\beta}) < \infty \qquad (A2)$$

for some $\alpha \in (0,1)$ and $\beta > 0$ holds.

Finally, $\sum_{n=1}^{\infty} c_n < \infty$ holds, because $E(X_1^{(n)}|\mathcal{A}_0) + E_{t_n}(X_1^{(n)}|\mathcal{A}_0) \leq E(X_1|\mathcal{A}_0) + \varepsilon$ is true uniformly P-almost everywhere for sufficiently large n, since by replacing X_i by $X_i - E(X_i|\mathcal{A}_0)$ we may assume without loss of generality $E(X_1|\mathcal{A}_0) = 0$ P-a.e., which implies $E(X_1^{(n)}|\mathcal{A}_0) + E_{t_n}(X_1^{(n)}|\mathcal{A}_0) \leq \varepsilon$ for sufficiently large n because of assumption (A1). Therefore, we have proved the following

Theorem.

Let $(X_n)_{n=1,2,...}$ be a sequence of identically distributed random variables with $E(X_1^2) < \infty$, which are conditionally independent relative to a sub-σ-algebra $\mathcal{A}_0$. Then the following assertions hold:

a) The assumption (A1) implies for any $\alpha \in (0,1)$ the existence of a sequence $(k_n)_{n=1,2,\ldots}$ of positive real numbers tending to infinity, such that for the corresponding truncated random variables

$X_i^{(n)} = (X_i - E(X_i|\mathcal{A}_0))I_{\{|X_i - E(X_i|\mathcal{A}_0)| \leq k_n\}}$, $i = 1,2,\ldots,n$, the inequality

$$P\left(\left\{\frac{\sum_{i=1}^{n} X_i^{(n)}}{n} - E(X_1^{(n)}|\mathcal{A}_0) > \varepsilon\right\}\right) \leqq E\left(\exp - \left(\frac{n^\alpha}{8} YI_{A_0}\right)\right)$$ holds for any given $\varepsilon > 0$ and for n sufficiently large ($Y = \mathrm{Var}(X_1|\mathcal{A}_0)$, $A_0 = \{Y > 0\}$).

b) The assumptions (A1) and (A2) imply $P\left(\left\{\sum_{i=1}^{n}(X_i - X_i^{(n)}) > 0\right\}\right) \leq \frac{\text{const.}}{n^{1+\delta}}$ for some $\delta > 0$ and all natural numbers n.

<u>Corollary.</u>

If (A1) holds and $\mathrm{Var}(X_1|\mathcal{A}_0)$ is P-a.e. bounded away from zero on A_o, the $\lim_{n\to\infty} P\left(\left\{\sum_{i=1}^{n}(X_i^{(n)} - E(X_1^{(n)}|\mathcal{A}_0))/n > \varepsilon\right\}\right)^{\frac{1}{n^\alpha}} = 0$ holds.

<u>Proof:</u> An upper bound for the corresponding probability is P-ess sup $\exp\left(-\frac{\lambda^2}{8} YI_{A_0}\right)$ for arbitrary $\lambda > 0$ if we choose for t_n instead of $\frac{1}{k_n^{4+\kappa}}$ the real number $\frac{1}{\lambda k_n^{4+\kappa}}$. Therefore, $\lambda \to \infty$ leads to the assertion of the corollary above.

<u>Remarks.</u>

1. If one wishes to replace α by 1 and the $X_i^{(n)}$ by the original

random variables X_i in the corollary above, then this can be done, if the moment generating function $E(e^{tX_1})$ exists for some $t > 0$ and the measurable function $\inf_{t>0} \{e^{-t(E(X_1|\mathcal{A}_0)+\varepsilon)} E(e^{tX_1}|\mathcal{A}_0)\}$ is P-a.e. bounded. The P-essential supremum of this function is in this case the exact rate of convergence for any $\varepsilon > 0$ (cf. Bártfai (1980)).

2. According to Erdős (1949) $P\left(\left\{\sum_{i=1}^{n} (X_i - E(X_1)) > n\,\varepsilon\right\}\right) \leq a_n + b_n + c_n$ holds, in the i.i.d. case, where $a_n = P\left(\left\{\sum_{i=1}^{n} (X_i^{(n)} - E(X_1^{(n)}) > n\frac{\varepsilon}{2}\right\}\right)$, $b_n = P(\{X_i - E(X_1) > n\frac{\varepsilon}{2}$ for at most one $i \in \{1,\ldots,n\}\})$, $c_n = P(\{X_i - E(X_1) > k_n$ for at most two distinct $i \in \{1,\ldots,n\}\})$, $X_i^{(n)} = (X_i - E(X_1))I_{\{|X_i - E(X_1)| \leq k_n\}}$, $i = 1,\ldots,n$, $b_n \leq E(\frac{2}{\varepsilon}(X_1+1)^2)$, $c_n \leq \binom{n}{2}(P(\{X_1 - E(X_1) \geq k_n\}))^2$. Majorizing now a_n according to the lemma above, leads to an upper bound for a_n of the type $\frac{1}{n^\alpha}$ (up to a constant) if we put $\frac{n}{k_n^2} = \ln(n^\alpha)$ for some $\alpha > 1$, whereas for the c_n we get an upper bound of the type $\frac{(\ln n)^{2m}}{n^{2m-2}}$ (up to a constant),

where m is a positive real number satisfying $E(|X_1|^m) < \infty$, which leads to the condition $m > \frac{3}{2}$, if the corresponding upper bound for c_n should be of the type $\frac{1}{n^{1+\delta}}$ for some $\delta > 0$ (up to a constant). The case of conditionally independence relative to a sub-σ-algebra $\mathcal{A}_0$ leads also to an upper bound for the c_n of the type $\frac{1}{n^{1+\delta}}$ for some $\delta > 0$ (up to a constant), if in addition $E(X_1^4) < \infty$ holds. The same is true for the a_n, if a_n is majorized according to Erdős (1949), i.e. complete convergence of the sample mean to the conditional expectation $E(X_1|\mathcal{A}_0)$ holds for a sequence $(X_n)_{n=1,2,...}$ of identically distributed random variables with $E(X_1^4) < \infty$, which are conditionally independent relative to a sub-σ-algebra $\mathcal{A}_0$.

REFERENCES

1. Asmussen, S. and Th. G. Kurtz: Necessary and sufficient conditions for complete convergence in the law of large numbers. Ann. Prob. 8 (1980), 176-182.
2. Bártfai, P.: Remarks on exchangeable random variables. Publ. Math. Debrecen 27 (1980), 143-148.
3. Erdős, P.: On a theorem of Hsu and Robbins. Ann. Math. Statisti. 20 (1949), 296-299.
4. Gut, A.: On complete convergence in the law of large numbers for subsequences. Ann. Prob. 13 (1985), 1286-1291.
5. Hsu, P.L. and H. Robbins: Complete convergence and the law of large numbers. Proc. Nat. Acad. Sci. 33 (1947), 25-41.
6. Lukács, E.: Stochastic Convergence. D. C. Heath and Company, 1968.
7. Stout, W. F.: Almost Sure Convergence. Academic Press, New York, 1974.

Detlef Plachky
Institut für Mathematische Statistik,
Universität Münster
Einsteinstr. 62, D-4400 Münster

Proc. of the 5th Pannonian Symp. on Math. Stat., Visegrád, Hungary 1985
W. Grossmann, J. Mogyoródi, I. Vincze, W. Wertz, eds

DER RANG EINER KLASSE MULTIDIMENSIONALER ZUFÄLLIGER PROZESSE

T. POGÁNY

Sombor

In the paper we consider of a family of weakly stationary multidimensional stochastic processes with continuous time, where the coordinates are linear combinations with complex coefficients of pairwise orthogonal stationary, scalar processes (or those of their mean square derivatives).
For the correlated case, one can only prove rank $(X(t)) \leq \dim(X(t))$ in Satz and Theoreme, see Cramér (1961).

AMS subject classification: 60G10

Sei $\{Y(t), t\in R\}$ ein multidimensionaler zufälliger Prozess mit Koordinaten der Form $\sum_{j=1}^{k} b_j X_i^{(j)}(t)$, wo $X_i(t)$, $i = \overline{1,k}$ ein stationärer in weitem Sinne zufälliger Prozess ist, doch $X_i^{(j)}(t)$ j-te "in medio" Ableitung des zufälligen Prozesses $X_i(t)$ ist, $X_i^{(0)}(t) \equiv X(t)$. Die Koeffizienten b_j, $j = \overline{1,n}$ sind beliebige komplexe Zahlen und egsistiert wenigstens ein nichtnullter b_j, zum Beispiel b_0. Wenn $X_i(t)$, $i = \overline{1,k}$ paarweise orthogonal sind, bestimmen wir den Rang $Y(t)$.

<u>SATZ.</u> Sei $X(t) = (X_1(t), \ldots, X_{k}(t))$ ein multidimensionaler zufälliger Prozess mit $m_r, r = \overline{1,k}$ "in medio" Ableitungen nach ihrer Koordinaten, welche unkorrelierte zufällige Prozesse sind. Sei

$$Y(t) = \left(\sum_{j=1}^{m_1} a_1^{(j)} X_1^{(j)}(t), \ldots, \sum_{j=1}^{m_k} a_k^{(j)} X_k^{(j)}(t)\right),$$

und

$$(a_1^{(0)},\dots,a_k^{(m_k)}) \in \mathbb{C}^{k+\sum_1^k m_j}, (a_1^{(0)},\dots,a_k^{(0)}) \neq 0,$$

dann ist rang $(Y(t)) = k$.

BEWEIS. Der Prozess $Z^{(j)}(t)$ ist untergeordnet dem Prozess $Z^{(k)}(t)$ für $j > k$. So gilt, daß der Hilbertsraum H_s der s-te Koordinate des Prozesses $Y(t)$, die Linearkombination $\sum_{j=0}^{m_s} a_s^{(j)} X_s^{(j)}(t)$, $1 \leq s \leq k$, enthalte. Weiter, H_r enthalte der Ausdruck $\sum_{j=0}^{m_r} a_r^{(j)} X_s^{(j)}(t)$, $1 \leq r \leq k$; $s \neq r$. Wegen der Orthogonalität der Prozesses $X_s(t)$ und $X_r(t)$, die Orthogonalität $H_r \perp H_s$, ist in Kraft. Natürlich, jetzt folgt, daß rang $(Y(t)) = \#\{H_1,\dots,H_k\} = k$ ist. Q.E.D.

Eine mögliche Verallgemeinheitung des vorherigen Ergebnisses kann man bekommen wenn die Koordinaten des multidimensionalen Prozesses $X(t)$ lineare Kombinationen der Form $\sum_{j,k} a_{jk} X_j^{(k)}(t)$ enthalten. Diese Generalisation basiert sich auf der Tatsache, daß sich der Rang des Prozesses mit Koordinaten der Form $\sum_j b_j X_i^{(j)}(t)$ nicht ändert, wenn

1) sich der Reihenordnung der Koordinaten ändert,
2) in $\sum b_i X_i^{(j)}(t)$, $i \in J \subset N$, bzw. J eine beliebige geordnete Untermenge der Menge der natürlichen Zahlen.
3) in $\sum b_i X_i^{(j)}(t)$ egsistiert wenigstens ein nichtnullter Koeffizient.

Die nächste Theoreme gibt uns die bestmögliche Generalisation nebst der bisherigen Vorausstzungen.

THEOREME. Sei $\{X(t) = (X_1(t),\dots,X_n(t)),\ t \in R\}$ ein stationärer in weitem Sinne, multidimensionaler zufälligen Prozess mit orthogonalen koordinaten, welche mittelquadratische Ableitungen der Potenz $m_i, i = \overline{1,n}$ haben. Sei weiter

$$Y(t) = \left(\sum_{J_1} \sum_{K_1} a_{jk}^{(1)} X_j^{(k)}(t), \ldots, \sum_{J_n} \sum_{K_n} a_{jk}^{(n)} X_j^{(k)}(t)\right)$$

wo $J_b = \{j_b, \ldots, r_b\}$, $K_b = \{k_b, \ldots, m_b\}$ beliebige, geordnete Untermengen der N sind, und

1) $\text{rang}(X(T)) = n$,

2) $a_{jk}^{(i)} \in C$, $i = \overline{1,n}$,

3) $(\forall i \in \{1\ldots,n\})(\exists j)(j \in J_i)(a_{j0}^{(i)} \neq 0)$,

4) $1 \leq \min_{1\leq b\leq n} (j_b) \leq \max_{1\leq b\leq n} (r_b) \leq n$.

Dann ist $\text{rang}\,(Y(t)) = \#\,\{J_b | J_b \leq J_c;\ b \neq c \text{ und } b,c = \overline{1,n}\}$.

<u>BEWEIS.</u> Suchen wir die matrische Spektraldichte $f_Y(\lambda) = (f_{bc}(\lambda))_{nxn}$ mittels Bochner-Khintchine's Theoreme aus der Korrelationsmatrix $K_Y(t) = (K_{bc}(t))_{nxn}$.

$$K_{bc}(t) = E \sum_{J_b, K_b} a_{jk}^{(b)} X_j^{(k)}(t)\, \overline{\sum_{J_c, K_c} a_{jk}^{(c)} X_j^{(k)}(0)}$$

$$= \sum_{J_b, J_c} \sum_{K_b, K_c} a_{jk}^{(b)} \overline{a_{in}^{(c)}}\; E\, X_j^{(k)}(t)\, \overline{X_i^{(n)}(0)}.$$

Aus der Orthogonalität der Koordinaten X(t) folgt daß

$$E\, X_j^{(k)}(t)\, \overline{X_i^{(n)}(0)} = \delta_{ij}\, K_{X_j^{(k)} X_i^{(n)}}(t) .$$

Darum ist wegen $\max(j_b, j_c) \leq j_{bc} \leq \min(r_b, r_c)$; $J_{bc} = \{j_{bc}, \ldots, r_{bc}\}$,

$$K_{bc}(t) = \sum_{J_{bc}} \sum_{K_b, K_c} a_{jk}^{(b)} \overline{a_{jn}^{(c)}} \int_R e^{it\lambda} (i\lambda)^k \overline{(i\lambda)^n}\, f_j(\lambda)\, d\lambda$$

wo $f_j(\lambda)$ die Spektraldichte des Prozesses $X_j(t)$ ist.

$$f_{bc}(\lambda) = \sum_{J_{bc}} f_j(\lambda) \sum_{K_b} a_{jk}^{(b)} (i\lambda)^k \sum_{K_c} \overline{a_{jn}^{(c)} (i\lambda)^n}. \qquad (1)$$

Natürlich, $f_{bb} = \sum_{J_b} f_j(\lambda) | \sum_{K_b} a_{jk}^{(b)} (i\lambda)^k |^2 \geq 0$ wegen 3), so daß die Diagonaselementen $f_x(\lambda)$ nichtnegativ sind.

Bestimmen wir jetzt den Rang des Prozesses Y(t). Voraussetzen wir daß rang(X(t)) = n für jede lineare Kombination

$$\sum_{J_b} \sum_{K_b} a_{jk}^{(b)} X_j^{(k)}(t) .$$

Sei dann $J_1 = J_2, J_i \neq J_m$, $i \neq m$; $i,m \in \{x,3,\dots,n\}$, $x \in \{1,2\}$. Wir bestimmen den Hauptminor M_2 aus der oberen linken Ecke der Matrix $f_x(\lambda)$. Wegen (1):

$$|M_2| = \begin{vmatrix} f_{11}(\lambda) & f_{12}(\lambda) \\ f_{21}(\lambda) & f_{22}(\lambda) \end{vmatrix} = (f_{11}f_{22} - f_{21}f_{22})(\lambda)$$

$$= \left(\sum_{J_1} f_j(\lambda)\right)^2 | \sum_{K_1} a_{jk}^{(1)} (i\lambda)^k |^2 | \sum_{K_2} a_{jk}^{(2)} (i\lambda)^k |^2 - \left(\sum_{J_{12}} f_j(\lambda)\right) \text{x}$$

$$\text{x} \left(\sum_{J_{21}} f_j(\lambda)\right) | \sum_{K_1} a_{jk}^{(1)} (i\lambda)^k |^2 | \sum_{K_2} a_{jk}^{(2)} (i\lambda)^k |^2$$

$$= | \sum_{K_1} a_{jk}^{(1)} (i\lambda)^k |^2 | \sum_{K_2} a_{jk}^{(2)} (i\lambda)^k |^2 \left[\left(\sum_{J_1} f_j(\lambda)\right)^2 - \right.$$

$$\left. - \sum_{J_{12}} f_j(\lambda) \sum_{J_{21}} f_j(\lambda) \right] .$$

Aus $J_1 = J_2 = J_{12} = J_{21}$ folgt $|M_2| = 0$. (Ähnlich ist $|M_3| = \dots = |M_n| \equiv$ $\equiv |f_X(\lambda)| = 0$, wenn gilt $J_1 = \dots = J_n$). Aber in unserem fall ist

$|M_i| \neq 0, i = \overline{3,n}$; und rang $(Y(t)) = n-1$. Also

$$\text{rang}(Y(t)) = \# \{J_b | J_b \nsubseteq J_c,\ b \neq c; b,c = \overline{1,n}\}.$$

Q.E.D.

ANMERKUNG. Die Theoreme gibt eine genügende Bedingung daß der Rang des Prozesses Y(t) - dessen Koordinaten aus X(t) herstammen und unkorreliert sind - maximal sein soll, oder er möge den voraus bestimmen Wert r_0, $1 \leq r_0 \leq n$ annehmen.

Die Theoreme ist gleichzeitig die Antwort auf das Problem der Konstruktion eines stationären zufälligen Prozesses mit vorausbestimmten Rang, aus Koordinates des Prozesses X(t), so daß beide gleicher Dimension sind. Natürlich, diese Frage kann man auch trivialisch lösen. Wenn X(t) ein Prozess aus der Theorem ist, dann ist rang$(Y(t)) = r_0$, eine voraus bestimmte Zahl r_0 aus $\{1,\ldots,n\}$ und

$$Y(t) = (X_1(t),\ldots,X_{r_0}(t),\underbrace{X_{r_0}(t),\ldots,X_{r_0}(t)}_{n-r_0})$$

dann ist dim(Y(t)) auch egal n.

REFERENCES

Cramér,H. (1961) On the Structure of Purely Non-deterministic Stochastic Processes, Ark.Math. 4., 2-3, 249-266.

Pogány,T. (1983) O rangu višedimenzionalnog slucajnog procesa stacionarnim koordinatama, Mat.Vesnik, 35, 1, 65-68.

Yaglom,A.M. (1962) Stationary Random Functions, Dover Publ.,NY.

Tibor Pogány
25000 Sombor
Fruškogorska A-5/I-1
Yugoslavia

Proc. of the 5th Pannonian Symp. on Math. Stat., Visegrád, Hungary 1985
W. Grossmann, J. Mogyoródi, I. Vincze, W. Wertz, eds

AUTOPHAGE AND ALLELOPHAGE PROBABILITY DISTRIBUTIONS

G.J. SZÉKELY

Budapest

We introduce some new types of probability distribution functions and prove theorems on their continuity, absolute continuity and infinite divisibility.
AMS subject classification: 60 E05

1. INTRODUCTION

Definition 1. The *menu-card* of a probability distribution function (pdf) F is the set of pdf's G such that

$$F(x) = F(ax) * G(x) \qquad (x\in R) \tag{1.1}$$

holds for some $a>1$ ($*$ denotes the convolution).

Remark 1. If the existence of $G=G_a$ is supposed for every $a>1$ then (by definition) F is self-decomposable. This type of pdf's was introduced and analysed by P. Lévy and Hinčin. Self-decomposable pdf's are known to be infinitely divisible (see e.g. Lukacs (1970)), absolutely continuous or degenerate (Fisz-Varadarajan (1963)), and unimodal (Yamazato (1978)). We shall see that the continuity of F is implied by the existence of a nondegenerate pdf on the menu-card of F.

Definition 2. The pdf F is *autophage* (eats itself as the cannibals or the Samoyedes) if there exists a $b>1$ such that $F(bx)$ is on the menu-card of $F(x)$, i.e. if

$$F(x) = F(ax) * F(bx) \qquad (x\in R) \tag{1.2}$$

holds for some $a>1$ and $b>1$.

Definition 3. The pdf's F and G are allelophage if G is on the menu-card of F and F(bx) is on the menu-card of G for some $b>1$.

In this paper we shall prove that autophage and allelophage pdf's are infinitely divisible, and autophage pdf's are absolutely continuous or degenerated.

2. EXAMPLES

(i) If g is the characteristic function of a pdf G from the menu-card of an exponential pdf F with characteristic function

$$f(t) = (1-it/\lambda)^{-1} \qquad (0<\lambda<\infty)$$

then

$$g(t) = (1-it/\lambda)^{-1}/(1-it/\mu)^{-1} = \lambda/\mu+(1-\lambda/\mu)(1-it/\lambda)^{-1} \quad (0<\lambda<\mu<\infty)$$

i.e. G is the mixture of a degenerate and an exponential pdf. Thus the menu-card of exponential pdf's consists of "almost exponential" pdf's.

The following example shows that given a nondegenerate G the pdf of F is not necessarily infinitely divisible nor absolutely continuous (see Theorem(i) below).

Let G be the pdf of the random variable X with distribution $Pr(X=0) = Pr(X=1) = 1/2$. If $a=2$, then F is uniformly distributed on the interval $[0,1]$, thus F is not infinitely divisible. The same G with $a=3$ gives a singular pdf F.

(ii) All stable pdf's are obviously autophage. For the construction of a non-stable autophage pdf we shall use the observation (see Theorem (ii) below) that all autophage pdf's are infinitely divisible. If the spectral functions of an infinitely diviseble pdf are

$M(x) = 2^{-[\log_2|x|]}$ $(x<0)$ and $N(x)=-2^{-[\log_2 x]}$ $(x>0)$

where [ ] denotes the integer part, then the corresponding pdf F satisfies, obviously, the equation (1.2), i.e. it is autophage but not stable.

(iii) If F and G are allelophage then

$F(x) = F(ax) * G(x)$ and $G(x) = F(bx) * G(cx)$

for some $a,b,c>1$. Allelophage pdf's are also infinitely divisible. Their spectral measures satisfy the following equations

$M(x) = M(ax) + L(x)$ and $L(x) = M(bx) + L(cx)$ $(x<0)$.

(A similar system of equations holds for $x>0$ but, of course, we may restrict ourselves to the negative half-line).

In the simplest case $a=b=c=2$. We get

$$M(x) = M(2x) + \sum_{k=1}^{\infty} M(2^k x).$$

If $M(2x) = \gamma M(x)$ then $M(x) = (\gamma+\frac{1}{1-\gamma} -1)\, M(x)$. Thus $\gamma = \frac{3-\sqrt{5}}{2}$ gives a pair of spectral measures of allelophage pdf's:

$$M(x) = \left(\frac{3-\sqrt{5}}{2}\right)^{[\log_2|x|]} \quad \text{and} \quad L(x) = \left(\frac{3-\sqrt{5}}{2}\right)^{[\log_2|x|]+1/2}$$

([ ] denotes the integer part; $\left(1-\left(\frac{3-\sqrt{5}}{2}\right)\right)^2 = \frac{3-\sqrt{5}}{2}$).

In this example $L(x) \neq M(bx)$, whatever $b>1$ is, since in the formula of L the exponent of $\frac{3-\sqrt{5}}{2}$ is not an integer. This means that $G(x) \neq F(bx)$, i.e. F is not autophage. Thus there exist allelophage pdf's that are not autophage.

3. MAIN RESULTS

Theorem

(i) If the menu-card of F contains a nondegenerate pdf then F is continuous.

(ii) If F is autophage, or more generally, if

$$F(x) = F(a_1x) * F(a_2x) * \dots * F(a_nx) \tag{3.1}$$

holds for some $a_1>1$, $a_2>1$, ..., $a_n>1$, then F is infinitely diviseble and absolutely continuous (if not degenerate). Allelophage pdf's are also infinitely divisible.

(iii) If F is autophage, or more generally, if (3.1) holds and F is not stable then $\log a_i/\log a_j$ $(i,j=1,2, \ldots, n)$ are rational numbers. Conversely: if all these quotients are rational numbers then one can construct non-stable pdf's satisfying (3.1)

In this paper we do not prove (iii). In fact it is a simple generalization of Borbély-Ruzsa-Székely (1984).

4. PROOF OF THE THEOREM

Proof of (i). The iteration of (1.1) gives

$$F(x) = G(x) * G(ax) * G(a^2x) * \ldots$$

By Hengartner-Theodorescu (1980) Th. 2.2.14, if the maximal jump of $G_1, G_2, \ldots$ is $p_1, p_2, \ldots$ then the maximal jump of $G_1 * G_2 * \ldots$ is not greater than $C\left(\sum_{j=1}^{\infty}(1-p_j)\right)^{-1/2}$ (C is a finite positive constant). But the maximal jump of $G(x)$, $G(ax)$, $G(a^2x)$, ... is the same, thus F is continuous if G is not degenerate.

Ramark 2. In the language of autoregressive schemes Theorem (i) can be restated as follows: for nondegenerate noise the pdf of a first order autoregressive sequence is always continuous.

Ramark 3. As T. F. Móri observed the following generalisation also holds. If h is an arbitrary one-to-one $R \to R$ function and

$$Y = h(Y) + X \quad \text{in distribution}$$

for idependent nondegenerate random variables X and Y, then the pdf of Y is continuous.

Proof of (ii). The iterative application of (3.1) shows that F can be decomposed into the convolution of pdf's corresponding to infinitesimally small random variables. The same decomposition holds for allelophage pdf's F. Thus Hinčin's classical limit theorem implies the infinite divisibility of F.

The absolute continuity of F satisfying (3.1) will be proved by showing the absolute integrability of its characteristic function f. (3.1) implies

$$f(t) = \prod_{j=1}^{n} f(\alpha_j t) \qquad (0<\alpha_j<1) \qquad (4.1)$$

where $\alpha_j = a_j^{-1}$ $(j=1,2, \ldots , n)$.
Denote by λ the unique positive finite number such that

$$\sum_{j=1}^{n} \alpha_j^{\lambda} = 1.$$

Since F is infinitely divisible its characteristic function f does not take the value 0. Let $g(t) = |t|^{-\lambda}\log|f(t)|$ $(t\neq 0)$. The equation (4.1) gives

$$g(t) = \sum_{j=1}^{n} \alpha_j^{\lambda} \; g(\alpha_j t)$$

i.e. g(t) is a convex linear combination thus the continuity of g implies the existence of a number $\alpha=\alpha(t)$ such that

$$0 < \min \alpha_j < \alpha < \max \alpha_j < 1 \quad \text{and } g(t) = g(\alpha t).$$

A repeated application of this observation shows that if g takes a value at $t_o\neq 0$ then it takes the same value at the points of a sequence $\{t_n\}$ which converges to 0. Thus g is bounded away from 0 in any interval $[c,d]$ $0<c<d<\infty$ $c/d < \min \alpha_j$ or $-\infty<c<d<0$, $|d/c|< \min \alpha_j$, otherwise g would take the value 0 in $[c,d]$ (by continuity) and thus there would exist a sequence $\{t_n\}$ converging to 0 such that

$$g(t_n) = 0 \quad \text{i.e.} \quad |f(t_n)| = 1 \qquad n = 1,2, \ldots$$

implying that f is a degenerate pdf. Thus there exists a $w>0$ such that $|g(t)|> w$ $\quad t\in [c,d]$.
By the considerations above the same is true for any t, $|t|>b>0$ where b is any positive number. Thus

$$|f(t)| < \exp\{-w|t|^{\lambda}\} \qquad (|t|>b>0).$$

Here the integral of the right-hand side is finite therefore F is absolutely continuous.

5. OPEN PROBLEMS

(i) Suppose the menu-card of F contains a continuous pdf G. Does this imply the absolute continuity of F?

(ii) If F is not autophage but there exists a G such that F and G are allelophage then

$$F(x) = F(a_1x) * F(a_2x) * \ldots \tag{5.1}$$

where $1<a_1<a_2< \ldots$, $\lim_{n\to\infty} a_n = \infty$. This equation differs from (3.1) only in the number of convolution factors: here it is infinite. In the language of characteristic functions we get the analogue of (4.1) as follows

$$f(t) = \prod_{j=1}^{\infty} f(\alpha_j t)$$

$0<\alpha_j<1$, $\lim_{j\to\infty} \alpha_j=0$. If we try to apply the argument of the proof of Theorem (ii) then it breaks down in the step:

$0 < c/d \leq \inf \alpha_j = 0 \qquad$ (or $|d/c|\leq 0$).

In spite of this difficulty I conjecture that all (nondegenerate) allelophage pdf's are absolutely continuous.

(iii) Prove the analogue of Theorem (iii) for allelophage pdf's. Here the essence of difficulties is the following. Let $f:(0,\infty)\to(0,\infty)$ be a real monotone increasing function satisfying the following ("conditional Cauchy") equation

$$f(x) = \sum_{j=1}^{\infty} f(\alpha_j x), \qquad \alpha_j>0 \qquad \sum_{j=1}^{\infty} \alpha_j = 1$$

or the equation

$$\sum_{j=1}^{n} f(\alpha_j x) = \sum_{k=1}^{m} f(\beta_k x)$$

$$\alpha_j > 0, \quad \beta_k > 0 \quad \sum_{j=1}^{n} \alpha_j = \sum_{k=1}^{m} \beta_k = 1.$$

Under what conditions on α_j (and β_k) follows that $f(x) = cx$ for some $c>0$? We can only handle the case when $n=1$ and m is finite (or $m=1$ and n is finite). For $n=1$, $m=2$ see Borbély-Ruzsa-Székely (1984). A consequence of this special result is Theorem (iii).

REFERENCES

Borbély, A. - Ruzsa, I. Z. - Székely, G.J. (1984): A note on stable distributions, Preprint No. 54 Math. Inst. Hungar. Acad. Sci. Budapest.

Fisz, M. - Varadarajan, V. S. (1963): A condition for absolute continuity of infinitely divisible distribution functions, Z.f. Wahrscheinlichkeitstheorie 1, 335-9.

Hengartner, W. - Theodorescu, R. (1980): Concentration Functions, Academic Press, New York.

Lukacs, E. (1970): Characteristic Functions (2nd ed.) Griffin, London.

Yamazato, M. (1978): Unimodality of infinitely divisible distribution functions of class L, Ann. Probab. 6. 523-531.

Gábor J. Székely
Department of Probability Theory and Statistics
Eötvös Loránd University
Budapest, Muzeum krt. 6-8.
H - 1088

Proc. of the 5th Pannonian Symp. on Math. Stat., Visegrád, Hungary 1985
W. Grossmann, J. Mogyoródi, I. Vincze, W. Wertz, eds

ON A PROBABILISTIC CHARACTERIZATION THEOREM

I. VINCZE

Budapest

Some twenty years ago A. Rényi and the author formulated a conjecture concerning entire functions. Recently Miles and Willianson have solved the problem in affirmative sense, given a joint characherization of the Poisson law and the gamma distributions. The present note is devoted to some related mathematical problems.
AMS subject classification: 62E10, 60E05

1. INTRODUCTION

Let

$$f(t) = 1 + a_1 t + a_2 t^2 + \dots \qquad a_i > 0 \tag{1.1}$$

be an entire function satisfying the relations

$$\int_0^\infty \frac{a_n t^n}{f(t)}\, dt = 1, \qquad n=0,1,2,\dots \tag{1.2}$$

Rényi and Vincze [3] formulated the conjecture that the only entire solution of (1.1) and (1.2) is $f(t) = e^t$. The first result in this direction was given by R. R. Hall and J. H. Williamson [2] proving among others that from (1.1) and (1.2)

$$\log f(t) = t + \mathcal{O}\left(\frac{t}{\log t}\right), \quad t \to \infty$$

follows. In the paper by W. K. Hayman and I. Vincze [4] - using elementary tools - the following statement is proved: there exists a constant $A \geq 19$, for which

$$e^{t - A\sqrt{t}} \leq f(t) \leq e^{t + A\sqrt{t}}, \qquad t > 0, \tag{1.3}$$

corresponding to $\log f(t) = t + \mathcal{O}(\sqrt{t}),\ t \to \infty$.

Starting from the relation (1.3) J. Miles and J. Williamson [5] solved the problem in affirmative sense: $f(t) = e^t$ *is the only solution of* (1.1) *and* (1.2). Besides (1.3) a lot of effort was needed to the final result.

This way a new characterization of the exponential function e^t is obtained, at the same time a joint characterization of the Poisson law and of the gamma-distributions.

The aim of the present note is to establish some mathematical problems which arose during the investigation of the topic.

2. A RELATED PROBLEM

Multiplying relation (1.2) by λ^n and summarizing with respect to n we obtain

$$\int_0^\infty \frac{f(\lambda t)}{f(t)}\,dt = \frac{1}{1-\lambda}, \quad 0 \leqq \lambda < 1 \tag{2.1}$$

wich condition is equivalent to (1.2). Let now $f(t)$ be an entire function

$$f(t) = a_0 + a_1 t + a_2 t^2 + \ldots + a_0 = 1, \quad a_i > 0, \tag{2.2}$$

in connection of which the notations

$$\int_0^\infty \frac{a_n t^n}{f(t)} = \varphi_n, \quad n = 0,1,2,\ldots$$

and

$$\int_0^\infty \frac{f(\lambda t)}{f(t)}\,dt = \varphi_0 + \varphi_1\lambda + \varphi_2\lambda^2 + \ldots = \Phi_f(\lambda), \quad 0 \leqq \lambda < \lambda_0 \tag{2.3}$$

will be used. We assume that the power series $\Phi_f(\lambda)$ converges in an interval $[0, \lambda_0)$, $\lambda_0 < \infty$.

Denoting by E_f the set of entire functions having the same self-transform $\Phi_f(\lambda)$ one can pose the *conjecture: the only member of* E_f *is the initial* $f(t)$ *itself*. This means that for any entire function $g(t) = b_0+b_1t + \ldots, b_i>0$, for which

$$\int_0^\infty \frac{b_n t^n}{g(t)}\, dt = \int_0^\infty \frac{a_n t^n}{f(t)}\, dt, \quad n= 0,1,2, \ldots$$

hold, $g(t) = f(t)$ is valid. An affirmative answer indicates the following statement.

Proposition 2.1. Whenever $f(t)$ and $g(t)$ are elements of E_f for which the supremum or infinum of

$$\alpha(t) = \log \frac{g(t)}{f(t)}, \qquad t>0$$

is taken on for a certain value $t = t_0 > 0$, then $g(t) = f(t)$.

Remark: This statement and its proof was suggested by [5], but Theorem 4 of [4] along with its proof is of the same nature.

Proof of Proposition 2.1. We can write $g(t) = f(t)e^{\alpha(t)}$. Let

$$-\beta_1 \doteq \inf_{t\geq 0} \alpha(t) \leq 0 = \alpha(0) \leq \sup_{t\geq 0} \alpha(t) = \beta_2$$

where $0 \leq \beta_1,\beta_2$. Let us consider the case when there exists a value t_0 for which $\alpha(t_0) = -\beta_1$. Now $\beta_1\geq 0$; when $\beta_1 = 0$ then trivially $f(t) = g(t)$. Suppose $\beta_1 > 0$; then

$$\frac{\varphi_n}{b_n} = \int_0^\infty \frac{t^n}{g(t)}\, dt = \int_0^\infty \frac{t^n}{f(t)}\, e^{-\alpha(t)} dt \leq \int_0^\infty \frac{t^n}{f(t)}\, dt\, e^{\beta_1} = \frac{\varphi_n}{a_n}\, e^{\beta_1}$$

for $n= 1,2, \ldots$ and sign $<$ is valid for $n=0$. Hence $a_n \leq b_n e^{\beta_1}$ and for all $n>0$ $b_n t^n \geq a_n t^n e^{-\beta_1}$; now

$$g(t_0) = \sum_0^\infty b_n\, t_0^n > f(t_0)e^{-\beta_1} = f(t_0)e^{\alpha(t_0)} = g(t_0)$$

yielding a contradiction. - The same consequence has the case $\alpha(t_0) = \beta_2$.

Miles and Williamson [5] came to contradiction in the same elegant way when the supremum or the infimum of $\alpha(t)$ is not attained for a certain value t_0; to come to this result they had to get over a lot of difficulties.

From Proposition 2.1 follows the validity of the following

<u>Corollary 2.1.</u> Whenever $f(t)$ and $g(t)$ are elements of E_f and

$$\log \frac{g(t)}{f(t)} = o(1),\ t \to \infty,$$

then $g(t) = f(t)$.

3. ON MULTIPLY MONOTONE FUNCTIONS

The original problem led to a solution $f(t) = e^t$, the reciprocal of which is completely monotone, i.e.

$$(-1)^k \frac{d^k}{dt^k} \frac{1}{f(t)} > 0, \qquad k = 0,1,2, \ldots$$

The problem posed in § 2. seems to be more difficult as $1/f(t)$ need not be even convex.

A function $\varphi(t)$ positive on the semiaxis $(0,\infty)$ will be called here k-times monotone when it has derivatives up to the k'th order and the relations

$$(-1)^m \frac{d^m}{dt^m} \varphi(t) > 0, \qquad m = 1,2, \ldots k$$

hold. In the original problem the reciprocals of the coefficients a_n form a moment sequence:

$$\frac{1}{a_n} = \int_0^\infty \frac{t^n}{f(t)}\, dt, n=0,1,2, \ldots$$

for which the moment inequalities

$$\left(\frac{1}{a_n}\right)^2 < \frac{1}{a_{n-1}} \cdot \frac{1}{a_{n+1}} \quad \text{i.e. } a_n^2 > a_{n-1}a_{n+1}, \; n = 1,2, \ldots \qquad (3.1)$$

are valid. In [7] the following lemma is proved:

Lemma. Let $f(t) = a_0 + a_1 t + \ldots$ be a power series with positive coifficients and with a radius of convergence $R>0$. Whenever the ralations $a_n^2 > a_{n-1}a_{n+1}$, $n=1,2, \ldots$ hold, then the function $1/f(t)$ is strictly convex in $[0,R)$, i.e.

$$\frac{d^2}{dt^2} \frac{1}{f(t)} > 0, \; 0 \leqq t < R.$$

Making use of this lemma to the original problem, in [6] the following conclusion of (2.1) was drawn:

$$\left(\frac{n+1}{a_n}\right)^2 < \frac{n}{a_{n-1}} \frac{n+2}{a_{n+1}}, \text{ i.e. } a_n^2 > a_{n-1}a_{n+1}\left(1 + \frac{1}{n(n+2)}\right),$$

$$n = 1,2, \ldots$$

Whenever higher order monotony of $1/f(t)$ is assumed, more information concerning the coefficients of $f(t)$ can be derived. In [1] it was proved that when $1/f(t)$ is completely monotone, from (1.1) and (2.1) $f(t) = e^t$ follows.

Returning to the convexity property of $1/f(t)$ when (3.1) holds, the following problem can be posed: *what kind of simple coefficient-relation guarantees the higher order monotonicity of* $1/f(t)$. In [6] a necessary condition is given, which is - in a little bit more general form - the following: A necessary condition for the function $\varphi(t)$ having moments of all order and derivatives up to the k-th order to be k-times monotone, is that for the moments

$$\mu_n = \int_0^\infty t^n \varphi(t)\,dt, \quad n = 0, 1,2, \ldots$$

the sequence

$$\left(\frac{1}{n} + \frac{1}{k}\right) \frac{\mu_n}{\mu_{n-1}}, \quad n = 1,2, \ldots \tag{3.2}$$

should be monotonically increasing. In the original problem from the k-times monotonicity of $1/f(t)$ the property (3.2) for the moments $\frac{1}{a_n} = \int_0^\infty \frac{t^n}{f(t)} dt$, i.e. for the coefficients follows. Another consequence of the above mentioned lemma is the following.

Corollary 3.1. Let $f(t) = a_0 + a_1 t + a_2 t^2 + \ldots$ be a power series with positive coefficients and with radius of convergence $R>0$. Whenever the relations $a_n^2 > a_{n-1}a_{n+1}$, $n=1,2,\ldots$ hold, then the functions

$$\frac{1}{f^{(k)}(t)}, \qquad k = 1,2, \ldots$$

are in $[0,R)$ strictly convex. Here the notation

$$f^{(k)}(t) = \frac{d^k}{dt^k} f(t) = k!\, a_k + \frac{(k+1)!}{1!} a_{k+1} t + \ldots + \frac{(k+n+1)!}{(n+1)!} a_{k+n+1} t^n + \ldots$$

is used.

This statement is true because from the relations $a_n^2 > a_{n-1}a_{n+1}$ and $n^2 > (n-1)(n+1)$, $(n\, a_n)^2 > (n-1)a_{n-1}(n+1)a_{n+1}$ follows, which settles for $f'(t)$ and in general the relations

$$\left(\frac{(k+n)!}{n!}\right)^2 > \frac{(k+n-1)!}{(n-1)!} \frac{(k+n+1)!}{(n+1)!},$$

$k = 1,2, \ldots$, $n = 1,2, \ldots$ hold, which completes the proof of Corollary 3.1.

Acknowledgement:

Author is indebted to the referee for his valuable comments.

REFERENCES

1. Hall, R. R. and Vincze, I. (1981). On a simultaneous characterization of the Poisson law and the gamma distribution. In: Analytical Methods in Probability Theory (Oberwolfach, 1980), pp. 54-59. Lecture Notes in Mathematics, 861, Springer, Berlin.

2. Hall, R. R. and Williamson, J. H. (1976). On a certain functional equation. J. London Math. Soc. (2) 12, 133-136.

3. Hayman, W. K. (1967). Research Problems in Function Theory. Athlone, London.

4. Hayman, W. K. - Vincze, I. (1978). A problem on entire functions. In: Complex Analysis and its Applications, pp. 591-594. Nauka, Moscow.

5. Miles, J. - Williamson, J. (1984). A Characterization of the Exponential Function. - preprint. To be published in J. London Math. Soc.

6. Vincze, I. (1982). On a certain class of multiply monotone and completely monotone functions. In: Functions, series, operators (Colloquia Math. Soc. J. Bolyai 35) pp. 1259-1266. Budapest.

7. Vincze, I. (1984). Contribution to a characterization problem. In: Statistics and Probability (Proc. of the 3rd Pannonian Symp. on Math. Stat. Visegrád (Hungary), Akadémiai Kiadó (Budapest) - D. Reidel Publ. Co. (Dordrecht), 1982. pp. 353-361.

István Vincze
Mathematical Institute of the
Hungarian Academy of Science
Budapest, Hungary H-1364, POB 127.

Proc. of the 5th Pannonian Symp. on Math. Stat., Visegrád, Hungary 1985
W. Grossmann, J. Mogyoródi, I. Vincze, W. Wertz, eds

ON THE ARITHMETICAL PROPERTIES OF THE MULTIPLICATIVE STRUCTURE OF PROBABILITY DISTRIBUTION FUNCTIONS

A. ZEMPLÉNI

Budapest

In this paper we discuss the two-dimensional multiplicative structure of probability distribution functions.
Sufficient conditions are given for a p.d.f. to be max-irreducible. Using it we prove that they form a dense set with respect to the weak topology.
Finally we deal with the problem of decomposition into irreducible factors and give examples for max-antiireducible p.d.f.-s.
AMS subject classification: 60E05

1. INTRODUCTION

<u>Definition</u>. Let the multiplication of the p.d.f.-s F and G be

$$F\cdot G(x) = F(x)G(x). \tag{1}$$

It is obvious, that F·G is also a p.d.f. The multiplication is an associative, commutative and continous operation /with respect to the weak topology/. So $(\mathfrak{F},\cdot)$ is a commutative topological semigroup.

The multiplication has an important meaning in probability theory: let X and Y be independent random variables with p.d.f.-s F and G. Then the p.d.f. of max(X,Y) is exactly F·G. This meaning gives the importance of this structure.
We shall use some notations:
Let $\underline{x},\underline{y} \in \mathbb{R}^k$. $\underline{x}\vee\underline{y} := (\max(x_1,y_1), \ldots, \max(x_k,y_k))$ is the maximum of the vectors.
For $S_1,S_2 \subset \mathbb{R}^k$ let $S_1 \vee S_2 := \{\underline{x}\vee\underline{y}: x\in S_1, \underline{y}\in S_2\}$

For a p.d.f. F let us denote the corresponding distribution by P_F. We shall often consider the support of a P_F:

$$\text{supp } P_F = \{\underline{x}\in \mathbb{R}^k: P_F(S(\underline{x},\varepsilon))>0 \text{ for any } \varepsilon>0\}.$$

supp P_F is obviously a closed set in $\mathbb{R}^k$ (if F is a k-dimensional p.d.f.).
Let $T= [x_o,x_1)\times[y_o,y_1)$ be a rectangle. Then $P_F(T) = F(x_1,y_1) + F(x_o,y_o) - F(x_1,y_o) - F(x_o,y_1)$. We simplify the notation using F_{11} instead of $F(x_1,y_1)$ etc. if it is not confusing.
In the one-dimensional case all $F\in\mathfrak{F}$ is infinitely divisible, because $\sqrt[n]{F}$ is also a p.d.f. So in the further parts we shall consider the two-dimensional case. (1) is even now a good definition of F·G, because

$$F(\underline{x})\cdot G(\underline{x}) = P(\max(\underline{X},\underline{Y})<\underline{x})$$

holds, where $\underline{X}$ and $\underline{Y}$ are independent random vectors with p.d.f.-s F and G, hence F·G is also a p.d.f. In this case not all p.d.f.-s are infinitely divisible because of the condition

$$P_F(T) = F_{11}+ F_{00}-F_{10}-F_{01}\geq 0.$$

A.A. Balkema and S.I. Resnick characterized the infinitely divisible distributions in [1].

Here we shall examine other arithmetical properties, but at first we mention a simple statement for the structure of $(\mathfrak{F},\cdot)$.

Proposition. In $(\mathfrak{F},\cdot)$ there are no units and associated elements. The proof is trivial.

Definition. A decomposition of the p.d.f. F having the form

$$F = G\cdot H$$

is called effective if $F\nmid G$ and $F\nmid H$ /othervise it is called noneffective/.

Lemma 1. $F \in \mathfrak{F}$ has non-effective decomposition if and only if

$$\text{supp } F \subset [x_1, +\infty) \times [x_2, +\infty) \quad \text{for an } \underline{x} = (x_1, x_2) \in \mathbb{R}^2,$$

and then the decomposition has a form

$$F = F \cdot G \tag{2}$$

for a G such that $G(\underline{x}+0) = 1$.

Proof. (2) is obviously a non-effective decomposition. Conversely, let $F = G \cdot H$ be a non-effective decomposition. Then F is associated with one of the factors of the right-hand side. We can suppose $F|G$, then the proposition mentioned above implies $F = G$. Hence $H(\underline{x}) = 1$ if $F(\underline{x}) > 0$. (3)

Suppose indirectly, that there is no $(x_1, x_2) \in \mathbb{R}^2$ for which $\text{supp} F \subset [x_1, +\infty) \times [x_2, +\infty)$. Then there is a sequence $(\underline{x}^{(n)})_{n=1}^{\infty}$ such that $F(\underline{x}^{(n)}) > 0$ and

$$x_1^{(n)} \to -\infty \text{ or } x_2^{(n)} \to -\infty$$

hold. Then (3) implies $H(\underline{x}^{(n)}) = 1$ which is a contradiction.

2. IRREDUCIBLE PROBABILITY DISTRIBUTION FUNCTIONS

In this part we shall give examples for irreducible distributions in $\mathfrak{F}$ and prove some properties of their set $\mathfrak{I}$.

Lemma 2. Let $F \in \mathfrak{F}$ such that there is a $T = [x_o, x_1) \times [y_o, y_1)$ rectangle with the properties

(i) $P_F(T) = 0$

(ii) $P_F(S_i) > 0$ $(i=1,2)$ where

$S_1 = (-\infty, x_o) \times [y_o, y_1)$

and $S_2 = [x_o, x_1) \times (-\infty, y_o)$.

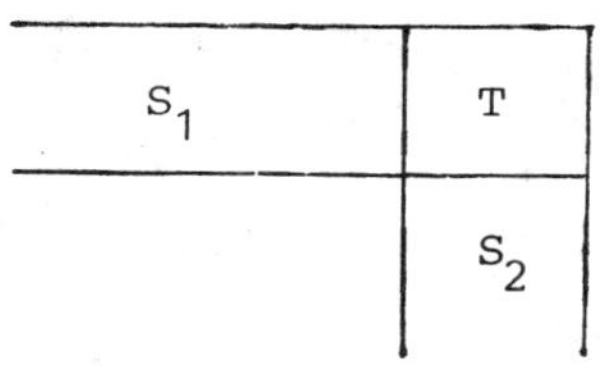

Suppose that $F = G \cdot H$ is a decomposition of F. Then one of the components is constant on T /i.e. supp $P_G \cap (S_1 \cup T \cup S_2) = \emptyset$ or supp $P_H \cap (S_1 \cup T \cup S_2) = \emptyset$/.

<u>Proof:</u> G and H are p.d.f.-s so

$$F_{11} + G_{00} \geqq G_{01} + G_{10}$$

$$H_{11} + H_{00} \geqq H_{01} + H_{10}.$$

Multiplying these equations and considering $F = G \cdot H$ we get

$$F_{00}+F_{11}+G_{11}H_{00}+G_{00}H_{11} \geqq F_{01}+F_{10}+G_{10}H_{01}+G_{01}H_{10} \quad . \quad (4)$$

$P_F(T) = 0$ implies that $F_{11}+F_{00} = F_{01}+F_{10}$. Substracting the double of this equalitiy from (4) and using again that $F=G\cdot H$, we get

$$(G_{11}-G_{00})(H_{11}-H_{00}) \leqq (G_{10}-G_{01})(H_{10}-H_{01}). \quad (5)$$

The left-hand side of (5) is nonnegative, so we can suppose, that

$$G_{10} > G_{01} \quad \text{and} \quad H_{10} > H_{01} \quad (6)$$

hold. Now we can finish the proof indirectly: if $G_{11}>G_{00}$ and $H_{11} > H_{00}$ hold, then (5) and (6) implies that

$$G_{11} = G_{10},\ G_{00} = G_{01},\ H_{11} = H_{10},\ H_{00} = H_{01}.$$

Hence $G_{11}H_{11}=F_{11} = G_{10}H_{10}=F_{10}$ but it contradicts to (ii). So one of the equations $G_{11}=G_{00}$, $H_{11}=H_{00}$ is true, which was to be proved.

<u>Theorem 1.</u> Suppose that given $F\in\mathcal{F}$ one can construct a sequence of rectangles $T_n = [x_1^{(n)}, x_2^{(n)}) \times [y_1^{(n)}, y_2^{(n)})$ /$n\in\mathbb{N}$/ for which the following properties hold.

(i) T_n and T_{n+1} have exactly two common vertices

(ii) at least one of the limit relations $x_2^{(n)} \to -\infty$ and $y_2^{(n)} \to -\infty$ $(n \to \infty)$ is true,

(iii) $P_F(T_n) = 0$ and for the corresponding stripes

$$P_F(S_i^{(n)}) > 0 \quad (i = 1,2,\ n \in \mathbb{N}).$$

Then F is irreducible.

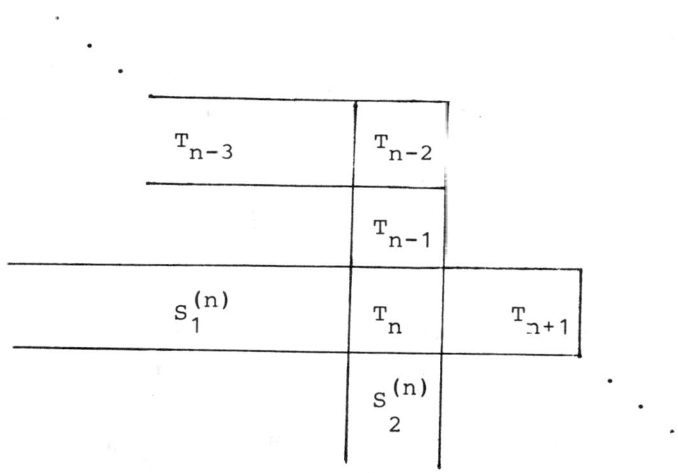

Proof. Suppose that $F = G \cdot H$.
Applying lemma 2. for the rectangle T_1 we get
$G(x_2^{(1)}, y_2^{(1)}) = G(x_1^{(1)}, y_1^{(1)})$ (or the same for H).

$G(x_2, y_2) > 0$ is obvious. We can apply lemma 2. for each rectangle T_n. Suppose that on T_{k-1} and on T_k not the same p.d.f. is constant. Then $G(\underline{x}_1) = G(\underline{x}_2)$ and $H(\underline{x}_1) = H(\underline{x}_2)$ also hold, so $F(\underline{x}_1) = F(\underline{x}_2)$ but it contradicts to (iii).

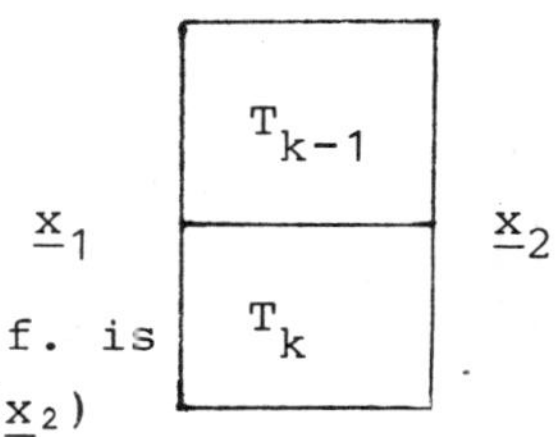

So the only possible case is that one of the components is constant on each T_n /suppose that it is G/. (7) implies that this constant is positive, then considering (ii) we get a contradiction.
The corollary below is a simple consequence of theorem 1.

Corollary. An $F \in \mathcal{F}$ for which $\mathrm{supp} P_F \subset e$ where e is a line with negative steepness and supp P_F is not bounded is max-irreducible.

Remarks.

1. In the corollary the boundedness is not necessary: one can compute that if supp $P_F \subset \bigcup_{k=1}^{N} \bigcup_{i=1}^{2} S_i$ and we omit the assumption (iii) in theorem 1. then F is still effective irreducible.

2. The theorem above proved the irreducibility under assumptions for supp P_F. So they can be written in a form that these sets are irreducible according to the $\vee$ (maximum) operation. This is similar to results for the classical convolutional case /see e.g. [3]/.
Here it is well-known that to each closed set H one can construct a p.d.f. F such that supp P_F=H and F is irreducible.
We shall see below in part 3. that in our case this is not true: there exist sets H for which supp P_F=H implies the max-antiirreducibility of F. Having seen these properties one may think that the properties of a p.d.f. in our case depend only on the properties of its support, but not even this is true.

Example. Let $H_1 = \{(2k+1, 5-(2k+1)) : k=0,1,2,\}$, $H_2 = \{(2k, 5-2k : k=0,1,2\}$, and denote $H_1 \vee H_2$ by H.

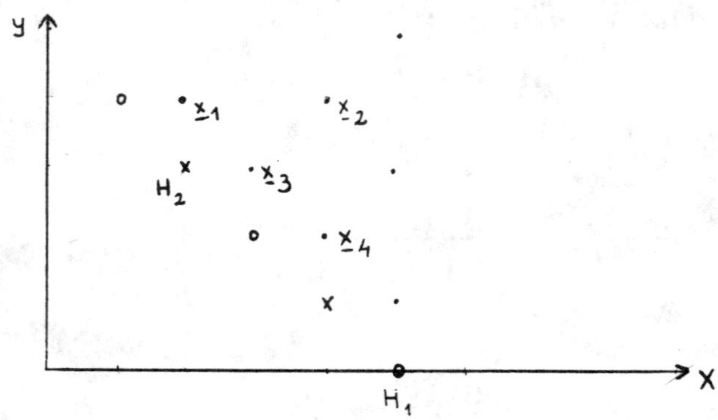

One can easily show that in each decomposition

$$H = H_1 \vee H_2$$

the points $\underline{x}_i$ /i=1,...,4/ can have only the same decomposition as in $H_1 \vee H_2$. Hence /see lemma 3. in part 3/ we get that any F for which

$$\text{supp } P_F = H \text{ and } \frac{P_F(\underline{x}_2)}{P_F(\underline{x}_1)} \neq \frac{P_F(\underline{x}_4)}{P_F(\underline{x}_3)}$$

hold is irreducible, but obviously its support is a reducible set.

The corollary gives a lot of irreducible p.d.f.-s. Using it we can prove the next theorem.

Theorem 2. In $(\mathcal{F},\cdot)$ the irreducible p.d.f.-s form a dense set.

Proof. Let $F \in \mathcal{F}$ be an aribitrary p.d.f. and $\varepsilon_n \searrow 0$. Let e_n be the line which satisfies the equality

$$x + y = -n$$

and $G_n \in \mathcal{F}$ such that supp $G_n \subset e_n$ and supp G_n is not bounded. Let

$$\tilde{F}_n := \begin{cases} F(x,y) & \text{if } x>-n+1 \text{ and } y>-n+1 \\ 0 & \text{otherwise} \end{cases}$$

and denote $(1-\varepsilon_n)\tilde{F}_n + \varepsilon_n G_n$ by F_n^*. $F_n^* \in \mathcal{F}$ is obvious and the corrollary gives that F_n^* is an irreducible p.d.f.

$$|F_n^*(\underline{x}) - F(\underline{x})| \leqq \varepsilon_n \text{ if } x>-n+1 \text{ and } y>-n+1$$

so $F_n^* \to F$ in the weak topology which proves our theorem.

[2] deals with the convolution-irreducible p.d.f.-s and it is proved that their set is also a dense one and G_δ.

3. REDUCIBLE, ANTIIRREDUCIBLE PROBABILITY DISTRIBUTION FUNCTIONS

In the previous part we have dealt with irreducible p.d.f.-s. Here we shall give a sufficient condition for a p.d.f. F to be reducible, then we give examples for antiirreducible distributions.

Theorem 3. Let $F \in \mathcal{F}$ such that supp P_F is bounded from below /i.e. there is an $\underline{x}_o \in \mathbb{R}^2$ for which $\underline{y} > \underline{x}_o$ holds for each $\underline{y} \in$ supp P_F/.

If there exists a constant $c>0$ such that for each rectangle $T = [x_1, x_2) \times [y_1, y_2)$ and for the stripes $S_1 = (-\infty, x_1) \times (y_1, y_2)$ and $S_2 = [x_1, x_2) \times (-\infty, y_1)$

$$P_F(S_1) P_F(S_2) \leqq c P_F(T) \tag{8}$$

holds, then F is reducible.

Proof. Let $G \in \mathcal{F}$ such that supp $P_G < \underline{x}_o$.
Let $0<\alpha<1$ and consider the p.d.f. $\alpha F + (1-\alpha) G$.
We shall prove that for suitable α

$$\alpha F + (1-\alpha) G \mid F$$

holds.
Let us denote $\dfrac{F}{\alpha F + (1-\alpha) G}$ by H.

It is obvious that H is monotonic in its arguments /G=1 holds where H>0/. So we have only to prove that $P_H(T) \geqq 0$ for an arbitrary rectangle T. We can suppose $G_{oo}=1$ because F>0 is the only interesting part of T. $P_H(T) \geqq 0$ means that

$$\frac{F_{11}}{\alpha F_{11} + 1 - \alpha} + \frac{F_{oo}}{\alpha F_{oo} + 1 - \alpha} \geqq \frac{F_{01}}{\alpha F_{01} + 1 - \alpha} + \frac{F_{10}}{\alpha F_{10} + 1 - \alpha}$$

This is equivalent to

$$\frac{2\alpha F_{11}F_{00}+(1-\alpha)(F_{11}+F_{00})}{(\alpha F_{11}+1-\alpha)(\alpha F_{00}+1-\alpha)} \geqq \frac{2\alpha F_{10}F_{01}+(1-\alpha)(F_{01}-F_{10})}{(\alpha F_{01}+1-\alpha)(\alpha F_{10}+1-\alpha)} .$$

Multiplying both sides by the denominators we get the equivalent inequality

$$2\alpha^2(1-\alpha)F_{11}F_{00}(F_{01}+F_{10})+(1-\alpha)\alpha^2(F_{11}+F_{10})F_{01}F_{10}+2(1-\alpha)^2\alpha F_{11}F_{00}+$$

$$+2\alpha^3F_{00}F_{01}F_{10}F_{11}+(1-\alpha)^3(F_{11}+F_{00})+(1-\alpha)^2\alpha(F_{11}+F_{00})(F_{01}+F_{10}) \geqq$$

$$\geqq 2\alpha^3F_{10}F_{01}F_{11}F_{00}+2\alpha^2(1-\alpha)F_{10}F_{01}(F_{11}+F_{00})+2\alpha(1-\alpha)^2F_{10}F_{01}+$$

$$+(1-\alpha)\alpha^2(F_{01}+F_{10})F_{11}F_{00}+(1-\alpha)^2\alpha(F_{01}+F_{10})(F_{11}+F_{00})+(1-\alpha)^3(F_{10}+F_{10}).$$

Hence rearranging we get the form

$$\left(\frac{1-\alpha}{\alpha}\right)^2 \cdot P_F(T)+2\left(\frac{1-\alpha}{\alpha}\right)\left[F_{11}F_{00}-F_{01}F_{10}\right]+F_{11}F_{00}(F_{10}+F_{01}) \geqq \tag{9}$$

$$\geqq F_{01}F_{10}(F_{11}+F_{00}).$$

Let us denote $\frac{1-\alpha}{\alpha}$ by α^*. To prove (9) we consider our assumption (8). It implies that

$$(F_{00}+P_F(S_1))(F_{00}+P_F(S_2))-(P_F(S_1)+P_F(S_2)+F_{00}+P_F(T))F_{00}<cP_F(T)$$

which is equivalent to

$$F_{01}F_{10}-F_{11}F_{00}<cP_F(T). \tag{10}$$

Thus

$$(F_{01}F_{10}-F_{11}F_{00})(F_{11}+F_{00}) < 2cP_F(T). \tag{11}$$

The inequality

$$F_{11}F_{00}(F_{11}+F_{00}-F_{10}-F_{01}) \leqq P_F(T) \tag{12}$$

is obvious. Multiplying (10) by a positive β and add it to (11) and (12) we get

$$(2c+c\beta+1)P_F(T) \geqq \beta(F_{01}F_{10}-F_{11}F_{00}) + (F_{01}F_{10}-F_{11}F_{00})(F_{11}+F_{00}) +$$

$$+F_{11}F_{00}(F_{11}+F_{00}-F_{01}-F_{10}) = \beta(F_{01}F_{10}-F_{11}F_{00})+F_{01}F_{10}(F_{11}+F_{00})-$$

$$- F_{11}F_{00}(F_{10}+F_{01}) \tag{13}$$

for arbitrary $\beta>0$. Let $\alpha^* = 2c+1$ and $\beta=2\alpha^*$. Then (13) is the same as (9) which was to be proved.

<u>Remark</u> We have seen in the proof that $P_F(S_1)P_F(S_2) \leqq c(T)$ implies

$$F_{01}F_{10}-F_{11}F_{00} \leqq cP_F(T).$$

The structure of this condition is very similar to the necessary and sufficient condition of infinite divisibility

$$F_{10}F_{01} \leqq F_{11}F_{00} \qquad ([1]).$$

<u>Corollary 2.</u> Let F be absolutely continous, supp P_F bounded. Let us denote the density of F by f. Suppose, that

$$0 < m \leqq f(\underline{x}) \leqq M < \infty$$

holds for any $\underline{x} \in$ supp P_F. Then F is reducible.

<u>Proof:</u> $P_F(T) \geqq \lambda(T)m$ is obvious.

Let us denote the diameter of supp P_F by d and let $T= [x_0,x_1) \times [y_0,y_1)$.
Then

$$P_F(S_1) \leqq d(y_1-y_0)M \quad \text{and}$$
$$P_F(S_2) \leqq d(x_1-x_0)M.$$

Hence $P_F(S_1)P_F(S_2) \leq d^2M^2\lambda(T)$ so for $c = \frac{M^2}{m}d^2$ we can apply theorem 3.

Lemma 3. Let $F,G \in \mathcal{F}$. Let us denote supp P_F by Z_1 and supp P_G by Z_2. Then supp $FG = \overline{Z_1 \vee Z_2}$.

Proof.

1. If $\underline{z}$ $\overline{Z_1 \vee Z_2}$ then $d(\underline{z}, \overline{Z_1 \vee Z_2}) = \eta > 0$ so $P_{FG}(S(\underline{z},\eta)) = 0$.

2. Let $\underline{z} \in \overline{Z_1 \vee Z_2}$. This implies, that for any $\varepsilon > 0$ there is a $\underline{z}_\varepsilon \in Z_1 \vee Z_2$ such that $d(\underline{z}, \underline{z}_\varepsilon) < \frac{\varepsilon}{2}$. $\underline{z} = \underline{x} \vee \underline{y}$ where $\underline{x} \in Z_1$ and $\underline{y} \in Z_2$. Hence $P_{FG}(S(\underline{z},\varepsilon)) \geqq P_{FG}(S(\underline{z}_\varepsilon, \frac{\varepsilon}{2})) \geqq$
$\geqq P_F(S(\underline{x}, \frac{\varepsilon}{4})) \cdot P_G(S(\underline{y}, \frac{\varepsilon}{4})) > 0$, so $\underline{z} \in$ supp P_{FG}.

Theorem 4. Let $F \in \mathcal{F}$ such that supp P_F is not bounded from below and for any $\underline{x} \neq \underline{y} \in$ supp P_F one of the inequalities

$$\underline{x} < \underline{y} \quad \text{and} \quad \underline{y} < \underline{x}$$

hold, and $|\text{supp } P_F \cap e| < 1$ where e is a line parallel to one of the axes. Then F is antiirreducible.

Proof. Using lemma 3. one can easily show that for a decomposition $F = GH$ supp $P_F \supset$ supp P_G and supp $P_F \supset$ supp P_H hold and none of them is bounded from below. Hence each divisor of F has the properties of the theorem. Such a G is obviously infinitely divisible, because $\sqrt[n]{F}$ is also a p.d.f. /one of the equalities $F_{01} = F_{00}$ or $F_{10} = F_{00}$ always hold/.

Suppose indirectly that F has a completely reducible divisor G /i.e. $G = \prod_{j=1}^{\infty} I_j$ where all I_j-s are irreducible/.

I_1 is also irreducible so there is an $\underline{x}_0 = (x_1^{(0)}, x_2^{(0)}) \in$ supp I_1 such that $\underline{x}_0 \notin$ supp P_F so $d(\underline{x}_0$, supp $P_F) > 2\varepsilon$ for an $\varepsilon > 0$.

$$P_G(S(\underline{x}_0,\varepsilon)) \geqq \prod_{j=2}^{\infty} I_j(x_1^{(0)} - \varepsilon, x_2^{(0)} - \varepsilon) \cdot P_{I_1}(S(\underline{x}_0,\varepsilon)). \qquad (14)$$

$\prod_{j=2}^{\infty} I_j | F$ so $\prod_{j=2}^{\infty} I_j(\underline{x}) \geqq F(x)$ for any $\underline{x} \in R^2$.

Hence (14) can be written in the form

$$P_G(S(\underline{x}_0,\varepsilon)) \geq F(x_1^{(0)}-\varepsilon, x_2^{(0)}-\varepsilon)\cdot P_{I_1}(S(\underline{x}_o,\varepsilon)) > \eta > 0.$$

But we have seen that it is impossible for a $G|F$. So we completed our proof.

4. AN OPEN PROBLEM

Hincin's decomposition theorem for the convolution semigroup of p.d.f.-s is well known /[3]/. It states that for any p.d.f. F

$$F = I_1 \ldots I_n \ldots J. \tag{15}$$

where all I_j-s are irreducible and J is antiirreducible. There are also some more general results /[4] [5]/. But the theorems proved in these papers cannot be applied to $(\mathfrak{F},\cdot)$ because the set of the divisors of an $F\in\mathfrak{F}$ is not always compact. To prove it we consider the theorem 3. There we showed that for a class of p.d.f.-s

$$\alpha F + (1-\alpha)G|F$$

where the only condition for G was that $G(\underline{x}) = 1$ where $F(\underline{x})>0$. Let $(G_n)_{n=1}^{\infty}$ be a sequence of p.d.f.-s such that

$$G_n((-n,n)) = 1.$$

It is obvious that $\alpha F + (1-\alpha)G_n \overset{W}{\to} F^*$ where $\lim\limits_{\substack{x_1\to-\infty\\ x_2\to-\infty}} F^*(x_1,x_2) = 1-\alpha > 0$

so the set $\{\alpha F+(1-\alpha)G_n\}$ cannot be a part of a compact set. This negative property causes that one cannot easily prove the existence of a decomposition having the form (15). It is still an open problem.

REFERENCES

1. A.A. Balkema, S.I. Resnick: Max-infinite divisibility, J. Appl. Prob. 14 (1977) 2. 309-319.
2. K. R. Parthasarathy, R.R. Rao, S.R. Varadhan: On the cathegory of indecomposable distributions on topological groups, Trans. Amer. Math. Soc. 102, 2 (1962) 200-217.
3. Ju. V. Linnik, I. V. Ostrowski: "Decomposition of Random Variables and Vectors", Trans. of. Math. Monographs Vol. 48, Amer. Math. Soc. Providence R.I., 1977 (Russian ed., Nauka, Moscow, 1972).
4. D.G. Kendall: Delphic semigroups, infinitely divisible regenarative phenomena, and the arithmetic of p-functions, Z. Wahrsch. Verw. Gebiete 9 (1968), 163-195.
5. I.Z. Ruzsa, G.J. Székely: Theory of decomposition in semigroups, Advances in Math. 56, 1 (1985), 9-27.

András Zempléni
Eötvös Loránd Univ.
Department of Probability Theory
and Mathematical Statistics
Budapest, H-1088
Muzeum krt. 6-8.

PART B

STATISTICS AND APPLICATIONS

Proc. of the 5th Pannonian Symp. on Math. Stat., Visegrád, Hungary 1985
W. Grossmann, J. Mogyoródi, I. Vincze, W. Wertz, eds

PERIODIC AUTOREGRESSION WITH EXOGENOUS VARIABLES AND EQUAL VARIANCES

J. ANDĚL

Prague

The periodic autoregressive process with non-vanishing mean and with exogenous variables is investigated in the paper. The statistical analysis is based on the Bayes approach with a vague prior densitiy. The modus of the posterior density is used as an estimate of the parameters of the model. Some tests of hypotheses are derived.

AMS subject classification: 62M10

1. INTRODUCTION

Let $\{Y_t\}$ be a white noise with $EY_t = 0$ and $\text{Var } Y_t = \sigma^2 > 0$. A process $\{X_t\}$ is called autoregressive, if it is generated by the relation

$$X_t = b_1 X_{t-1} + \dots + b_n X_{t-n} + Y_t \tag{1.1}$$

where $b = (b_1, \dots, b_n)'$ is a vector of autoregressive parameters.

If we analyse a seasonal time series with a period p, it is possible to use a model of the autoregressive type (1.1) where the coefficients b_i are substituted by some periodic functions. More precisely, we assume that $X_1, \dots, X_n$ are given random variables and $X_{n+1}, X_{n+2}, \dots$ are defined by

$$X_{n+(j-1)p+k} = \sum_{i=1}^{n} b_{ki} X_{n+(j-1)p+k-i} + Y_{n+(j-1)p+k} \tag{1.2}$$

for $k = 1,2,\dots,p$ and $j = 1,2,\dots$ Here we have a set of autoregressive vectors $b_k = (b_{k1}, \dots b_{kn})'$, $k = 1,\dots,p$. Pagano (1978) investigates some properties of estimates of parameters of process

(1.2) and shows that this process is tightly connected with a multivariate autoregressive process. Newton (1982) proposes to use this correspondence for estimating statistical characteristics of multivariate autoregressive processes. A statistical analysis of model (1.2) is described by Anděl (1983). If it is necessary to assume that the process $\{X_t\}$ has a non-vanishing mean, e.g. $EX_{n+(j-1)p+k} = \xi_k$ for $k = 1,\dots,p$, then we come to the model

$$X_{n+(j-1)p+k} = \mu_k + \sum_{i=1}^{n} b_{ki} X_{n+(j-1)p+k-i} + Y_{n+(j-1)p+k} \tag{1.3}$$

where

$$\mu_k = \xi_k - \sum_{i=1}^{n} b_{ki}\, \xi_{n+(j-1)p+k-i}$$

[see Anděl et al. (1985)]. Both the models (1.2) and (1.3) are called periodic autoregressions. If $\operatorname{Var} Y_t = \sigma^2$ does not depend on t, we have models with equal variances. If $\operatorname{Var} Y_{n+(j-1)p+k} = \sigma_k^2$ depends on k, we get models with periodic variances.

In this paper we assume that Y_t are i.i.d. $N(0,\sigma^2)$ variables and we generalize (1.3) in the following way. Let $X_1,\dots,X_n$ be independent of $Y_{n+1}, Y_{n+2},\dots$ and let X_t for $t > n$ be given by

$$X_{n+(j-1)p+k} = \mu_k + \sum_{i=1}^{n} b_{ki} X_{n+(j-1)p+k-i} + \sum_{s=1}^{S} \sum_{r=0}^{m_s} a_{ksr}\, \varphi_{s,n+(j-1)p+k-r} + Y_{n+(j-1)p+k} \tag{1.4}$$

where μ_k, b_{ki}, a_{ksr} and σ^2 are unknown parameters, n, S and m_s are given orders and $\varphi_{s,t}$ are known values of exogenous variables $\varphi_1,\dots,\varphi_S$. It would be possible to include μ_k into set of the parameters a_{ksr}, but because of their special meaning we prefer to express them separately. We assume that a realization $x_1,\dots,x_N$ of the variables $X_1,\dots,X_N$ is given, where N is a sufficiently large number. More precisely, we require throughout the paper that

$$N > n+p+\max\{n+p[m+(p-1)n],\ m+p[n+(p-1)m]\}.$$

For our statistical analysis of the model we use the Bayes approach in which we choose a vague prior density of the parameters. We assume that parameters μ_k, b_{ki}, a_{ksr} and $\log \sigma$ are distributed independently and have the improper rectangular distribution. This choice is discussed in detail by Zellner (1971), Chap. 2. It corresponds to the Jeffreys rule for calculating the invariant prior distribution and leads to the density with minimal information. This seems to be in a good agreement with real cases, because any specific prior knowledge about the parameters can be hardly assumed. Moreover, the estimates for μ_k, b_{ki} and a_{ksr} are identical with the maximum likelihood estimates and according to well known theorems about the Bayesian theory the asymptotic distribution of the estimates does not depend on the prior density under very general conditions. In the same time, the derivations are substantially simpler than a direct method for calculating asymptotic results and for this reason they were used also by other authors, for example in the intervention analysis [see Box and Tiao (1975)]. Our procedures are suitable for very large samples. They cannot be recommended in the cases where the length of the realization is small or where some more precise prior information about the parameters is available.

2. STATISTICAL ANALYSIS OF THE MODEL

Denote

$x = (x_1,\ldots,x_N)'$, $x_t^0 = (x_{t-1},\ldots,x_{t-n})'$, $m = m_1+\ldots+m_S$,

$\mu = (\mu_1,\ldots,\mu_p)'$, $b = (b_1',\ldots,b_p')'$, $a_{ks} = (a_{ks0},\ldots,a_{ksm_s})'$,

$a_k = (a_{k1}',\ldots,a_{kS}')'$, $a = (a_1',\ldots,a_p')'$, $\varphi_{st}^0 = (\varphi_{st},\ldots,\varphi_{s,t-m_s})'$,

$\theta_k = (\mu_k, b_k', a_k')'$, $\theta = (\theta_1',\ldots,\theta_p')'$,

$$w_{kj} = (1, x^{0\prime}_{n+(j-1)p+k}, \varphi^{0\prime}_{1,n+(j-1)p+k}, \dots, \varphi^{0\prime}_{S,n+(j-1)p+k})',$$

$\alpha_k = [(N-n-k)/p]+1$, where [] denotes the integer part,

$$\varkappa_k = \sum_{j=1}^{\alpha_k} x^2_{n+(j-1)p+k}, \quad q_k = \sum_{j=1}^{\alpha_k} x_{n+(j-1)p+k} w_{kj}, \quad Q_k = \sum_{j=1}^{\alpha_k} w_{kj} w'_{kj}.$$

We shall assume that the matrices $Q_1, \dots, Q_p$ are positive definite. This is satisfied with probability one if $\alpha_k > 1+n+m$.

Under our assumptions the conditional density of $X_{n+1}, \dots, X_N$, given $X_1 = x_1, \dots, X_n = x_n$, θ, σ, is

$$f(x_{n+1}, \dots, x_N | x_1, \dots, x_n, \theta, \sigma) = (2\pi)^{-(N-n)/2} \sigma^{-N+n} \times$$

$$\exp\left\{- \sum_{k=1}^{p} \sum_{j=1}^{\alpha_k} z^2_{kj}/(2\sigma^2)\right\}$$

where

$$z_{kj} = x_{n+(j-1)p+k} - \mu_k - b'_k x^0_{n+(j-1)p+k} - \sum_{s=1}^{S} a'_{ks} \varphi^0_{s,n+(j-1)p+k}.$$

Obviously, $z_{kj} = x_{n+(j-1)p+k} - \theta'_k w_{kj}$. Now, we apply the method described in Zellner (1971), Chap. 3 and 7. The following assertion is an easy consequence of the Bayes theorem.

<u>Theorem 2.1.</u> Let θ and σ have the prior density $\pi(\theta,\sigma) = \sigma^{-1}$ for $\sigma > 0$ and $\theta \in R_{(1+n+m)p}$ and vanishing otherwise and let (θ,σ) be independent of $(X_1, \dots, X_n)$. Then the posterior density of θ and σ is

$$g(\theta,\sigma|x) = c\sigma^{-N+n-1} \exp\left\{-\frac{1}{2\sigma^2} \sum_{j=1}^{p} [(\theta_k - \theta^*_k)' Q_k (\theta_k - \theta^*_k) + v_k]\right\} \tag{2.1}$$

for $\sigma > 0$ and zero otherwise, where $\theta^*_k = Q_k^{-1} q_k$, $v_k = \varkappa_k - \theta^{*\prime}_k q_k$.

Theorem 2.2. (i) Modus of the posterior density (2.1) is $\theta=\theta^*, \sigma^2=\sigma^{*2}= v/(N-n+1)$, where $v = v_1+\ldots+v_p$.

(ii) The marginal posterior density of θ is

$$g_1(\theta|x) = c\left[1 + v^{-1}\sum_{k=1}^{p}(\theta_k-\theta_k^*)'Q_k(\theta_k-\theta_k^*)\right]^{-(N-n)/2}.$$

(iii) The marginal posterior density of σ is

$$g_2(\sigma|x) = c\sigma^{-N+n-1+(1+n+m)p}e^{-v/(2\sigma^2)}, \quad \sigma > 0.$$

Proof. (i) is obvious, (ii) and (iii) follow from

$$g_1(\theta|x) = \int_0^\infty g(\theta,\sigma|x)\, d\sigma, \quad g_2(\sigma|x) = \int_{R^{(1+n+m)p}} g(\theta,\sigma|x)\, d\theta.$$

From here we can see that the posterior distribution of v/σ^2 is the $\chi^2_{N-n-(1+n+m)p}$.

The following two theorems follow from the general properties of the multivariate t-distribution [see Zellner (1971), Appendix B].

Theorem 2.3. Denote the elements of θ_k and θ_k^* by θ_{ki} and θ_{ki}^* $(i=1,\ldots,1+n+m)$, respectively. Let $q^{(k)ii}$ be the (i,i) th element of the matrix Q_k^{-1}. Then the posterior distribution of

$$T_{ki} = \left\{[N-n-(1+n+m)p]/(vq^{(k)ii})\right\}^{1/2}(\theta_{ki}-\theta_{ki}^*)$$

is the Student $t_{N-n-(1+n+m)p}$ distribution.

Theorem 2.4. The random variable

$$F = \frac{N-n-(1+n+m)p}{(1+n+m)p}\sum_{k=1}^{p}(\theta_k-\theta_k^*)'Q_k(\theta_k-\theta_k^*)$$

has the posterior $F_{(1+n+m)p,\ N-n-(1+n+m)p}$ distribution.

The statistics T_{ki} and F can serve for tests of fit in the cases when a hypothesis specifies a value of θ or a value of a given component θ_{ki}. The tests of hypotheses $\theta=0$ (independence) or $\theta_{ki}=0$ can occur most frequently. One of the most important questions is if θ_k really depends on k. If not, we can use the classical autoregressive model with exogenous variables. A test of the hypothesis $\theta_1=\ldots=\theta_p$ can be based on the following theorem.

<u>Theorem 2.5.</u> Let $\Delta_k=\theta_k-\theta_p-(\theta_k^*-\theta_p^*)$ for $k = 1,\ldots,p-1$ and $\Delta = (\Delta_1',\ldots,\Delta_{p-1}')'$. Let
$H = \mathrm{Diag}(Q_1,\ldots,Q_{p-1}) - (Q_1,\ldots,Q_{p-1})'Q^{-1}(Q_1,\ldots,Q_{p-1})$,
where $Q=Q_1+\ldots+Q_p$. Then the posterior distribution of

$$F_\Delta = \frac{N-n-(1+n+m)p}{(p-1)(1+n+m)v}\,\Delta'H\Delta$$

is the $F_{(p-1)(1+n+m),\ N-n-(1+n+m)p}$.

<u>Proof.</u> Put $\Delta_p = \theta_p-\theta_p^*$. The posterior density of $\Delta_1,\ldots\Delta_p$ is

$$g_3(\Delta_1,\ldots,\Delta_p|x)=c\left\{1+v^{-1}\left[\Delta_p'Q_p\Delta_p+\sum_{k=1}^{p-1}(\Delta_k+\Delta_p)'Q_k(\Delta_k+\Delta_p)\right]\right\}^{-(N-n)/2}.$$

If we denote

$$h=\sum_{k=1}^{p-1}Q_k\Delta_k,\qquad G=\sum_{k=1}^{p-1}\Delta_k'Q_k\Delta_k-h'Q^{-1}h$$

then the density g_3 can be written in the form

$$c\left[1+v^{-1}G+v^{-1}(\Delta_p+Q^{-1}h)'Q(\Delta_p+Q^{-1}h)\right]^{-(N-n)/2}.$$

The marginal posterior density of Δ is

$$g_4(\Delta|x) = c(1+v^{-1}\Delta' H\Delta)^{-(N-2n-m-1)/2}$$

and from here we get the assertion.

Since $\theta_k=(\mu_k, b_k', a')'$, it is also convenient to introduce μ_k^*, b_k^* and a_k^* by $\theta_k^*=(\mu_p^*, b_k^{*\prime}, a^{*\prime})'$. Of course, the blocks of θ_k^* correspond to those of θ_k.

<u>Theorem 2.6.</u> The marginal posterior density of θ_k is

$$g_5(\theta_k|x) = c[1+v^{-1}(\theta_k-\theta_k^*)' Q_k(\theta_k-\theta_k^*)]^{-[N-n-(p-1)(1+n+m)]/2}$$

for $k = 1,\dots,p$.

<u>Proof.</u> The formula for g_5 follows from Theorem 2.2 (ii). Denote $Q_k^{-1}=(Q_k^{ij})_{i,j=1}^3$, where Q_k^{11}, Q_k^{22} and Q_k^{33} are 1x1, nxn and mxm blocks, respectively.

<u>Theorem 2.7.</u> (i) The marginal posterior density of μ is

$$g_6(\mu|x) = c[1+v^{-1}\sum_{k=1}^{p}(\mu_k-\mu_k^*)^2/Q_k^{11}]^{-[N-n-p(n+m)]/2}.$$

(ii) The marginal posterior density of b is

$$g_7(b|x) = c\left[1+v^{-1}\sum_{k=1}^{p}(b_k-b_k^*)'(Q_k^{22})^{-1}(b_k-b_k^*)\right]^{-[N-n-p(1+m)]/2}.$$

(iii) The marginal posterior density of a is

$$g_8(a|x) = c\left[1+v^{-1}\sum_{k=1}^{p}(a_k-a_k^*)'(Q_k^{33})^{-1}(a_k-a_k^*)\right]^{-[N-n-p(1+n)]/2}.$$

<u>Proof.</u> Theorem is a consequence of Theorem 2.2 (ii).

If we wish to test that μ_k, b_k and a_k do not depend on k separately, we can use one of the next three theorems. For example, in this way we can come to the conclusion that μ_k depends on k whereas the hypothesis that b_k and a_k do not depend on k cannot be rejected. All the next theorems follow from Theorem 2.7 and from results summarized by Zellner (1971), Appendix B. The quadratic forms used in the assertions are positive definite, which is ensured by Theorem 2.2 in Anděl (1983). The proofs are similar to that of Theorem 2.5 and they are omitted.

Theorem 2.8. Let $\Delta_k = \mu_k - \mu_p - (\mu_k^* - \mu_p^*)$ and $Q^{11} = (Q_1^{11})^{-1} + \ldots + (Q_p^{11})^{-1}$. Then the posterior distribution of the statistic

$$F_{\Delta,\mu} = \frac{N-n-(1+n+m)p}{(p-1)v} \left\{ \sum_{i=1}^{p-1} (\Delta_i^2 / Q_i^{11}) - (Q^{11})^{-1} \left[\sum_{i=1}^{p-1} (\Delta_i / Q_i^{11}) \right]^2 \right\}$$

is the $F_{p-1, N-n-(1+n+m)p}$.

Theorem 2.9. Let $\Delta_k = b_k - b_p - (b_k^* - b_p^*)$ and $Q^{22} = (Q_1^{22})^{-1} + \ldots + (Q_p^{22})^{-1}$. Then the posterior distribution of the statistic

$$F_{\Delta,b} = \frac{N-2n-p[1+m+(p-1)n]}{np(p-1)v} \left[\sum_{i=1}^{p-1} \Delta_i' (Q_i^{22})^{-1} \Delta_i - \sum_{i=1}^{p-1} \sum_{j=1}^{p-1} \Delta_i' (Q_i^{22})^{-1} (Q^{22})^{-1} (Q_j^{22})^{-1} \Delta_j \right]$$

is the $F_{np(p-1), N-2n-p[1+m+(p-1)n]}$.

Theorem 2.10. Let $\Delta_k = a_k - a_p - (a_k^* - a_p^*)$ and $Q^{33} = (Q_1^{33})^{-1} + \ldots + (Q_p^{33})^{-1}$. Then the posterior distribution of the statistic

$$F_{\Delta,a} = \frac{N-n-m-p[1+n+(p-1)m]}{mp(p-1)v} \left[\sum_{i=1}^{p-1} \Delta_i' (Q_i^{33})^{-1} \Delta_i - \right.$$

$$\sum_{i=1}^{p-1}\sum_{j=1}^{p-1} \Delta_i'(Q_i^{33})^{-1}(Q^{33})^{-1}(Q_j^{33})^{-1}\Delta_j\Big]$$

is the $F_{mp(p-1),N-n-m-p[1+n+(p-1)m]}$.

For example, if we wish to test the hypothesis that μ_k does not depend on k, we have $\mu_1=\ldots=\mu_p$. Then the test is based on $F_{\Delta,\mu}$ with $\Delta_k = \mu_p^* - \mu_k^*$, since $\mu_k = \mu_p$ holds under the hypothesis. Applications of Theorems 2.9 and 2.10 are quite analogous.

REFERENCES

Anděl, J. (1983) Statistical analysis of periodic autoregression, Apl. mat. 28, 364-385.

Anděl, J., Rubio, A. and Insua, A. (1985) On periodic autoregression with unknown mean, Apl. mat. 30, 126-139.

Box, G. E. P. and Tiao, G. C. (1975) Intervention analysis with applications to economic and environmental problems, J. Amer. Statist. Assoc. 70, 70-79.

Newton, H. J. (1982) Using periodic autoregression for multiple spectral estimation, Technometrics 24, 109-116.

Pagano, M. (1978) On periodic and multiple autoregression, Ann. Statist. 6, 1310-1317.

Zellner, A. (1971) An Introduction to Bayesian Inference in Econometrics, Wiley, New York.

Jiří Anděl
Department of Statistics
Charles University
Sokolovská 83
186 00 Prague 8
Czechoslovákia

Proc. of the 5th Pannonian Symp. on Math. Stat., Visegrád, Hungary 1985
W. Grossmann, J. Mogyoródi, I. Vincze, W. Wertz, eds

ON STATE ESTIMATION OF DYNAMICAL SYSTEM PERTURBATED BY SMALL NOISE

G. APOYAN, YU. KOTOYANTS

Yerevan

In this paper the problem of state estimation of dynamical system by the observations of perturbated system states is considered. It is proved that the state of perturbated system itself is nonimprovable in some sense estimate of the state as the noise tends to zero.

AMS subject classification: 62G05

1. INTRODUCTION

Let the state $X(t_o)$ of dynamical system

$$\frac{dX(t)}{d\ t} = S(X(t)),\ X(0) = x,\ 0 \le t \le T \tag{1}$$

be required to estimate by the observations $X_\varepsilon = (X_\varepsilon(t),\ 0 \le t \le T)$ or perturbated system

$$dX_\varepsilon(t) = S(X_\varepsilon(t))dt + \varepsilon dw(t),\ X_\varepsilon(0) = x_o\ 0 \le t \le T. \tag{2}$$

It is naturally to consider the state $X_\varepsilon(t_o)$ itself as an estimate and the following question arises: is it possible to construct some other estimate of $X(t_o)$ which is better than this one? We are interested in the properties of estimates as $\varepsilon \to 0$. It corresponds to the presence of small Gaussian noise in the system (1). In this paper proved that in certain minimax sense the estimate $X_\varepsilon(t_o)$ of $X(t_o)$ is optimal. So the observations $X_\varepsilon(t)$, $t \ne t_o$ are superfluous. We use the minimax bound of L. Le Cam and minimax result of W. Millar. Note that the function $S(x)$ is not known to the observer, so we have a nonparametrical problem. Parametrical problems for this scheme of observations were considered by Yu. Kutoyants (1984).

2. ASSUMPTIONS AND AUXILIARY RESULTS

We assume that the conditions

$$|S(x) - S(y)| \le L|x - y|, \quad |S(x)| \le L(1 + |x|) \tag{3}$$

which guarantee the existence and uniquiness of the solution of equation (2) are fulfilled. Denote $\Theta(L)$ the set of functions which satisfy (3) and introduce $\Theta = \bigcup_{L>0} \Theta(L)$.

Observer knows that $S \in \Theta$ and the functions $S(x)$ is positive and continuously differentiable. From these conditions it follows that uniformly on $t \in [0,T]$ with probability 1

$$\lim_{\varepsilon \to 0} \varepsilon^{-1}[X_\varepsilon(t) - X(t)] = Z(t)$$

and the process $Z(t)$ has the differential

$$dZ(t) = S'(X(t))\, Z(t)\, dt + dw(t), \quad Z(0) = 0 . \tag{4}$$

The solution of equation (4) could be written in the form

$$Z(t) = S(X(t))\, W(F(t)), \quad F(t) = \int_0^t \frac{dv}{S(X(v))^2} ,$$

where $W(\cdot)$ is some Wiener process.

To characterize the merit of estimates we introduce a loss function $l(x)$, $x \in C[0,T] = C$ by the formula

$$l(x) = g(|x|),$$

where $|x| = \sup_{0 \le t \le T} |x(t)|$ and $g(x)$ is increasing on $[0,\infty)$ function,

$$0 \le g(u) \le C \exp\{u^{2-\gamma}\}, \quad u \ge 0,$$

where $C > 0$ and $\gamma \in (0,2]$.

We recall that the procedure $b(\cdot,\cdot)$ is a Markov kernel on $(C,\mathcal{B})/(C,\mathcal{B})$ ($\mathcal{B}$-σalgebra of Borel sets in C) if for each $x \in C$, $b(x,\cdot)$ is a probability on $(C,\mathcal{B})$, and for each $A \in \mathcal{B}$, $b(\cdot,A)$

is $\mathcal{B}$ -measurable. Note that the observation and decision spaces coincide in our problem. The procedure $b(\cdot,\cdot)$ defines a rule of an estimate construction.

3. MAIN RESULT

Fix the positive function $S_o(\cdot)\in\Theta(L)$ and corresponding dynamical system

$$\frac{dX_o(t)}{dt} = S_o(X_o(t)),\ X_o(0) = x_o \qquad 0\le t\le T$$

and introduce the family of functions

$$S(x,h,\varepsilon) = S_o(x) + \varepsilon h(x),$$

where $h\in\Theta(L)$. Then denote $X_o^h(t)$ and $X_\varepsilon^h(t)$ the solutions of equations

$$\frac{dX_o^h(t)}{dt} = S_o(X_o^h(t)) + \varepsilon h(X_o^h(t)),\ X_o^h(0) = x_o,$$

$$dX_\varepsilon^h(t) = [S_o(X_\varepsilon^h(t)) + \varepsilon h(X_\varepsilon^h(t))]dt + \varepsilon dw(t),\ X_\varepsilon^h(0) = x_o \quad (5)$$

The measure induced in the $(C,\mathcal{B})$ by the process (5) is denoted by $P_h^{(\varepsilon)}$.

Theorem.

a) $$\lim_{L\uparrow\infty}\lim_{\varepsilon\to 0}\inf_{b(\cdot,\cdot)}\sup_{h\in\Theta(L)}\int\int l(\varepsilon^{-1}(X-X_o^h))\, b(X_\varepsilon^h,dX)\, dP_h^{(\varepsilon)} =$$

$$= E\, l(S_o(X\ (\cdot))\, W(F(\cdot)))$$

b) $$\lim_{L\uparrow\infty}\lim_{\varepsilon\to 0}\sup_{h\in\Theta(L)}\int l(\varepsilon^{-1}(X_\varepsilon^h - X_o^h))\, dP_h^{(\varepsilon)} =$$

$$= E\, l(S_o(X_o(\cdot))\, W(F(\cdot)))\ .$$

Proof. Let $h\in\Theta$. The family of measures $\{P_h^{(\varepsilon)},\ h\in\Theta\}$ has the likelihood ratio

$$\frac{dP_h^{(\varepsilon)}}{dP_o^{(\varepsilon)}} = \exp\{\int_0^T h(X_\varepsilon^o(t))\, dw(t) - \frac{1}{2}\int_0^T h(X_\varepsilon^o(t))^2\, dt\}$$

with $P_o^{(\varepsilon)}$ probability 1.

Introduce also the family of processes

$$dZ^h(t) = [S_o'(X_o(t))\, Z^h(t)+h(X_o(t))]dt+dw(t),\quad Z^h(0)= 0 \tag{6}$$

and denote $\{P_h,\ h\in\Theta\}$ corresponding family of measures.

For each $h\in\Theta$ we have convergence

$$\lim_{\varepsilon\to 0}\ \sup_{0\le t\le T}\ |\overset{o}{X}_\varepsilon(t) - X_o(t)| = 0$$

with $P_o^{(\varepsilon)}$ probability 1. This allows to prove the convergence of vectors $\{dP_h^{(\varepsilon)}/dP_o^{(\varepsilon)},\ h\in\Theta_o\}$ in distribution under $P_o^{(\varepsilon)}$ to vector $\{dP_h/dP_o,\ h\in\Theta_o\}$, where Θ_o is an arbitrary finite collection of $h\in\Theta$. From this it follows that the sequence of experiments $\{P_h^{(\varepsilon)},\ h\in\Theta\}$ converge to the experiment $\{P_h,\ h\in\Theta\}$ and according to the minimax bound of L. Le Cam (1972) (see also W. Miller (1981), theorem 3.1.1.) we obtain

$$\lim_{L\uparrow\infty}\ \varliminf_{\varepsilon\to 0}\ \inf_b\ \sup_{h\in\Theta(L)}\ \iint l(\varepsilon^{-1}(X - X_o^h))\, b(X_\varepsilon^h,dX)\, dP_h^{(\varepsilon)} \ge$$

$$\ge \lim_{L\uparrow\infty}\ \inf_b\ \sup_{h\in\Theta(L)}\ \iint l(Y - q(h))\, b(Z^h,\, dY)\, dP_h$$

where

$$q(h) = S_o(X_o(t)) \int_a^{X_o(t)} \frac{h(y)\, dy}{S_o(y)^2}\ .$$

Let us introduce the Hilbert space

$$H = \{h(x) : \int_a^b \frac{h(x)^2}{S_o(x)}\, dx < \infty\},\ a = x_o,\ b = X_o(T)$$

and note that the set Θ is dense in H, so (see Millar (1981), §8.2)

$$\lim_{L\uparrow\infty}\ \inf_b\ \sup_{h\in\Theta(L)}\ \iint l(Y - q(h))\, b(Z^h,\, dY)\, dP_h =$$

$$= \inf_b\ \sup_{h\in H}\ \iint\ l(Y - q(h))\, b(Z^h,\, dY)\, dP_h\ .$$

Let us introduce Hilbert space H^*, Banach space B and one-to-one map $\tau: H^* \to B$ such, that (τ, H^*, B) is an abstract Wiener space. $\{Q_h, h \in H\}$ is its Gaussian shift experiment.

<u>Theorem</u> (Miller (1981)) If $\int l(x)\, dQ_0(x) < \infty$, then

$$\inf_b \sup_{h \in H^*} \int\int l(x - \tau h)\, b(y, dx)\, Q_h(dy) = \int l(x)\, Q_0(dx).$$

The solution $Z^h(t)$ of (6) could be written in the form

$$Z^h(t) = S_0(X_0(t)) \int_0^t \frac{h(X_0(v))}{S(X_0(v))} dv + S_0(X_0(t))\, W(F(t)) .$$

Introduce a new variable $x = X_0(t)$, then the process $Y^h(x) = S_0(x^{-1})\, Z^h(t(x))$ has the representation

$$Y^h(x) = \int_a^x \frac{h(y)}{S_0(y)^2} dy + W\left(\int_a^x \frac{dy}{S_0(y)^3}\right) \quad , x \in [a,b].$$

Hence the measures $\{Q_h, h \in H\}$ induced by this process in $(C[a,b], \mathcal{B}[a,b])$ define a Gaussian shift experiment for the abstract Wiener space $(\tau, H, C[a,b])$ where mapping is

$$\tau h(x) = \int_a^x \frac{h(y)dy}{S_0(y)^2} \quad .$$

Introducing the new variable $Y^h(x)$ instead of $Z^h(t)$ and using Millar's theorem we estate

$$\inf_b \sup_h \int\int l(Y - q(h))\, b(Z^h, dY)\, dP_h =$$

$$= \inf_b \sup_h \int\int l(Y - \tau h)\, b(Y^h, dY)\, dQ_h =$$

$$= \int l(x)\, dQ_0(x) = E\, l(S_0(X_0(\cdot))\, W(F(\cdot))) .$$

So we have proved the inequality $\geq$ in the (a) statement of theorem. Equality would follow from this one and (b). To prove (b) note that the process $Z_\varepsilon^h(t) = \varepsilon^{-1}[X_\varepsilon^h(t) - X_0^h(t)]$ by the mean value theorem has the representation

$$Z^h_\varepsilon(t) = \int_0^t S'_o(\tilde{X}^h_\varepsilon(v))\, Z^h_\varepsilon(v)\, dv + \int_0^t [h(X^h_\varepsilon(v)) - h(X^h_o(v))]dv +$$

$$+ w(t), \quad Z^h_\varepsilon(0) = 0, \quad 0 \le t \le T \quad . \qquad (7)$$

By the condition $S_o(x) \in \Theta(L)$ the function $|S'_o(\cdot)|$ is bounded and uniformly on $h \in \Theta(L)$

$$\lim_{\varepsilon \to 0} \sup_{0 \le t \le T} |X^h_\varepsilon(t) - X^h_o(t)| = 0.$$

This allows to prove the convergence

$$\lim_{\varepsilon \to 0} \sup_{0 \le t \le T} |Z^h_\varepsilon(t) - Z(t)| = 0$$

uniformly on $h \in \Theta(L)$.

From (7) using standart arguments we derive the inequallity

$$\sup_{0 \le t \le T} |Z^h(t)| \le K \sup_{0 \le t \le T} |w(t)|, \quad K > 0$$

which guarantee the uniform integrability of the $l(Z^h_\varepsilon(\cdot))$ because

$$\sup_{h \in \Theta(L)} P\{ \sup_{0 \le t \le T} |Z^h(t)| > N\} \le P \sup_{0 \le t \le T} |w(t)| > N K^{-1}\} \le$$

$$\le \exp\left\{- \frac{N^2}{2T K^2}\right\} \quad .$$

Hence (b) of our theorem is proved.

REFERENCES

Kutoyants, Yu.A. (1984). Parameter Estimation for Stochastic Processes, Heldermann, Berlin.

Le Cam, L. (1972). Limits of experiments, Proc. 6th Berkeley Symposium, I.

Millar, W. (1981). The minimax principle in asymptotic statistical theory, Lecture Notes in Math. Nc 976, Springer, Berlin.

Yurii Kutoyants
Yerevan State University
375049 Yerevan, Mravian I.
USSR

Proc. of the 5th Pannonian Symp. on Math. Stat., Visegrád, Hungary 1985
W. Grossmann, J. Mogyoródi, I. Vincze, W. Wertz, eds

COMPARISON OF SOME RELIABILITY ESTIMATORS IN THE EXPONENTIAL CASE UNDER RANDOM CENSORSHIP

J. HURT

Prague

Alternative parametric estimators of the reliability function of an exponential distribution based on randomly censored data are studied. The emphasis is put on their asymptotic properties.

AMS subject classification: 62F10, 62N05, 62E20

1. INTRODUCTION

A useful model for analysing biometrical and reliability data is the model of random censorship. Suppose that $X_1,\dots,X_n$ are independent identically distributed random variables with the distribution function F and $T_1,\dots,T_n$ are independent identically distributed random variables which are independent of X_i's and possess the distribution function G. In the model of random censorship X_i's, the random variables in question, often represent times to failure, while T_i's are time censors, G being a nuisance distribution. Under random censorship one can only observe the pairs

$$(W_1,I_1),\dots,(W_n,I_n), \tag{1.1}$$

where

$$W_j = \min(X_j,T_j),$$
$$I_j = 1 \text{ if } X_j \leq T_j \quad (j^{th} \text{ observation is uncensored})$$
$$\quad = 0 \text{ if } X_j > T_j \quad (j^{th} \text{ observation is censored at time } T_j),$$
$$j = 1,\dots,n.$$

Thus (1.1) are independent identically distributed random vectors. Let X_1 and T_1 have the Lebesgue densities f and g, respectively. Then (W_1, I_1) has the density

$$h(w,i) = \left\{f(w)[1-G(w)]\right\}^{i}\left\{g(w)[1-F(w)]\right\}^{1-i}, \qquad (1.2)$$

$$w \in R, \quad i = 0,1,$$

with respect ot Lebesgue times counting product measure. Usually both F and G contain unknown parameters. These are often estimated by the method of maximum likelihood. Denote $U = \{j: I_j = 1\}$, $C = \{j: I_j = 0\}$. If there is no functional dependence between parameters of F and G, it is sufficient to maximize the sub-likelihood function

$$L_F = \prod_{j \in U} f(X_j) \prod_{j \in C} [1 - F(T_j)]$$

only to obtain maximum likelihood estimators of the parameters of F.

An attractive model connecting F and G is the Koziol-Green model (see [5]) in which $1 - G = (1 - F)^{\beta}$ for some $\beta \geqq 0$. Note that $\beta = 0$ corresponds to no censoring.

In this paper we assume that the X_i's are exponentially distributed:

$$F(x) = 1 - e^{-x/\theta}, \quad x > 0, \qquad (1.3)$$

$$= 0 \qquad \text{otherwise},$$

θ being an unknown parameter. Here we employ the Koziol-Green model so that the T_i's are also exponentially distributed with

$$G(t) = 1 - e^{-t/\omega}, \quad t > 0,$$
$$= 0 \qquad \text{otherwise}.$$

Put

$$\frac{1}{\delta} = \frac{1}{\theta} + \frac{1}{\omega} \quad .$$

Then we can express the density of (W_1, I_1) as

$$h(w,i) = (\delta/\Theta)^{i} (\delta/\omega)^{n-i} \delta^{-n} e^{-w/\delta}, \quad w \geqq 0,\ i = 0,1. \tag{1.4}$$

Therefore W_1, I_1 are independent random variables, W_1 possesses the exponential distribution with $P(I_1 = 1) = \delta/\theta$. Denote $I = I_1 + \ldots + I_n$, $W = W_1 + \ldots + W_n$. From (1.4) it follows that the likelihood function corresponding to (1.1) is $(\delta/\theta)^{I}(\delta/\omega)^{n-I}\delta^{-n}e^{-W/\delta}$. We see that (W,I) is a sufficient statistic for (θ,ω). The maximum likelihood estimator of θ is $\hat{\theta} = W/I$ if $I>0$ and undefined if $I = 0$. Its properties have been studied in [3].

The reliability corresponding to (1.3) taken at mission time $x \geqq 0$ is

$$R(x) = e^{-x/\theta} \quad .$$

Without any loss of generality, after a proper scaling we can restrict ourselves to $x = 1$ so that we shall study

$$R = e^{-1/\theta} \quad .$$

In [3] asymptotic properties of the maximum likelihood estimator of R have been studied. In this paper we give a more detailed analysis of that estimator which will enable us to suggest an alternative estimator with an asymptotically smaller bias. Such kind of estimators are important also for systems analysts, see [6].

2. PRELIMINARIES

We shall use the following notation:

$$p = \delta/\theta, \quad \overline{W} = W/n, \quad \overline{I} = I/n, \quad Z = 1/W, \quad Y = IZ, \quad Z' = 1/\overline{W},$$

$$g_i = g_i(Y,Z') = Ye^{-iY}(\tfrac{1}{2}Z'-1),$$

$$O_j = O(n^{-j}) \quad \text{for } n \to \infty.$$

The maximum likelihood estimator of R is

$$R_1 = e^{-Y}.$$

The main tool of our analysis will be Theorem 5 from [4]. Here we state it without proof.

Theorem 1. Let $\{\underline{X}(t)\}_{t \in T}$ be an r-dimensional stochastic process and $g = g(\underline{x},t)$ be a real function defined on $R_r \times T$. Assume that

a) for all $t \in T$ g is (q+1) times totally differantiable with respect to x_i's in the interval $K = X_{i=1}^{r} [\theta_i - \delta, \theta_i + \delta]$ where $\delta > 0$ is independent of t and $\underline{\theta} = (\theta_1, \ldots, \theta_r)$ is a vector parameter,

b) there exist integers p,u, $p \geq q+1$, $0 \leq u \leq p$ such that

$$|g(\underline{x},t)| \leq C_1 + C_2 \frac{\max_{1 \leq i \leq r} E|X_i(t) - \theta_i|^{q+1}}{\max_{1 \leq i \leq r} E|X_i(t) - \theta_i|^p} \sum_{i=1}^{r} |x_i|^u$$

for all $\underline{x} \in R_r$, $t \in T$, where C_1 and C_2 are independent of $\underline{x}$ and t,

c) all the derivatives of g up to the order q+1 are bounded on $K \times T$. Let us denote

$$D^j g(m_1, \ldots, m_r; \underline{\theta}) = (\partial^j g / \partial y_1^{m_1} \ldots \partial y_r^{m_r})_{\underline{y} = \underline{\Theta}}$$

If $\max_{1 \leq i \leq r} E|X_i(t) - \theta_i|^{q+1} \to 0$ as $t \to t_0$ then

$$Eg(\underline{X}(t),t) = g(\underline{\theta},t) + \sum_{j=1}^{q} \frac{1}{j!} \sum_{\substack{m_1+\ldots+m_r=j \\ m_1,\ldots,m_r \geq 0}} \ldots \sum D^j g(m_1, \ldots, m_r; \underline{\theta}) \times$$

$$E(X_1(t)-\theta_1)^{m_1} \ldots (X_r(t)-\theta_r)^{m_r} + O(\max_{1 \leq i \leq r} E|X_i(t)-\theta_i|^{q+1}).$$

If $\max_{1\leq i\leq r} E|X_i(t)-\theta_i|^{q+2} \to 0$ as $t \to t_0$ then

var $g(\underline{X}(t),t) =$

$$\sum_{\substack{j=1\\ j+k\leq q+1}}^{q} \sum_{k=1}^{q} \frac{1}{j!k!} \sum_{\substack{m_1+\ldots+m_r=j\\ m_1,\ldots,m_r\geq 0}} \cdots \sum \; \sum_{\substack{n_1+\ldots+n_r=k\\ n_1,\ldots,n_r\geq 0}} \cdots \sum D^j g(m_1,\ldots,m_r;\underline{\theta}) \times$$

$$D^k g(n_1,\ldots,n_r;\underline{\theta}) \operatorname{cov}((X_1(t)-\theta_1)^{m_1}\ldots(X_r(t)-\theta_r)^{m_r},$$

$$(X_1(t)-\theta_1)^{n_1}\ldots(X_r(t)-\theta_r)^{n_r}) + 0(\max_{1\leq i\leq r} E|X_i(t)-\theta_i|^{q+2}).$$

By means of this theorem we are able to give asymptotic expansions both of the expected value and the mean squared error (MSE) of reliability estimates with remainder terms $0(n^{-3})$. Then we make a comparison. We need moments of I, Z, and Y. Since I possesses a binomial distribution with parameter p we have

$$EI = np$$
$$EI^2 = n(n-1)p^2+np$$
$$EI^3 = n(n-1)n-2)p^3+3n(n-1)p^2+np$$
$$EI^4 = n(n-1)(n-2)(n-3)p^4+6n(n-1)(n-2)p^3+7n(n-1)p^2+np.$$

Obviously W possesses a gamma distribution with the parameter δ so that $EZ^k = EW^{-k}$ exist for $k < n$ and

$$EZ^k = \delta^{-k}\Gamma(n-k)/\Gamma(n).$$

Particularly,

$$EZ = (\delta(n-1))^{-1}$$
$$EZ^2 = (\delta^2(n-1)(n-2))^{-1}$$
$$EZ^3 = (\delta^3(n-1)(n-2)(n-3))^{-1}$$
$$EZ^4 = (\delta^4(n-1)(n-2)(n-3)(n-4))^{-1}.$$

Taking into account the independence of I and Z we obtain after some algebra

$$EY = \theta^{-1}(1+n^{-1}+n^{-2})+O_3$$
$$EY^2 = \theta^{-2}+n^{-1}(2\theta^{-2}+\theta^{-1}\delta^{-1})+n^{-2}(4\theta^{-2}+3\theta^{-1}\delta^{-1})+O_3$$
$$EY^3 = \theta^{-3}+n^{-1}(3\theta^{-3}+3\theta^{-2}\delta^{-1})+n^{-2}(9\theta^{-3}+15\theta^{-2}\delta^{-1}+\theta^{-1}\delta^{-2})+O_3$$
$$EY^4 = \theta^{-4}+n^{-1}(4\theta^{-4}+6\theta^{-3}\delta^{-1})+n^{-2}(16\theta^{-4}+42\theta^{-3}\delta^{-1}+7\theta^{-2}\delta^{-2})+O_3$$

and

$$E(Y-1/\theta) = \theta^{-1}(n^{-1}+n^{-2})+O_3$$
$$E(Y-1/\theta)^2 = n^{-1}\theta^{-1}\delta^{-1}+n^{-2}(2\theta^{-2}+3\theta^{-1}\delta^{-1})+O_3$$
$$E(Y-1/\theta)^3 = n^{-2}(6\theta^{-2}\delta^{-1}+\theta^{-1}\delta^{-2})+O_3$$
$$E(Y-1/\theta)^4 = 3n^{-2}\theta^{-2}\delta^{-2}+O_3.$$

Lemma 1. We have

$$Eg_i(Y,Z') = e^{-i/\theta}\left[\frac{1}{2\theta\delta} - \frac{1}{\theta} + \frac{1}{n}\left(-\frac{1}{\theta} + \frac{i}{\theta^2} + \frac{2i+3}{2\theta\delta} - \frac{i^2+2i}{2\theta^2\delta} - \frac{i}{2\theta\delta^2} + \frac{i^2}{4\theta^2\delta^2}\right)\right] + O_2.$$

Proof. The function $g_i(y,z) = ye^{-iy}(z/2-1)$ obviously fulfils the assumptions of Theorem 1. To achieve the remainder term O_2 it suffices to take $q = 2$ because the higher moments of $Y-1/\theta$ and $Z'-1/\delta$ are actually O_2. So we have

$$Eg_i(Y,Z') = g_i(1/\theta,1/\delta) + (\partial g_i/\partial y)E(Y-1/\theta) + (\partial g_i/\partial z)\,E(Z'-1/\delta)+.5(\partial^2 g_i/\partial y^2)E(Y-1/\theta)^2 + (\partial^2 g_i/\partial y\partial z)E((Y-1/\theta)(Z'-1/\delta)) + .5(\partial^2 g_i/\partial z^2)E(Z'-1/\delta)^2+O_2$$

where all the derivatives should be taken at point $(1/\theta,1/\delta)$. The last derivative vanishes and $E(Z'-1/\delta) = n^{-1}\delta^{-1}+O_2$, $E((Y-1/\theta)(Z'-1/\delta)) = E((\bar{I}Z'-1/\theta)(Z'-1/\delta)) = n^{-1}\theta^{-1}\delta^{-1}+O_2$. After substitution and some calculations we get the assertion. QED

3. ESTIMATION OF THE HAZARD RATE

Results of the previous section enable us to give an unbiased estimator of the hazard rate $\lambda = 1/\theta$. A natural estimator of λ is

$$\tilde{\lambda} = I/W = Y.$$

Immediately we see that

$$E\,\tilde{\lambda} = \frac{I}{\theta}\,\frac{1}{n-1}$$

so that

$$\hat{\lambda} = \frac{n-1}{n}\,Y \tag{3.1}$$

is an unbiased estimator of the hazard rate $1/\theta$.

<u>Theorem 2.</u> For $n \to \infty$

$$n^{1/2}(\hat{\lambda}-\lambda) \xrightarrow{D} N(0,(\theta\delta)^{-1}).$$

<u>Proof.</u> Due to the independence of I and W the proof is obvious.

4. ESTIMATION OF THE RELIABILITY FUNCTION

In this section we derive the asymptotic properties of the maximum likelihood estimator R_1 and use them to construct an alternative estimator with smaller bias.

<u>Theorem 3.</u> For $n \to \infty$

$$n^{1/2}(R_1-R) \xrightarrow{D} N(0,R^2(\theta\delta)^{-1}), \tag{4.1}$$

$$ER_1 = R\left[1 + n^{-1}\left(-\frac{1}{\theta} + \frac{1}{2\theta\delta}\right) + n^{-2}\left(-\frac{1}{\theta} + \frac{1}{\theta^2} + \frac{3}{2\theta\delta} - \frac{1}{\theta^2\delta} - \frac{1}{6\theta\delta^2} + \frac{1}{8\theta^2\delta^2}\right)\right] - O_3, \tag{4.2}$$

$$\operatorname{var} R_1 = R^2\left[n^{-1}\frac{1}{\theta\delta} + \right. \tag{4.3}$$

$$\left. n^{-2}\left(\frac{1}{\theta^2} + \frac{3}{\theta\delta} - \frac{5}{\theta^2\delta} - \frac{1}{\theta\delta^2} + \frac{3}{2\theta^2\delta^2}\right)\right] + O_3,$$

$$\operatorname{MSE}(R_1) = R^2\left[n^{-1}\frac{1}{\theta\delta} + \right. \tag{4.4}$$

$$\left. n^{-2}\left(\frac{2}{\theta^2} + \frac{3}{\theta\delta} - \frac{6}{\theta^2\delta} - \frac{1}{\theta\delta^2} + \frac{7}{4\theta^2\delta^2}\right)\right] + O_3$$

hold.

Proof. For (4.1) see [3]. Put $g(y) = e^{-y}$ in Theorem 1. It is sufficient to take $q = 4$ for the calculation of ER_1 and $q = 3$ for that of $\operatorname{var} R_1$. Using formulas from Section 2 we get (4.2) and (4.3). The expansion (4.4) follows from the identity

$$\operatorname{MSE}(R_1) = \operatorname{var} R_1 + (ER_1 - R)^2. \qquad \text{QED}$$

On the basis of (4.2) we can now eliminate the bias of the first order. We simply substract the estimate of it from R_1 and obtain

$$R_2 = R_1 - \frac{1}{n}R_1 Y(.5Z' - 1) = R_1 - \frac{1}{n}g_1(Y,Z'). \tag{4.5}$$

Theorem 4. For $n \to \infty$

$$n^{1/2}(R_2 - R) \overset{D}{\to} N(0, R^2(\theta\delta)^{-1}), \tag{4.6}$$

$$ER_2 = R\left[1 + n^{-2}\left(-\frac{1}{\theta\delta} + \frac{1}{2\theta^2\delta} + \frac{1}{3\theta\delta^2} - \frac{1}{8\theta^2\delta^2}\right)\right] + O_3, \tag{4.7}$$

$$\operatorname{var} R_2 = \operatorname{MSE}(R_1) + O_3 =$$

$$R^2\left[(n\theta\delta)^{-1} + n^{-2}\left(\frac{1}{\theta^2} + \frac{1}{\theta\delta} - \frac{2}{\theta^2\delta} + \frac{1}{2\theta^2\delta^2}\right)\right] + O_3.$$

Proof. It is easy to see that $n^{-1/2}g_1(Y,Z')$ tends to zero in probability. Therefore, the limiting distributions of R_1 and R_2 are the same. To derive (4.7) we point out that $ER_2 = ER_1 - n^{-1}Eg_1(Y,Z')$. Now we make use of Theorem 3 and Lemma 1. Finally we have

$$\operatorname{var} R_2 = \operatorname{var} R_1 - 2n^{-1}\operatorname{cov}(R_1, g_1(Y,Z')) + n^{-2}\operatorname{var} g_1(Y,Z'). \tag{4.9}$$

Write

$$\operatorname{cov}(R_1, g_1(Y,Z')) = Eg_2(Y,Z') - ER_1 Eg_1(Y,Z') =$$

$$n^{-1}\left(\frac{1}{\theta\delta} - \frac{3}{2\theta^2\delta} - \frac{1}{2\theta\delta^2} + \frac{1}{2\theta^2\delta^2}\right) + O_2$$

by Lemma 1 and Theorem 3. It is not difficult to show that $\operatorname{var} g_1(Y,Z') = O_1$, so that the last term in (4.9) is O_3. Hence the result.

QED

5. COMPARISON OF THE ESTIMATES

From the asymptotic expansions for MSE it follows that both the estimates are (strongly) asymptotically efficient. For that reason we make use of asymptotic deficiency. The asymptotic expansion of the MSE of the estimates are of the form

$$\operatorname{MSE}(R_i) = \frac{a}{n} + \frac{b_i}{n^2} + o(n^{-2})$$

so that the asymptotic deficiency of R_1 with respect to R_2 is

$$d_{12} = \frac{b_1 - b_2}{a}$$

(see [1]). More specifically,

$$d_{12} = \frac{\delta}{\theta} + 2 - \frac{4}{\theta} - \frac{1}{\delta} + \frac{5}{4\theta\delta} . \tag{5.1}$$

In reliability analysis mainly the behaviour of the distribution tales at higher mission times is of interest. Under our agreement, lower values of θ correspond to higher mission times. If we wish to estimate the reliability function at mission time equal to the percentile of order q we find the corresponding θ_q from

$$1 - \exp(-1/\theta_q) = q$$

or

$$\theta_q = -1/\ell n(1-q).$$

Further $k = \delta/\theta$ is the expected proportion of uncensored observations. For the uncensored sample $k = 1$. Thus d_{12} expressed as a function of θ and k becomes

$$d_{12} = 2 + k - \frac{4}{\theta} - \frac{1}{k\theta} + \frac{5}{4k\theta^2} \quad . \tag{5.2}$$

It is obvious that in the tails the term $5/(4k\theta^2)$ prevails. For small values of θ, the greater proportion of censored data, the higher d_{12} is so that the usual maximum likelihood estimator is disadvantageous when estimating the tails and when large number of data is censored. For greater vatues of θ and slight censoring R_1 is preferable. Table 1 shows d_{12} for some values of k and θ.

6. CONCLUDING REMARKS

Here we only mention further possibilities in estimation of the reliability function. Miller [7] gave a formal Taylor expansion to obtain the first order bias of the maximum likelihood estimator of the reliability function from an uncensored sample. He proposed then a correction to the maximum likelihood estimator. If we employ an analogous method for randomly censored data we obtain with a help of (4.2)

$$R_3 = R_1/[1 + n^{-1}Y(.5Z' - 1)] \quad .$$

Handling this estimator is a little bit more complicated.

Table 1

q	k	1	4/5	2/3	1/2	1/5
	θ_q					
0.5	1.443	0.13	-0.01	-0.24	-0.46	-1.03
0.9	0.434	-1.89	-1.00	-0.06	1.93	14.61
0.95	0.334	-0.76	1.09	3.02	6.96	31.33
0.99	0.217	6.48	11.76	17.10	27.89	93.30
0.999	0.145	28.11	41.09	54.14	80.35	238.26

Finally we give a Bayes estimator of R . Suppose that $\Lambda = 1/\theta$ possesses the gamma a priori distribution with parameters a,p so that Λ has the density

$$q(\lambda) = \frac{a^p}{\Gamma(p)} e^{-a\lambda} \lambda^{p-1}, \quad \lambda \geqq 0$$

$$= 0 \qquad \text{otherwise.}$$

Then the a priori distribution of $R = e^{-\Lambda}$ has the density

$$s(r) = \frac{a^p}{\Gamma(p)} r^{a-1}(-\ln r)^{p-1}, \quad 0 < r < 1 \tag{6.1}$$

$$= 0 \qquad \text{otherwise.}$$

From (1.4) and (6.1) we get the a posteriori distribution of R given (1.1):

$$\rho(r|W_1,I_1,\ldots,W_n,I_n) = \frac{(a+W)^{I+p}}{\Gamma(I+p)} r^{a+W-1}(-\ln r)^{I+p-1}, \tag{6.2}$$

$$0 < r < 1$$

If we take the expectation of R with respect to (6.2) we obtain a Bayes estimator

$$R_4 = \left(\frac{W+a}{W+a+1}\right)^{I+p} .$$

REFERENCES

1. Hodges jr., J. L. and Lehmann, E.L. (1970) Deficiency, Ann. Math. Statist., 41, 783-801.

2. Hurt,J. (1976) On estimation of reliability in the exponential case, Apl.Mat., 21, 263-272.

3. Hurt, J. (1982) On estimation of reliability in the exponential case under censorship, Proc. of the 3rd Pannonian Symp. on Math. Stat., Visegrád, Hungary, 115-124 (J. Mogyoródi, I. Vincze, W.Wertz, eds), Akadémiai Kiadó, Budapest.

5. Hurt, J. (1986) Asymptotic expansions for moments of functions of stochastic processes and their applications, Statistics and decisions, 4, NO. 2/3.

6. Koziol, J.A. and Green, S.B. (1976) A Cramér-von Mises statistic for randomly censored data, Biometrika, 63, 465-474.

7. Leitl, R. (1982) Systems diagnostics (In Czech), Technical report ÚVZÚ Skoda Plzeň.

8. Miller, jr., R.G. (1981) Survival analysis, Wiley, New York.

J. Hurt
Department of Statistics
Charles University
Sokolovská 83
186 00 Praha 8
Czechoslovakia

Proc. of the 5th Pannonian Symp. on Math. Stat., Visegrád, Hungary 1985
W. Grossmann, J. Mogyoródi, I. Vincze, W. Wertz, eds

A NEW GOODNESS-OF-FIT TEST

P. KOSIK K. SARKADI

Budapest

A new goodness-of-fit test is proposed for testing uniformity. Monte Carlo experiments indicate that this test is superior in power to its competitors.
AMS subject classification: 62F03

1. INTRODUCTION

Let $X_1 \leq \ldots \leq X_n$ the order statistics of a random simple sample drawn from a population with distribution function F(x).

For testing the hypothesis that the distribution F(x) is uniform in (0,1), Lohrding (1973) proposed the following test statistics:

$$T_4 = c_4 \max_{1 \leq i \leq n} |d_i|$$

$$T_5 = c_5 \sum_{i=1}^{n} |d_i|$$

$$T_6 = c_6 \sum_{i=1}^{n} d_i^2$$

where

$$d_i = \frac{X_i - \frac{i}{n+1}}{i(n-i+1)} \qquad (1)$$

and c_4, c_5, c_6 are constants. The above tests are weighted version of the tests of Kolmogorov, Finkelstein-Schafer and Cramér-Mises, respectively. In the way of weighting, they slightly differ from the well-known Anderson-Darling test. In deriving the weights, Anderson and Darling started from the variance of the empirical distribution function and Lohrding from those of the order statistics.

The Monte-Carlo results of the same author indicate that the proposed modification improves power in general.

In this paper we compare Lohrding's tests with further competitive tests. In our experiment we employ the alternative distributions of Stephens (1974).

In the same time, we propose a new test: the Lohrding type modification of the Watson test. For the Watson test, we refer to Pearson and Hartley (1972, p.118), for its good power properties to Stephens (1974). The new test statistic is the following:

$$J = n^2 \sum_{i=1}^{n} d_i^2 - n\left(\sum_{i=1}^{n} d_i\right)^2$$

where d_i is defined by (1).
Section 3 compares this test and the tests of Lohrding with those found powerful by Stephens (1974).

2. CRITICAL VALUES

The critical values of the new test have been determined by Monte-Carlo method. Ten thousand samples of size n from the rectangular distribution were generated for values of n equal to 5,10,40 and 70. For other sample sizes the critical values were determined by interpolation and extrapolation. They are given in Table 1. If J exceeds the critical value the hypothesis on the uniformity of the sampled population is to be rejected.

For the other tests, the critical values tabulated in Pearson and Hartley (1972, p. 359) and in Lohrding (1973) were taken.

3. POWER

Table 2 gives results based on 1000 Monte-Carlo samples,

drawn from given distributions and tested for uniformity. The table gives the percentage of samples declared significant by various test statistics. Results marked with asterisk are reprinted from Stephens (1974), the rest comes from our experiment. The test is at the 10% level.

The distribution functions are the following (k=1.5,2),cf. Stephens (1974, p. 733)

$$\text{Ak:}\quad F(x) = 1-(1-x)^k \qquad 0\leq x\leq 1$$

$$\text{Bk:}\quad F(x) = \begin{cases} 2^{k-1}x^k & 0\leq x\leq 0.5 \\ 1-2^{k-1}(1-x)^k & 0.5<x\leq 1 \end{cases}$$

$$\text{Ck:}\quad F(x) = \begin{cases} 0.5-2^{k-1}(0.5-x)^k & 0\leq x\leq 0.5 \\ 0.5+2^{k-1}(x-0.5)^k & 0.5<x\leq 1 \end{cases}$$

The tests are the following: Cramér-Mises test (CM), Anderson-Darling test (AD), Watson test (WA), Lohrding's T_4,T_5 and T_6-tests and the J-test.

We do not include tests which are dominated in power in the light of past results. For these, we refer to Stephens (1974), Prosini (1983) and Lohrding (1973).

In the opinion of the present authors, Table 2 indicates that the J-test is the best if overall performance is considered. Watson test is the second one.

4. TESTING FOR NORMALITY AND EXPONENTIALITY

We began to perform experiments concerning the versions of the J-test for testing normality and exponentiality, respectively. This section reports on our first results.

The unknown parameters have been estimated and the statistics modified in the customary way, see Pearson and Hartley (1972, pp.117-118).

The critical values of the J-test were again determined by Monte-Carlo method, based on 2000 samples in both cases.

The empirical powers are based on 1000 samples. In both cases, the sample size is 10 and the size of the test is 10%.

In the J-test for normality, the apprioximation

$$\phi(x) \approx 1 - \frac{1}{2}(1 + \sum_{j=1}^{4} c_j x^j)^{-4} \qquad (x \geq 0)$$

was used for the standardized normal distribution function, cf. Abramovitz and Stegun (1964, p. 932). Here c_1=0.196854, c_2=0.115194, c_3=0.000344 and c_4=0.019527.

Table 3 gives the empirical power of the J-test for normality. For comparison those of the Anderson-Darling and Shapiro-Wilk tests are given, since they are the best ones in the light of the studies of Stephens (1974) and Dyer (1974). Figures marked with asterisk are reprinted form Stephens (1974).

The table shows that the J-test for normality is slightly inferior to the two other tests in overall performance. The differences are, however not significant.

Table 4 gives the empirical power of the J-test for exponentiality. In addition, the table contains the corresponding data for the tests T_4 and T_5 of Lohrding, which seem the best ones in the light of the report of Csörgő et al. (1975). The alternative distributions are: χ^2-distributions with 1 and 4 degrees of freedom, respectively; logarithmic normal distribution with shape parameter 1 and half normal distribution. Figures marked with asterisk are reprinted from Csörgő et alii (1975).

Further experiments are in progress in both cases.

Table 1 Critical values of the J test.

Sample size	Size of the test in percentage 10	5
5	0.408	0.499
10	0.599	0.741
15	0.695	0.881
20	0.748	0.931
25	0.782	0.973
30	0.806	1.00
35	0.823	1.02
40	0.836	1.04
50	0.851	1.05
60	0.862	1.06
70	0.870	1.07
80	0.876	1.07
90	0.880	1.08
100	0.883	1.08

Table 2 Power comparisons. Testing for uniformity.

Alternative distribution	Sample size	Tests CM	AD	Wa	T_4	T_5	T_6	J
A 1.5	40	70*	74	43*	72	76	77	91
A 2	10	60*	58*	35*	39	59	45	46
B 1.5	40	22*	26	61*	58	31	40	71
B 2	10	7*	6*	44*	49	29	37	28
C 1.5	40	32*	38	63*	15	30	30	56
C 2	10	21	31	39	16	24	22	53

Table 3 Power comparisons. Testing for normality.

Alternative distribution	Empirical power AD	W	J
Exponential	41*	43*	39
Double exponential	16*	14*	15
Cauchy	62*	59*	58

Table 4 Power comparisons. Testing for exponentiality.

Alternative distribution	Empirical power		
	T_4	T_5	J
X_1^2	37*	45*	30
X_4^2	33*	34*	32
Lognormal (1)	22*	17*	24
Half normal	16*	20*	18

REFERENCES

Abramovitz, M. and Stegun, I. (1964) *Handbook of Mathematical Functions*, Dover, New York.

Csörgő, M., Seshadri, V. and Yalovsky, M. (1975) Application of characterizations in the area of goodnes of fit. *Statist. Distrib. in Scientific Work*, Vol.2, 79-90 G.P.Patil, Skotz, J.Ord, eds Reidel, Dordrecht.

Dyer, R.(1974) Comparison of tests for normality with a cautionary note. Biometrika 61, 185-189.

Frosini, B.V. (1983) Distributions and power of a new goodnes-of--fit statistic. Statistica 43, 389-413.

Lohrding, R.K. (1973) Three Kolmogorov-Smirnov-type one-sample test with improved power properties. J. Statist. Comput. Simulation 2, 139-148.

Pearson, E.S. and Hartley, H.O.(1972) Biometrika Tables for Statisticians, Vol.2, University Press, Cambridge.

Stephens, M.A. (1974) EDF statistics for goodnes of fit and some comparisons. J. Amer. Statist. Assoc. 69, 730-737.

Mathematical Institute of the
Hungarian Academy of Siences
Budapest, POB. 127
H-1364, Hungary

Proc. of the 5th Pannonian Symp. on Math. Stat., Visegrád, Hungary 1985
W. Grossmann, J. Mogyoródi, I. Vincze, W. Wertz, eds

SOME STATISTICAL CONSEQUENCES OF THE GEOMETRY OF NONLINEAR REGRESSION

A. PÁZMAN

Bratislava

The uniqueness and the probability density of the non-linear least square estimates are considered.
AMS subject classification: 62J02

1. INTRODUCTION

Let

$$y_i = \eta_{x_i}(\theta) + \varepsilon_i; \quad (i=1,\dots,N) \tag{1}$$

be a nonlinear regression model. Here $x_1,\dots,x_N$ are design points, $\varepsilon = (\varepsilon_1,\dots,\varepsilon_N)'$ is the vector of randon errors, $y = (y_1,\dots,y_N)'$ is the vector of observed data, $\theta = (\theta_1,\dots,\theta_m)'$ is the vector of unknown parameters, and $\eta_{x_i}:\Theta\to R$ are known functions, defined and twice continuously differentiable on an open set $\Theta \subset R^m$. We suppose that $E(\varepsilon_i)=0$ and that the covariance matrix K with entries $\{K\}_{ij} = \text{cov}(\varepsilon_i,\varepsilon_j)$ is nonsingular.

When the design points $x_1,\dots,x_N$ are fixed, the dependence of $\eta_{x_i}(\theta)$ on x_i is of no importance, and instead of (1) we prefer a simpler notation

$$y = \eta(\theta) + \varepsilon \tag{2}$$

where $\eta(\theta) = (\eta_{x_1}(\theta),\dots,\eta_{x_N}(\theta))'$.

The set

$$\xi = \{\eta(\theta) : \theta \in \Theta\}$$

(called the "solution locus" in [2]) is geometrically a "surface" (a C^2-manifold) contained in R^N. The regression model (2) is a linear model if ξ is a linear manifold.

We shall suppose that for every $\theta \in \Theta$ the vectors of the first order derivatives $\partial\eta(\theta)/\partial\theta_1,\ldots,\partial\eta(\theta)/\partial\theta_m$ are linearly independent, and that the mapping $\eta:\Theta\to\xi$ is one-to-one.

Denote by

$$\langle a,b\rangle = a'K^{-1}b, \quad \|a\| = \langle a,a\rangle^{1/2}$$

an inner product and the corresponding norm in R^N. The least-square (= L. S.) estimate of θ is given by

$$\hat{\theta} = \hat{\theta}(y) = \operatorname{Arg}\min_{\theta\in\Theta}\|y - \eta(\theta)\|^2. \tag{3}$$

It is usually computed by solving first the normal equations

$$\langle y - \eta(\theta), \frac{\partial\eta(\theta)}{\partial\theta_i}\rangle = 0, \quad (i = 1,\ldots,m) \tag{4}$$

(which correspond to $\partial\|y - \eta(\theta)\|^2/\partial\theta_i = 0$; $(i = 1,\ldots,m)$).

Besides numerical difficulties arising when solving (3) and (4), we meet problems of statistical origin: how probable is that (3) has a unique solution, what is the probability distribution of $\hat{\theta}$ etc. The insight into the geometry of the model (2) gives some answers.

2. THE GEOMETRY OF THE MODEL

As said, the set ξ is a manifold imbedded in R^N. The tangent hyperplane to ξ at a point θ is equal to the set

$$T_\theta = \left\{\eta(\theta) + \sum_{i=1}^{m} \frac{\partial\eta(\theta)}{\partial\theta_i} (\gamma_i - \theta_i) : (\gamma_1,\ldots,\gamma_m) \in R^m\right\}.$$

According to (4), a point θ is a solution of (4) iff the residual vector $y - \eta(\theta)$ is orthogonal to the tangent hyperplane T_θ. The L. S. estimate $\hat{\theta}$ is the solution of (4) giving the shortest distance $\|y - \eta(\theta)\|$. We see that the Euclidean geometry of R^N (with the inner product $\langle .,.\rangle$) is closely connected with the L. S. estimation. This geometry induces a Riemannian geometry in ξ (or equivalently in the parameter space Θ) which is statistically meaningful. Reversely, the nonlinear manifold ξ induces a nonlinear parametrization of the sample space R^N.

2.1 The geometry of Θ (resp. of ξ)

In differential geometry, the Riemannian metric tensor of a manifold Θ which is imbedded in the Euclidean space by the mapping $\eta(.)$, is equal to

$$M_{ij}(\theta) = \left\langle \frac{\partial\eta(\theta)}{\partial\theta_i}, \frac{\partial\eta(\theta)}{\partial\theta_j} \right\rangle ; \quad (i,j = 1,\ldots,m, \ \theta \in \Theta).$$

But this is the well-known information matrix for the case that the errors ε_i are gaussian,

$$M_{ij}(\theta) = E_\theta\left\{- \frac{\partial^2 \ell n\ f(y|\theta)}{\partial\theta_i \partial\theta_j}\right\}$$

where

$$f(y|\theta) = \frac{1}{(2\pi)^{N/2}\det^{1/2}K} \exp\left\{- \frac{1}{2}[y-\eta(\theta)]'K^{-1}[y-\eta(\theta)]\right\} . \qquad (5)$$

When the variances of the errors ε_i are small, it is often used to approximate simply the model (2) by a linear model

$$E(y) = \eta(\theta^o) + \sum_{i=1}^{m} \frac{\partial\eta(\theta^o)}{\partial\theta_i}(\theta_i - \theta_i^o)$$

where θ^o is an adequately choosen parameter value. That means the tangent hyperplane T_{θ^o} is used instead of ξ. In such a case, approximately

$$\hat{\theta}_i = \theta_i^o + \sum_{j=1}^{m} \left\{M^{-1}(\theta^o)\right\}_{ij} < \frac{\partial\eta(\theta^o)}{\partial\theta_j}, y - \eta(\theta^o)>,$$

$$\mathrm{cov}(\hat{\theta}) = M^{-1}(\theta^o).$$

When the nonlinearity of ε is to be taken into account, besides the information matrix also measures of curvatures of ξ (resp. of Θ) must be used. They are all based on the second order derivatives of $\eta(.)$, but different approaches are possible here.

a) Take $w \in R^N$ a unit vector orthogonal to ξ at θ (i. e. $<w, \partial\eta/\partial\theta_i> = 0$; $(i = 1,\dots,m)$). The mxm matrix with entries

$$\left\langle w, \frac{\partial^2\eta(\theta)}{\partial\theta_i\partial\theta_j}\right\rangle \tag{6}$$

is called the second order form of ξ at the point $\eta(\theta)$ in the direction w. We use the second order form in the expression (14) for the probability density of $\hat{\theta}$.

b) Another possibility is to use geodesic curves on ξ, resp. on Θ, and to define curvatures at $\eta(\theta)$ through the curvatures of the geodesics. This has been a useful tool for proving statistical statements in [6] and [8].

c) Methods for correcting linear confidence regions for θ in sligtly nonlinear models need the use of measures of nonlinearity of the model (2) (cf. [2]) which are connected to "intrinsic" measures of curvature of Θ (cf. [1]).

2.2 The nonlinear coordinates in R^N

For every $\tilde{\theta} \in \Theta$ consider vector-valued functions $w_1(\theta),\dots$ $\dots, w_{N-m}(\theta)$ which are defined, orthonormal, and are ortogonal to T_θ at every point $\theta \in U_{\tilde{\theta}}$ (= a neighbourhood of $\tilde{\theta}$), and which have continuous first order derivatives $\partial w_i(\theta)/\partial\theta_j$. (The existence of such $w_i(\theta)$ follows from the implicite function theorem, cf. e. g. [7]).

Denote by $h\colon U_{\tilde\theta}\times R^{N-m}\to R^N$ the mapping

$$h(\theta,a) = \eta(\theta) + \sum_{i=1}^{N-m} a_i w_i(\theta). \tag{7}$$

This mapping is usefull when discussing the uniqueness of $\hat\theta$. Denote by $\mathcal{R}$ the set
$\mathcal{R} = \{y\in R^N\colon$ there is a unique L. S. estimate $\tilde\theta(y)\}$.
Local coordinates $\theta_1,\dots,\theta_m$, $a_1,\dots,a_{N-m}$ can be defined by the equation

$$y = h(\hat\theta,a)\Big|_{\hat\theta=\theta}\ .$$

A slightly different choice of local coordinates useful when considering the probability density of $\hat\theta$ is defined as follows: Take a fixed point $\bar\theta\in\Theta$. Denoty by P^θ the N×N matrix with entries

$$\{P^\theta\}_{ij} = \sum_{k,\ell,r} \frac{\partial\eta_i(\theta)}{\partial\theta_k}\{M^{-1}(\theta)\}_{k\ell}\frac{\partial\eta_r(\theta)}{\partial\theta_\ell}\{K^{-1}\}_{rj} \tag{8}$$

(the projector onto T_θ), and denote by $\Psi_{\bar\theta}(\theta)$ the vector

$$\Psi_{\bar\theta}(\theta) = P^\theta[\eta(\theta)-\eta(\bar\theta)] + \eta(\bar\theta). \tag{9}$$

Define a mapping $g\colon U_\theta\times R^{N-m}\to R^N$ by

$$g(\theta,b) = \Psi_{\bar\theta}(\theta) + \sum_{i=1}^{N-m} b_i w_i(\theta). \tag{10}$$

Local coordinates $\theta_1,\ldots,\theta_m$, $b_1,\ldots,b_{N-m}$ are defined by the equation

$$y = g(\hat{\theta},b)\Big|_{\hat{\theta}=\theta}. \tag{11}$$

3. STATISTICAL RESULTS

3.1 The uniqueness of the L. S. estimate

Proposition 1. The Lebesgue measure of the set of all $y \in R^N$ giving at least two L. S. estimate $\hat{\theta}(y)$ is zero.

We note, that we do not state that $\hat{\theta}(y)$ is unique for *every* $y \in R^N$.

The proof of Proposition 1 is presented in [5], and its generalization to the maximal likelihood estimation in curved exponential families is given in [7]. First, it can be proved, using the Sard theorem from differential geometry, that the set of singular points of the mapping h defined by (7) has zero Lebesgue measure. (This ensure the local uniquenes of $\hat{\theta}(y)$.) Then, using a simple version of the transversality theorem from differential geometry, the nonlocal uniqueness can be proved.

3.2. The probability distribution of the L. S. estimate

Let $\bar{\theta}$ be the true value of θ. We shall look for an expression approximating the probability density $p(\hat{\theta}|\bar{\theta})$ of the L. S. estimate $\hat{\theta}$.

For $r > 0$ denote

$$G_{\bar{\theta}}(r) = \{y \in R^N:\ \|y - \eta(\bar{\theta})\| \le r\}$$

$$A_{\bar{\theta}}(r) = \{\theta \in \Theta:\ \underset{y \in G_{\bar{\theta}}(r)}{\exists}\ \theta \text{ is a solution of } (4)\}$$

$$H_{\bar{\theta}}(r) = \{y \in R^N:\ \hat{\theta}(y) \in A_{\bar{\theta}}(r) \text{ et } \|y - \Psi_{\bar{\theta}}[\hat{\theta}(y)]\| \le r\}.$$

The set $G_{\bar{\theta}}(r)$ is a hypersphere centred at $\eta(\bar{\theta})$. $A_{\bar{\theta}}(r)$ is the set of possible solutions of the normal equations (4) when

$y \in G_{\bar\theta}(r)$. The set $H_{\bar\theta}(r)$ is a tube of radius r arround the "surface" $\{\Psi_{\bar\theta}(\theta) : \theta \in A_{\bar\theta}(r)\}$.

Obviously, $G_{\bar\theta}(r) \subset H_{\bar\theta}(r)$, since $\Psi_{\bar\theta}(\hat\theta)$ is the projection of the center of $G_{\bar\theta}(r)$ onto the set $\{y: \hat\theta=\hat\theta(y)\}$, and $\{y:\hat\theta=\hat\theta(y)\} \cap G_{\bar\theta}(r)$ is a sphere of radius $\le r$.

We make the following assumption:

(A) The number r is chosen so small and the point $\bar\theta$ is so distant from the boundary of Θ that for every $y \in H_{\bar\theta}(r)$ there is exactly one L. S. estimate $\hat\theta(y)$.

We shall compute the contribution of the sample points belonging to $H_{\bar\theta}(r)$ to the probability density of $\hat\theta$.

We introduce some notations:

$$\{Q_{\bar\theta}(\theta)\}_{ij} = \{M(\theta)\}_{ij} + [\eta(\theta)-\eta(\bar\theta)]'(I-P^\theta)\frac{\partial^2\eta(\theta)}{\partial\theta_i\partial\theta_j}, \tag{12}$$

$$\{D(b,\theta)\}_{ij} = -\sum_{\ell-1}^{N-m} b_\ell \left\langle w_\ell(\theta), \frac{\partial^2\eta(\theta)}{\partial\theta_i\partial\theta_j}\right\rangle \tag{13}$$

(compare with (6)),

$$\varphi^*(b) = \exp\left\{-\frac{1}{2}\sum_{i=1}^{N-m} b_i^2\right\}; \text{ if } \sum_i b_i^2 \le r^2,$$

$$= 0 \qquad ; \text{ if } \sum_i b_i^2 > r^2,$$

$$q(\theta|\bar\theta) = \frac{\det Q_{\bar\theta}(\theta)}{(2\pi)^{m/2}\det^{1/2}M(\theta)} \exp\left\{-\frac{1}{2}\|P^\theta[\eta(\theta)-\eta(\bar\theta)]\|^2\right\} \tag{14}$$

<u>Proposition 2</u> If $\varepsilon_1,\dots,\varepsilon_N$ are gaussian random variables and if the assumption (A) holds, then the contribution of samples belonging to $H_{\bar\theta}(r)$ to the probability density of $\hat\theta$ is proportional to

$$q(\hat\theta|\bar\theta) \int_{R^{N-m}} \det[I + D(b,\hat\theta)Q_{\bar\theta}^{-1}(\hat\theta)]\varphi^*(b)\, db \tag{15}$$

Proof. Take $g(\theta,b)$ as in (10). Denote

$$\nabla_\theta g = \left(\frac{\partial g}{\partial\theta_1},\ldots,\frac{\partial g}{\partial\theta_m}\right),\quad \nabla_b g = \left(\frac{\partial g}{\partial b_1},\ldots,\frac{\partial g}{\partial b_{N-m}}\right)$$

$$\nabla g = (\nabla_\theta g,\ \nabla_b g).$$

Evidently $\nabla_b g = (w_1,\ldots,w_{N-m})$. Denote the last matrix by W. According to (11) the joint density of $\hat\theta$ and b is equal to

$$f(g(\hat\theta,b)|\bar\theta)|\det \nabla g(\hat\theta,b)|$$

which in turn, using (5), is proportional to

$$\frac{\det \nabla g(\hat\theta,b)}{\det^{1/2} K}\exp\left\{-\frac{1}{2}\|\Psi_{\bar\theta}(\theta)-\eta(\bar\theta)\|^2-\frac{1}{2}\sum_i b_i^2\right\}.$$

Hence the required contribution to the density of $\hat\theta$ is proportional to

$$\int_{R^{N-m}} \det\nabla g(\hat\theta,b)\varphi^*(b)\,db\ \frac{\exp\left\{-\frac{1}{2}\|\Psi_{\bar\theta}(\hat\theta)-\eta(\bar\theta)\|^2\right\}}{\det^{1/2} K} \tag{16}$$

We can write

$$\begin{aligned}\frac{\det^2(\nabla g)}{\det K} &= \det\left[(\nabla'_\theta g,\ \nabla'_b g)K^{-1}\begin{pmatrix}\nabla_\theta g\\ \nabla_b g\end{pmatrix}\right]\\ &= \det\begin{pmatrix}\nabla'_\theta g\ K^{-1}\ \nabla_\theta g, & W'\ K^{-1}\ \nabla_\theta g\\ \nabla'_\theta g\ K^{-1}\ W, & I\end{pmatrix} \qquad (17)\\ &= \det\left[\nabla'_\theta g\ K^{-1}(I-WW'\ K^{-1})\nabla_\theta g\right]\end{aligned}$$

(cf. [3] chpt. II, eq. (II b)).

Obviously, the matrix $WW'K^{-1}$ is the orthogonal projector onto the linear space spanned by $w_1, \ldots, w_{N-m}$. Thus $I - WW'K^{-1}$ is equal to the orthogonal projector P^θ defined by (8). Setting the expression (8) into (17) we obtain

$$\frac{\det^2(\nabla g)}{\det K} = \frac{\det^2[\{\langle\frac{\partial g}{\partial\theta_i}, \frac{\partial\eta}{\partial\theta_j}\rangle\}_{i,j=1}^m]}{\det M} . \tag{18}$$

From the definition of $g(\theta,b)$ it follows that

$$\frac{\partial g}{\partial\theta_i} = \frac{\partial\Psi_{\bar\theta}}{\partial\theta_i} + \sum_{\ell=1}^{N-m} b_\ell \frac{\partial w_\ell}{\partial\theta_i} . \tag{19}$$

From

$$\langle w_\ell, \frac{\partial\eta}{\partial\theta_j}\rangle = 0$$

we obtain by differentiation with respect to θ_i that

$$\langle\frac{\partial w_\ell}{\partial\theta_i}, \frac{\partial\eta}{\partial\theta_j}\rangle = - \langle w_\ell, \frac{\partial^2\eta}{\partial\theta_i\partial\theta_j}\rangle.$$

From (9) it follows that

$$\langle\Psi_{\bar\theta} - \eta, \partial\eta/\partial\theta_i\rangle = 0,$$

and by differentiation we obtain using (12) that

$$\{Q_{\bar\theta}\}_{ij} = \langle\partial\eta/\partial\theta_i, \partial\eta/\partial\theta_j\rangle + \langle\eta - \Psi_{\bar\theta}, \partial^2\eta/\partial\theta_i\partial\theta_j\rangle$$

$$= \langle\partial\Psi_{\bar\theta}/\partial\theta_i, \partial\eta/\partial\theta_j\rangle.$$

Hence using (13) and (19) we obtain

$$\langle\frac{\partial g}{\partial\theta_i}, \frac{\partial\eta}{\partial\theta_j}\rangle = \{Q_{\bar\theta}\}_{ij} + \{D(b)\}_{ij}.$$

Thus, according to (18)

$$\det{}^2(\nabla g) = \frac{\det K \det{}^2 Q_{\bar{\theta}}}{\det M} \det{}^2[I + D(b)Q_{\bar{\theta}}^{-1}].$$

Setting this and the expression (9) into (16) we obtain the required formula (15). □

As shown in [6, 8], the term $q(\hat{\theta}|\bar{\theta})$ is the main part not only in the formula (15) but also in the approximation of the whole density $p(\hat{\theta}|\bar{\theta})$. In [8] the following inequality is proved for every Borel set B $A_{\bar{\theta}}(r)$

$$\left|\int_B p(\hat{\theta}|\bar{\theta})d\hat{\theta} - \int_B q(\hat{\theta}|\bar{\theta})d\hat{\theta}\right| \leq \int_B q(\hat{\theta}|\bar{\theta})d\hat{\theta}\cdot\left\{\frac{P_{N-m}}{P_N} - 1 + \frac{P_{N-m}}{P_N}\sum_{\ell=1}^{INT(m/2)} \frac{1}{r^{2\ell}}\binom{m}{2\ell}\prod_{i=0}^{\ell-1}(N-m+2i)\right\} \tag{20}$$

where INT(x) is the integer part of a number x, $P_\ell = P\{\chi_\ell^2 \leq r^2\}$, and χ_ℓ^2 is a χ^2-random variable with ℓ degrees of freedom.

We note that in the case that the model (2) is linear, the formula (14) for $q(\hat{\theta}|\bar{\theta})$ gives the well known density

$$q(\hat{\theta}|\bar{\theta}) = \frac{\det^{1/2} M}{(2\pi)^{m/2}} \exp\left\{-\frac{1}{2}(\hat{\theta}-\bar{\theta})'\, M(\hat{\theta}-\bar{\theta})\right\}.$$

REFERENCES

1. D.M. Bates - D.G. Watts (1980). Relative curvature measures of nonlinearity. J. Roy. Statist. Soc. Ser. B 40, 1-25.
2. E.M.L. Beale (1960). Confidence regions in nonlinear estimation. J. Roy. Statist. Soc. B 22, 41-88.
3. F.R. Gantmacher (1966). Matrix Theory (Russian). Nauka, Moscow.
4. J.Milnor (1963). Morse Theory. Princeton Univ. Press., Princeton, N. J.

5. A. Pázman (1984). Nonlinear least squares-uniqueness versus ambiguity. Math. Operationsforsch. u. Statist., ser. statist. 15, 323-336.
6. A. Pázman (1984). Probability distribution of the multivariate nonlinear least squares estimates. Kybernetika (Prague) 20, 209-230.
7. A. Pázman (1986). On the uniqueness of the M. L. Estimates in curved exponential families. Kybernetika (Prague), 22, 124-132.
8. A. Pázman (1985). On formulas for the distribution of nonlinear L. S. estimates. To appear in Statistics (MOS).

A. PÁZMAN
Mathematical Institute
ul. Obrancov Mieru
814 713 Bratislava
Czechoslovakia

Proc. of the 5th Pannonian Symp. on Math. Stat., Visegrád, Hungary 1985
W. Grossmann, J. Mogyoródi, I. Vincze, W. Wertz, eds

ON THE CONVERGENCE TO THE POISSON DISTRIBUTION IN REJECTIVE SAMPLING FROM A FINITE POPULATION

Z. PRÁŠKOVÁ

Prague

The convergence of a sample sum in rejective sampling from a finite population to the Poisson distribution is considered in the case that the size of the population as well as that of the sample tend to infinity. A modification of the method of characteristic functions developed by Franken (1964) for a sum of independent random variables is used. A similar result was obtained by Bikeliene (1973, 1980) for simple random sampling.
AMS subject classification: 62D05, 62E20, 60F05

1. INTRODUCTION

Let us consider a population of N real numbers $\{y_1,\dots,y_N\}$ and a sample of size n drawn from this population according to rejective sampling plan with parameters $p_1,\dots,p_N$, $0<p_i<1$, $\Sigma_{j=1}^N p_j = n$ (see Hájek (1964) for details). Let $S_{N,n}$ be the sample sum. In the next section we shall study the convergence of $S_{N,n}$ to the Poisson distribution in the case that $N\to\infty$, $n\to\infty$. We shall use the method of characteristic functions. First we shall give some lemmas.

Lemma 1.1. Let X,Y be integer-valued random variables, X with a distribution function (df) F and a characteristic function (cf) f, Y with a df G and a cf g. Then

$$\sup_x |F(x) - G(x)| \le \frac{1}{4\pi} \int_{-\pi}^{\pi} |f(t) - g(t)|\,|\sin \frac{t}{2}|^{-1}\,dt, \qquad (1.1)$$

$$\sup_k |P(X = k) - P(Y = k)| \le \frac{1}{2\pi} \int_{-\pi}^{\pi} |f(t) - g(t)|\,dt. \qquad (1.2)$$

Proof. For (1.1) see Tsaregradski (1958), (1.2) follows

immediately from the relation between the probability function of an integer-valued random variable and the corresponding cf.

□

Lemma 1.2. Let $b>0$ be a constant. Then

$$\int_0^{\pi} e^{-2b\sin^2(t/2)}dt \le \frac{\pi}{2}^{3/2} b^{-1/2} , \tag{1.3}$$

$$\int_0^{\pi} \sin^{2k}(t/2)e^{-2b\sin^2(t/2)}dt \le (k-1)!b^{-k}(1-e^{-2b}), \tag{1.4}$$

$k\ge 1$, integer,

$$\int_0^{\pi} \sin^{2k+1}(t/2)e^{-2b\sin^2(t/2)}dt \le k!\, b^{-(k+1)}(1-e^{-2b}), \tag{1.5}$$

$k\ge 0$, integer.

Proof. The inequalities (1.4) and (1.5) follow from Lemma 3 in Franken (1964). From the inequality $\sin t\ge 2t/\pi$ valid for $t\le\pi/2$ we get

$$\int_0^{\pi} e^{-2b\sin^2(t/2)}dt\le\int_0^{\pi} e^{-2bt^2\pi^{-2}}dt\le\int_0^{\infty} e^{-2bt^2\pi^{-2}}dt=(\frac{\pi}{2})^{3/2} b^{-1/2}. \;\square$$

Lemma 1.3. Let $0<p<\frac{1}{2}$, $q=1-p$. Then for any $\Theta \in \langle -\pi,\pi\rangle$

$$|q + p\, e^{i\Theta}| \le e^{-2pq\sin^2(\Theta/2)} , \tag{1.6}$$

$$|q + p\, e^{i\Theta}|^{-1} < (q - p)^{-1} . \tag{1.7}$$

Proof. We have

$$|q+pe^{i\Theta}|=(1-4pq\sin^2(\Theta/2))^{1/2}\le 1-2pq\sin^2(\Theta/2) \le$$

$$\le e^{-2pq\sin^2(\Theta/2)},$$

further

$$|q+pe^{i\Theta}|^{-1}=(1-4pq\sin^2(\Theta/2))^{-1/2}\le(1-4pq)^{-1/2}=(q-p)^{-1}.$$

□

Lemma 1.4. The cf of $S_{N,n}$ is

$$f_{N,n}(t) = (2\pi P_{N,n})^{-1} \int_{-\pi}^{\pi} e^{-i\Theta n} \prod_{j=1}^{N} (q_j+p_j\, e^{i(\Theta+ty_j)})d\Theta, \quad (1.8)$$

where

$$P_{N,n} = P(\sum_{j=1}^{N} I_j = n),$$

$I_1,\ldots,I_N$ are independent random variables, $P(I_j=1) = p_j$, $P(I_j=0) = 1 - p_j = q_j$, $j = 1,\ldots,N$.

Proof. It can be shown that the distribution of $S_{N,n}$ coincides with the conditional distribution of $\Sigma_{j=1}^{N} y_j I_j$ under the condition $\Sigma_{j=1}^{N} I_j = n$, where the I_j's are independent random variables, $P(I_j=1) = p_j = 1 - P(I_j=0)$. Hence, after some calculations we obtain

$$f_{N,n}(t) = (2\pi P_{N,n})^{-1} \int_{-\pi}^{\pi} e^{-i\Theta n}\, Ee^{i(\Theta\Sigma I_j+t\Sigma I_j y_j)}d\Theta =$$

$$= (2\pi P_{N,n})^{-1} \int_{-\pi}^{\pi} e^{-i\Theta n} \prod_{j=1}^{N}(q_j+p_j\, e^{i(\Theta+ty_j)})d\Theta . \qquad \square$$

Remark. Notice that the cf of $\Sigma_{j=1}^{N} I_j$ is

$$h_N(\Theta) = \prod_{j=1}^{N} (q_j + p_j\, e^{i\Theta}) \quad . \quad (1.9)$$

Consequently,

$$(2\pi P_{N,n})^{-1} \int_{-\pi}^{\pi} e^{-i\Theta n}\, h_N(\Theta)\, d\Theta = 1. \quad (1.10)$$

2. CONVERGENCE TO THE POISSON DISTRIBUTION

Let $\{y_1,\ldots,y_N\}$ be a population of nonnegative integers. Let g_N be the cf of Poisson distribution with the parameter

$$a_N = \sum_{j=1}^{N} y_j p_j \quad ,$$

that is

$$g_N(t) = \exp(a_N(e^{it} - 1)).$$

In the sequel we shall suppress the dependence on N for the sake of simplicity by omitting the subscript N from $f_{N,n}$, $P_{N,n}$, g_N, h_N, a_N. Denote

$$h(t,\Theta) = \prod_{j=1}^{N} (q_j + p_j\, e^{i(\Theta+ty_j)})$$

and put

$$\varphi(t,\Theta) = \frac{h(t,\Theta)}{h(\Theta)} \quad .$$

Obviously,

$$\varphi(t,\Theta) = \prod_{j=1}^{N} \varphi_j(t,\Theta) = \prod_{j=1}^{N} (q_j(\Theta) + p_j(\Theta)\, e^{ity_j}) \tag{2.1}$$

where

$$q_j(\Theta) = q_j(q_j + p_j\, e^{i\Theta})^{-1},\ p_j(\Theta) = 1 - q_j(\Theta) \quad . \tag{2.2}$$

Define

$$\psi(t,\Theta) = (g(t))^{-1}\ \varphi(t,\Theta) = e^{a(1-e^{it})}\ \varphi(t,\Theta) \quad .$$

Now, in accordance with (1.8) and (1.10) we can write

$$\begin{aligned} f_n(t) - g(t) &= (2\pi P_n)^{-1} \int_{-\pi}^{\pi} e^{-i\Theta n}(h(t,\Theta) - h(\Theta)g(t))d\Theta = \\ &= (2\pi P_n)^{-1} \int_{-\pi}^{\pi} e^{-i\Theta n}\ h(\Theta)g(t)(\psi(t,\Theta) - 1)\ d\Theta \quad . \end{aligned} \tag{2.3}$$

We shall prove the following lemma.

Lemma 2.1. Suppose that

$$\max_j p_j < \frac{1}{2} \quad .$$

Denote by

$$A = \sum_{j=1}^{N} y_j p_j (q_j - p_j)^{-1} ,$$

$$B_1 = \sum_{j=1}^{N} y_j p_j q_j (q_j - p_j)^{-2} ,$$

$$B_2 = \sum_{j=1}^{N} [y_j(y_j - 1)p_j + (y_j p_j)^2] (q_j - p_j)^{-2} .$$

Then for $|t| \leq \pi$, $|\Theta| \leq \pi$

$$|\psi(t,\Theta)-1| \leq 4e^{2(a+A)}(|\sin \tfrac{t}{2} \sin \tfrac{\Theta}{2}| B_1 + (\sin \tfrac{t}{2})^2 B_2). \qquad (2.4)$$

Proof. For a complex z consider functions

$$\hat{\varphi}_j(z,\Theta) = q_j(\Theta) + p_j(\Theta) z^{y_j} \qquad 1 \leq j \leq N$$

and

$$\hat{\psi}(z,\Theta) = e^{a(1-z)} \prod_{j=1}^{N} \hat{\varphi}_j(z,\Theta) = e^{a(1-z)} \hat{\varphi}(z,\Theta) .$$

Obviously,

$$\psi(t,\Theta) = \hat{\psi}(e^{it}, \Theta) \quad .$$

From the Taylor expansion of the function $\hat{\psi}$ we obtain

$$\psi(t,\Theta)-1 = \int_1^{e^{it}} \hat{\psi}'(z,\Theta)dz = (e^{it}-1) \int_0^1 \hat{\psi}'(1+\tau(e^{it}-1),\Theta)d\tau \quad . \qquad (2.5)$$

Recalling that $a = \Sigma y_j p_j$, we easily get

$$\hat{\psi}'(z,\Theta) = \frac{d}{dz}\hat{\psi}(z,\Theta) =$$

$$= e^{a(1-z)}\hat{\varphi}(z,\Theta)\sum_{j=1}^{N}[\hat{\varphi}_j'(z,\Theta) - p_j y_j \hat{\varphi}_j(z,\Theta)](\hat{\varphi}_j(z,\Theta))^{-1}. \quad (2.6)$$

Another Taylor expansion yields

$$\hat{\varphi}_j(z,\Theta) = \hat{\varphi}_j(1,\Theta) + (z-1)\,\beta_{j1}(\Theta) = 1 + (z-1)\,\beta_{j1}(\Theta), \quad (2.7)$$

where

$$|\beta_{j1}(\Theta)| \le |p_j(\Theta)|\; y_j$$

and similarly

$$\hat{\varphi}_j'(z,\Theta) = \hat{\varphi}_j'(1,\Theta) + (z-1)\beta_{j2}(\Theta) = p_j(\Theta) y_j + (z-1)\beta_{j2}(\Theta), \quad (2.8)$$

$$|\beta_{j2}(\Theta)| \le y_j(y_j-1)|p_j(\Theta)|\;.$$

Thus, from (2.7), (2.8), (2.2) and (1.7) we get for $|z|<1$

$$|\hat{\varphi}_j'(z,\Theta) - p_j y_j \hat{\varphi}_j(z,\Theta)| \le [2p_j q_j y_j |\sin\tfrac{\Theta}{2}| +$$

$$+ |z-1|(y_j(y_j-1)p_j + (y_j p_j)^2](q_j - p_j)^{-1} \quad (2.9)$$

Notice, that any complex z, $|z|<1$, can be written as $z=\rho e^{i\omega}$, $0<\rho<1$, $\omega \in (-\pi,\pi)$.

Hence

$$|q_j + p_j\, e^{i\Theta} z^{y_j}| = |q_j + p_j \rho^{y_j}\, e^{i(\Theta+\omega y_j)}| \ge q_j - p_j$$

and

$$|\varphi_j(z,\Theta)|^{-1} = |q_j + p_j e^{i\Theta}|\,|q_j + p_j e^{i\Theta} z^{y_j}|^{-1} \le (q_j - p_j)^{-1}\;. \quad (2.10)$$

Combining (2.6), (2.9) and (2.10) we get for $|z|<1$

$$|\hat{\psi}'(z,\Theta)| \le |e^{a(1-z)}|\,|\hat{\varphi}(z,\Theta)|\,[2B_1|\sin\tfrac{\Theta}{2}| + |z-1|B_2]\;.$$

Put

$$z = z_0 = 1 + \tau(e^{it} - 1), \qquad 0<\tau<1.$$

Then, utilizing (2.7), (2.2) and (1.7) we have

$$|\hat{\varphi}(z_0,\Theta)| = \prod_{j=1}^{N} |\hat{\varphi}_j(z_0,\Theta)| \le \prod_{j=1}^{N} (1+2\tau|\sin\frac{t}{2}||\, p_j(\Theta)|y_j) \le$$

$$\le \prod_{j=1}^{N} (1+2p_j y_j (q_j-p_j)^{-1}) \le e^{2A} \quad .$$

Since

$$|e^{a(1-z_0)}| = e^{2a\tau\sin^2(t/2)}$$

we can summarize that

$$|\hat{\varphi}'(1+\tau(e^{it}-1),\Theta)| \le e^{2a\tau\sin^2(t/2)+2A}[2B_1|\sin\frac{\Theta}{2}|+2B_2\tau|\sin\frac{t}{2}|]$$

and from here and from (2.5) we easily get (2.4). □

<u>Lemma 2.2</u>. Under the notation and the assumption of Lemma 2.1,

$$|f_n(t)-g(t)| \le c_1 e^{-2a\sin^2(t/2)+2(a+A)}[|\sin\frac{t}{2}|B_1 d^{-1/2}+c_2 B_2 \sin^2\frac{t}{2}]$$

holds for $|t|\le\pi$ and N,n sufficiently large ($N>2n$), where c_1, c_2 are positive constants,

$$d = \sum_{j=1}^{N} p_j(1 - p_j) \quad .$$

<u>Proof</u>. According to (2.3) we have

$$|f_n(t)-g(t)| \le (2\pi P_n)^{-1} \int_{-\pi}^{\pi} |h(\Theta)||g(t)||\psi(t,\Theta)-1|d\Theta = (2\pi P_n)^{-1} J(t).$$

From (1.6) and (1.9) we get

$$|h(\Theta)| \le e^{-2d\,\sin^2(\Theta/2)} , \tag{2.11}$$

further,

$$|g(t)| = |e^{a(e^{it}-1)}| = e^{-2a\sin^2(t/2)} . \qquad (2.12)$$

Hence, utilizing (2.11), (2.12) and (2.4) together with (1.3) and (1.5) we obtain that

$$J(t) \le c_1 e^{-2a\sin^2(t/2)+2(a+A)[|\sin\frac{t}{2}|d^{-1}B_1+c_2B_2d^{-1/2}\sin^2\frac{t}{2}]} .$$

It remains to estimate the probability P_n. It can be shown that $P_n(2\pi d)^{1/2}\to 1$ for $d\to\infty$ (see Hájek (1981), p. 62). In our case (recall that we have assumed $\Sigma p_j = n$ and $p_j<1/2$ for all j) we have

$$\frac{n}{2} = \frac{1}{2}\sum_{j=1}^{N} p_j < d = \sum_{j=1}^{N} p_j(1-p_j) < \frac{N}{4} .$$

Hence, for n and N sufficiently large there exist positive constants k_1, k_2 such that

$$k_1\, d^{-1/2} \le 2\pi P_n \le k_2 d^{-1/2}$$

which completes the proof. □

<u>Theorem 2.1</u>. Let $F_{N,n}$ be the df of $S_{N,n}$ and let G_N be the df of the Poisson random variable Y_N with the cf g_N considered above. Provided that max $p_j<1/2$ there exist positive constants c, c_1, c_2 such that

$$\sup_x |F_{N,n}(x) - G_N(x)| \le e^{2(A+a)}(c_1B_1d^{-1/2}\, a^{-1/2} + c_2B_2a^{-1})$$

and

$$\sup_k |P(S_{N,n} = k) - P(Y_N = k)| \le c\, e^{2(A+a)}(B_1d^{-1/2}a^{-1}+B_2a^{-1}).$$

<u>Proof</u>. The assertion follows from Lemma 1.1, Lemma 2.2 and Lemma 1.2. □

Let us consider a sequence of rejective sampling designs, the ν-th of which has the size n_ν and parameters $p_{\nu_1},\ldots,p_{\nu_{N_\nu}}$, and refers to a population $\{y_{\nu_1},\ldots,y_{\nu_{N_\nu}}\}$. Suppose that $n_\nu\to\infty$, $N_\nu\to\infty$ if $\nu\to\infty$. Let F_{N_ν,n_ν} and G_{N_ν} refer to the ν-th design.

<u>Theorem 2.2</u>. Suppose that the following conditions are fulfilled:

(i) $$\max_{1\le j\le N_\nu} p_{\nu_j} \le \frac{1}{4} \qquad \nu\ge 1 ,$$

(ii) $$\lim_{\nu\to\infty} \sum_{j=1}^{N_\nu} p_{\nu_j}\, y_{\nu_j} = \lambda > 0 ,$$

(iii) $$\lim_{\nu\to\infty} \max_{1\le j\le N_\nu} p_{\nu_j}\, y_{\nu_j} = 0 ,$$

(iv) $$\lim_{\nu\to\infty} \sum_{j=1}^{N_\nu} p_{\nu_j}\, y_{\nu_j}\,(y_{\nu_j} - 1) = 0 .$$

Then

$$\lim_{\nu\to\infty} \sup_x |F_{N_\nu,n_\nu}(x) - G_{N_\nu}(x)| = 0 ,$$

$$\lim_{\nu\to\infty} \sup_k |P(S_{N_\nu,n_\nu} = k) - P(Y_{N_\nu} = k)| = 0 .$$

<u>Proof</u>. Under the condition (i) we get for fixed ν that

$$A \le 2\Sigma p_{\nu_j}\, y_{\nu_j} ,$$

$$B_1 \le 4\Sigma p_{\nu_j}\, y_{\nu_j} ,$$

$$B_2 \le 4(\Sigma p_{\nu_j} y_{\nu_j}(y_{\nu_j} - 1) + \max p_{\nu_j} y_{\nu_j}\, \Sigma p_{\nu_j}\, y_{\nu_j}) ,$$

$$d \ge \frac{3}{4}\, n_\nu .$$

The rest of the proof follows from (ii) - (iv) and Theorem 2.1.

□

REFERENCES

1. Bikeliene, V. (1973). Limit theorems for simple random sampling from a finite population. (In Russian.) Litovsk. mat. sb. 13, No 1, 47-54.
2. Bikeliene, V. (1980). Sampling from a finite population. Poisson approximation. (In Russian.) In: Application of Probability Theory and Mathematical Statistics, 3, 111-129, Vilnius.
3. Franken, P. (1964). Approximation der Verteilungen von Summen unabhängiger nichtnegativer ganzzahliger Zufallsgrössen durch Poissonsche Verteilungen. Math. Nachr. 27, 303-340.
4. Hájek, J. (1964). Asymptotic theory of rejective sampling with varying probabilities from a finite population. Ann. Math. Statist. 35, 1491-1523.
5. Hájek, J. (1981). Sampling from a Finite Population (Edited by V. Dupač), Dekker, New York.
6. Tsaregradski, I. P. (1958). On a uniform approximation for the binomial distribution by infinitely divisible laws. (In Russian.) Teor. Verojatnost. i Primenen. 3, 470-474.

Zuzana Prášková
Charles University
Department of Statistics
Sokolovská 83
186 00 Prague
Czechoslovakia

Proc. of the 5th Pannonian Symp. on Math. Stat., Visegrád, Hungary 1985
W. Grossmann, J. Mogyoródi, I. Vincze, W. Wertz, eds

UNBIASED ESTIMATION AND LOCAL STRUCTURE

L. RÜSCHENDORF

Münster

We develop the importance of the local structure of the estimation model to the theory of unbiased estimation, especially, in the case of large nonparametric models. On one side this gives us an useful tool for the construction of MVUE, justifying already in the finite theory intuitively appealing construction methods, which are known from the asymptotic theory. On the other side our results show under general conditions, that UMVUE typically are asymptotically efficient and, furthermore, that asymptotically efficient estimators are obtained from MVUE for a linearized estimation functional combined with an adaption procedure.
AMS subject classification: 62G05

1. INTRODUCTION

In the theory of unbiased estimation many examples are known, where unbiased estimators are very bad and, more worse, where unbiased estimators do not exist (e.g. in many models involving nuisance paramters). Furthermore, for bounded loss functions the principle of MVUE is not effective as was demonstrated by Basu (1955). A beautiful presentation of the relation of unbiasedness with different estimation principles can be found in the recent book of Lehmann (1983).

On the other hand it was shown by Sharma (1973) and Portnoy (1977), that UMVUE in regular exponential families typically are asymptotically equivalent to the MLE and, therefore, are asymptotically efficient. So the bad behaviour in these cases is due to too small sample sizes. We want to show that this behaviour extends to a general class of estimation problems, especially in nonparametric models, and that on the other side MVUE lead in a natural way to a construction method for asymptotically efficient estimators.

In the asymptotic theory the importance of the local structure was impressively demonstrated in the LAN-theory of Le Cam and Hajek for parametric models, while for general models one can find a convincing and far reaching presentation in the recent book of Pfanzagl (1982). We want to show, that the local structure is already important in the finite unbiased theory, at least for large nonparametric models.

2. UNBIASED ESTIMATION AND TANGENT CONES

Let $\mathcal{P}$ be a class of probability measures on $(X,\mathcal{B})$ and $g:\mathcal{P} \to \mathbb{R}^1$ a function which we want to estimate assuming a quadratic loss. The leading idea of the following construction method of minimum variance unbiased estimators (MVUE) of g is that the local structure of $\mathcal{P}$ contains essential information on the estimation problem at least in large nonparametric models.

We shall concentrate in this paper on the first order tangent cones and, therefore, to linear estimation functionals g; nonlinear estimation functionals and the corresponding higher order tangent structure will be treated in a forthcoming paper. For $P \in \mathcal{P}$ let $T(P,\mathcal{P}) \subset L^2(P)$ denote the set of all *tangent vectors* at P in $\mathcal{P}$, i.e. the set of all $h \in L^2(P)$ such that there exists a path $(P_t)_{t\geq 0} \subset \mathcal{P}$ with

a) $P_0 = P$

b) $\left\| 2\left[\left(\frac{dP_t}{dP}\right)^{\frac{1}{2}} - 1\right] - th \right\| = o(t)$ (1)

c) $P_t\left\{\frac{dP_t}{dP} = \infty\right\} = o(t^2)$ for $t \to 0$

$\| \ \|$ denoting the norm in $L^2(P)$, $\frac{dP_t}{dP}$ the generalized Radon Nikodym derivative.

For the general theory of L^2-differentiation we refer to Pukelsheim (1981) and Witting (1985). The importance of L^2-differentiation and of the concept of tangent cones in asymptotic estimation theory was fully established by Pfanzagl (1982) where one can find many parametric and nonparametric examples for the calculation of $T(P,\mathcal{P})$.

If $(P_t) \subset \mathcal{P}$ and h, $\frac{dP_t}{dP} \in L^2(P)$, $t \geq 0$ such that

$$||\frac{dP_t}{dP} - 1 - th|| = o(t) \text{ and } P_t\left\{\frac{dP_t}{dP} = \infty\right\} = o(t^2) \qquad (2)$$

then $h \in T(P,\mathcal{P})$. We remark that derivatives in the sense of (2) have been considered by Barankin (1949) and Parzen (1959) in a parametric model in connection with unbiased estimation We extend this result to the general nonparametric case. Define

$$T^1(P,\mathcal{P}) = T(P,\mathcal{P}) \cup \{1\}; \qquad (3)$$

the constant 1 corresponding to the derivative 0 of the constant path $P_t = P$, $t \geq 0$. Furthermore let

$$D_g := \{d \in L^2(\mathcal{P});\ \int d\, dP = g(P),\ \forall\, P \in \mathcal{P}\} \qquad (4)$$

be the set of all unbiased estimators of g which are square integrable w.r.t. $\mathcal{P}$ and $D_0 = D_0(\mathcal{P})$ the corresponding class of estimators of zero

<u>Theorem 1</u>.

a) $D_0 \subset \bigcap_{P \in \mathcal{P}} T^1(P,\mathcal{P})^{\perp}$

where $T^1(P,\mathcal{P})^{\perp} = \{h \in L^2(P);\ \int hf\, dP = 0,\ \forall f \in T^1(P,\mathcal{P})\}$.

b) Let $H_p := cl\langle T^1(P,\mathcal{P})\rangle$ - the closure of the linear space generated by $T^1(P,\mathcal{P})$ in $L^2(P)$; if $d^* \in D_g \cap H_p$, then d^* is MVUE for g in P.

<u>Proof</u>. a) Let $h \in D_0$ and $f \in T(P,\mathcal{P})$, then there exists a path $(P_t) \subset \mathcal{P}$ with $P_0 = P$ and L^2-derivative f. The function $F(t) := \int h dP_t$, $t \geq 0$ is differentiable in 0 with $F'(0) = \int hf\, dP$ (cf. Pukelsheim (1981), Pfanzagl (1982), Prop. 5.1.5 and Witting (1985), Satz 1.179, 1.190). Since $F(t) = 0$, we obtain $0 = F'(0)$ i.e. h is orthogonal to f. Since by assumption h is orthogonal to 1 we obtain $D_0 \subset T^1(P,\mathcal{P})^{\perp}$.

b) Since by a) $D_0 \subset T^1(P,\mathcal{P})^{\perp}$ we get

$$H_P = cl\langle T^1(P,\mathcal{P})\rangle = (T^1(P,\mathcal{P})^{\perp})^{\perp} \subset D_0^{\perp} \quad .$$

Therefore, by the covariance method any $d^* \in D_g \cap H_P$ is MVUE for g in P. □

Corollary 1. If for all $P \in \mathcal{P}$ holds

$$H_P = L^2(P), \text{ then } D_0(\mathcal{P}) = \{0\} \tag{5}$$

i.e. the class $\mathcal{P}$ is L^2-complete. ({0} denotes the set of all functions which are equal to $0[\mathcal{P}]$.)

Remark 1. a) A model with the property (5) is called a full model. Typical examples of full models are robust models (e.g. variation neighbourhood models or e.g. absolutely continuous distributions on $(\mathbb{R}^k, \mathbb{B}^k)$ w.r.t. λ^k).

b) By Theorem 1 the following method is proposed: For $d \in D_g$ determine the projection $\tilde{d}$ on H_P. If $\tilde{d} \in D_g$, then $\tilde{d}$ is MVUE for g. If $\tilde{d} \in \bigcap_{P\in\mathcal{P}} H_P$, then $\tilde{d}$ is UMVUE for g. The chance to find such an element is especially good, if $T^1(P,\mathcal{P})$ is independent of $P \in \mathcal{P}$. This projection method turns out to be especially effective when projectioning an optimal estimator from a larger model down to the actual model, in this way obtaining a finite version of the projection method known from asymptotic statistics.

3. EXAMPLES

a) Linear restrictions

Let $f_i : (X,\mathcal{B}) \to (\mathbb{R}^1, \mathbb{B}^1)$, $1 \le i \le k$,

$$\mathcal{P}_i = \{P \in M^1(X,\mathcal{B});\ f_i \in L^2(P),\ \int f_i dP = 0\},\ 1 \le i \le k, \text{ and } \mathcal{P} = \bigcap_{i=1}^{k} \mathcal{P}_i .$$

Then for $P \in \mathcal{P}$

$$T(P,\mathcal{P}_i) = \langle f_i \rangle^{\perp} \quad . \tag{6}$$

Proof. For $h \in T(P,\mathcal{P}_i)$ let $(P_t) \subset \mathcal{P}_i$ be a path with $P_0=P$ and tangent vector h, then with $F(t) = \int f_i \, dP_t$ we obtain as in the proof of Theorem 1, $0 = F'(0) = \int f_i h \, dP$ i.e. $h \in \langle f_i \rangle^{\perp}$. If, conversely, $h \in \langle f_i \rangle^{\perp}$, let $h_t \in \langle f_i \rangle^{\perp}$ satify, $t\|h_t - h\| \to 0$ and $th_t \geq -1$ and define $P_t = (1+th_t)P$. Then (P_t) is a path in $\mathcal{P}_i$ with tangent vector h by (2), i.e. $h \in T(P,\mathcal{P}_i)$. □

As in Pfanzagl (1982), pg. 118, one now obtains

$$T(P,\mathcal{P}) = \bigcap_{i=1}^{k} T(P,\mathcal{P}_i) = \bigcap_{i=1}^{k} \langle f_i \rangle^{\perp} \quad . \tag{7}$$

Define for $c \in \mathbb{R}^k$ and $d \in D_g$

$$d_c = d + c^T f, \quad f:=(f_1,\dots,f_k)^T$$

and $G_P = (\int f_i f_j \, dP)_{1 \leq i,j \leq k}$ - the 'information' matrix of $\mathcal{P}$ in P.

Proposition 2.

a) $D_0 = \langle f_1,\dots,f_k \rangle$.

b) If $c^* \in \mathbb{R}^k$ solves the equation

$$(c^*)^T G_P + E_P df = 0, \tag{8}$$

then d_{c^*} is MVUE for g in P.

c) $d^*(x_1,\dots,x_n) := \frac{1}{n} \sum_{i=1}^{n} d_{c^*}(x_i)$ is MVUE for g(Q) in P^n w.r.t. $\mathcal{P}^n = \{P^n;\ P \in \mathcal{P}\}$.

Proof. a) From Theorem 1 and (7)

$$D_0 \subset \bigcap_{P \in \mathcal{P}} (T^1(P,\mathcal{P}))^{\perp} = \langle f_1,\dots,f_k \rangle$$ the converse inclusion is obvious.

b) By a) we obtain

$D_g = \{d_c;\ c \in \mathbb{R}^k\}$. So by the covariance method d_{c^*} is MVUE in P if and only if

$c^T E_P df + (c^*)^T G_P c = 0$ for all $c \in \mathbb{R}^k$, or, equivalently

$E_P df + (c^*)^T G_P = 0$.

c) The tangent space of P^n in $\mathcal{P}^n$ is

$$T(P^n, \mathcal{P}^n) = \{ \sum_{i=1}^n h(x_i);\ h \in T(P, \mathcal{P})\} .$$

Since $d^*(x) = \frac{1}{n} \sum_{i=1}^n d_{c^*}(x_i) \in D_g \cap H_{P^n}$

Theorem 1 implies its optimality in P^n.

□

We remark that $D_o(\mathcal{P}^n)$ has been characterized by Hoeffding (1977) as the set of all functions of the form $\{ \sum_{i=1}^k \sum_{j=1}^n f_i(x_j) h_{ij}(x_{(j)})$ where $x_{(j)} := (x_1, \ldots, x_{j-1}, x_{j+1}, \ldots, x_n)$. The formulation in terms of tangent cones allows many modifications. To give a concrete example: Let $X = \mathbb{R}^1$, $\mathcal{P}_o := \{P \in M^1(\mathbb{R}^1, \mathbb{B}^1)\}$; P symmetric around 0, f_i $L^2(P)$, $\int f_i dP = 0$, $1 \le i \le k\}$ then for $P^n \in \mathcal{P}_o^n$ we get

$$T(P^n, \mathcal{P}_o^n) = \{ \sum_{i=1}^n h(x_i) \in T(P^n, \mathcal{P}^n);\ h \text{ symmetric around } 0\}.$$

Therefore, defining

$$\tilde{d}(x_1, \ldots, x_n) := \frac{1}{2n} \sum_{i=1}^n (d_{c^*}(x_i) + d_{c^*}(-x_i)) \tag{9}$$

we obtain $\tilde{d} \in D_g \cap T^1(P^n, \mathcal{P}_o^n)$ and, therefore, by Theorem 1 $\tilde{d}$ is MVUE for g in P^n w.r.t. $\mathcal{P}_o^n$.

b) <u>G invariant distributions</u>

Let G be a finite group of measurable transformations of $(X, \mathcal{B})$ and let $\mathcal{P}$ be a large subset of all distributions which are invariant w.r.t. G, large meaning, that $\mathcal{P}$ has the same tangent cone as the full model has, i.e. for $P \in \mathcal{P}$

$$T(P, \mathcal{P}) = \{h \in L^2(P);\ h \circ g = h[P] \text{ for } g \in G,\ \int h\, dP = 0\} \tag{10}$$

Let for $f \in L^2(\mathcal{P})$, $g(P) = \int f dP$. Without any 'essential' restrictions on P the UMVUE of $g(P)$ after n independent observations would be

$$d_n(x) = \int f dP_{n,x} = \frac{1}{n} \sum_{i=1}^{n} f(x_i), \text{ where} \tag{11}$$

$P_{n,x}(B) = \frac{1}{n} \sum_{i=1}^{n} 1_B(x_i)$ is the empirical measure. According to our general idea we consider

$$\tilde{P}_{n,x}(B) = \frac{1}{n|G|} \sum_{i=1}^{n} \sum_{g \in G} 1_{g^{-1}(B)}(x_i) \quad , \tag{12}$$

the projection of $P_{n,x}$ on P.

<u>Proposition 3</u>. For $f \in L^2(P)$ is

$$\tilde{d}(x) = \int f d\tilde{P}_{n,x} = \frac{1}{n|G|} \sum_{i=1}^{n} \sum_{g \in G} f \bullet g(x_i) \text{ the UMVUE for} \tag{13}$$

$g(P^n) = \int f dP$.

<u>Proof</u>. Since $T^1(P^n, P^n) = \{ \sum_{i=1}^{n} h(x_i);\ h \in L^2(P),\ h \bullet g = h[P],\ g \in G\}$, $\tilde{d} \in D_g \cap T^1(P^n, P^n)$. Therefore, by Theorem 1, $\tilde{d}$ is UMVUE.

□

<u>Special case</u>. 1) If $(X, B) = (\mathbb{R}^1, \mathbb{B}^1)$, $G = \{id, s\}$, where $s(x) = -x$, $x \in \mathbb{R}^1$, then P is a large subclass of the distributions which are symmetric around 0. If $f = 1_{(-\infty, x_0]}$, then the UMVUE for $g(P^n) = F_P(x_0)$ is

$$\tilde{F}_{n,x}(x_0) = \frac{1}{2}[F_{n,x}(x_0) + 1 - F_{n,x}((-x_0)-)] =$$

$$= \frac{1}{2n} \sum_{i=1}^{n} [1_{(-\infty, x_0]}(x_i) + 1_{[-x_0, \infty)}(x_i)] \tag{14}$$

the symmetrized df.

2) If $(X, B) = (\mathbb{R}^k, \mathbb{B}^k)$, $G = S_k$ denotes the group of coordinate changes, then P is a large subclass of the set of symmetric

measures on $(\mathbb{R}^k, \mathcal{B}^k)$. If $f = 1_{(-\infty,x_0]} \times 1_{\mathbb{R}^{k-1}}$, then $g(P^n) = P_1(-\infty,x_0] = F_{P_1}(x_0)$ is the df of the marginal of P. The UMVUE is

$$F_{n,x}(x_0) = \frac{1}{n\cdot k} \sum_{i=1}^{n} \sum_{j=1}^{k} 1_{(-\infty,x_0]}(x_{ij}) \tag{15}$$

where $x = (x_1,\ldots,x_n)$, $x_i = (x_{i1},\ldots,x_{ik})$. We remark, that for the case that $\mathcal{P}$ denotes the set of all continuous symmetric probability measures one could also apply a completeness result of Smith (1969), pg. 35.

4. UNBIASED ESTIMATORS AND ASYMPTOTICALLY EFFICIENT ESTIMATORS

Consider an asymptotic estimation problem $(\mathcal{P}^n)_{n\in\mathbb{N}}$ with iid observations and let $g\colon \mathcal{P} \to \mathbb{R}^1$ be a differentiable functional which is to be estimated; differentiability of g means that for all $P \in \mathcal{P}$ there exists $g_P \in L^2(P)$ such that $\int g_P dP = 0$ and for all $\{P_t\}_{t\geq 0} \subset \mathcal{P}$ with $P_0=P$ and tangent vector h holds:

$$\lim_{t\to 0} \frac{g(P_t) - g(P)}{t} = \int g_P h dP . \tag{16}$$

$g_P = g(\cdot,P)$ is called *gradient of g* in P.

As is clear from the definition (16) a gradient is not uniquely determined. Call $\dot{g}_P(\cdot,P) = \dot{g}_P$ a *canonical gradient* if $\dot{g}_P$ is a gradient and if $\dot{g}_P \in cl\langle T(P,\mathcal{P})\rangle$; i.e. $\dot{g}_P$ is the (unique) projection of any g_P on $cl\langle T(P,\mathcal{P})\rangle$.

An estimator sequence is called asymptotically efficient (of first order) if it is as.median-unbiased and as. $N(0,\sigma^2(P))$ distributed, with $\sigma^2(P) = E_P(\dot{g}(\cdot,P))^2$. The typical as. eff. estimator sequences have a stochastic expansion of the form

$$\sqrt{n}(d_n - g(P)) = \frac{1}{\sqrt{n}} \sum_{i=1}^{n} \dot{g}(x_i,P) + o_{P^n}(1). \tag{17}$$

If (d_n^*) is a UMVUE sequence for g constructed according to Theorem 1, then (d_n^*) is as. efficient.

Proposition 4. Let $d_n^* \in D_g(P^n)$ satisfy

$$d_n^*(x) = \frac{1}{n} \sum_{i=1}^{n} h(x_i), \; h \in \bigcap_{P \in \mathcal{P}} H_P, \; n \in \mathbb{N} \;,$$

then (d_n^*) is asymptotically efficient.

Proof. Since $d_n^* \in D_g(P^n)$, $n \geq 1$ we have for $P \in \mathcal{P}$, $d_1^*(x) = g(P) + f_P(x) = h(x)$, i.e. $f_P \in cl\langle T(P,\mathcal{P})\rangle = H_P$. Therefore, $g(Q) = \int d_1^* dQ = g(P) + \int f_P dQ$ for all $Q \in \mathcal{P}$, implying, that $f_P = \dot{g}_P$ is the canonical gradient of g. By the CLT

$$\sqrt{n}(d_n^* - g(P)) \xrightarrow{D} N(0, \int (\dot{g}_P)^2 dP)$$

i.e. (d_n^*) is as. efficient.

□

Proposition 4 shows that the phenomenon detected by Sharma (1977) holds true in very general situations (at least for linear estimable functionals g). Loosely speaking one can say that for large nonparametric models UMVUE are typically as. efficient.

Let $\delta(Q,P) = \left\| \left(\frac{dQ}{dP}\right)^{1/2} - 1 \right\|$ denote the Hellinger distance ($\| \ \|$ in $L^2(P)$). $g_P \in L^2(P)$ is called a strong gradient of g in P w.r.t. δ if $\int g_P dP = 0$ and

$$g(Q) - g(P) = \int g_P dQ + o(\delta(Q,P)) \text{ for } Q \text{ with}$$
$$Q\left(\frac{dQ}{dP} = \infty\right) = o(\delta(Q,P)^2) \tag{18}$$

As in Pfanzagl (1982), pg. 66, if g_P is a strong gradient, then g_P is a gradient.

Since for nonlinear functionals there are no unbiased estimators (in our context) we consider for strongly differentiable g a linearization

$$\tilde{g}(Q) = g(P) + \int \dot{g}_P dQ \tag{19}$$

of g which approximates g(Q) for P near to Q w.r.t. δ. For linear functionals $g(Q) = \int f dQ$ as considered in section 2 we have $\tilde{g}(Q) = g(Q)$. By Theorem 1 the MVUE of $\tilde{g}$ in P^n is

$$d_n^*(x) = d_n^*(x,P) = g(P) + \frac{1}{n} \sum_{i=1}^{n} \dot{g}_P(x_i) \ . \tag{20}$$

We can now try to make the MVUE d_n^* to a global estimator by an *adaption-step* namely by replacing the unknown P by a 'good' estimator $P_n = P_n(\cdot,x)$, good means according to (18), that P_n should approach P w.r.t Hellinger distance at a certain order. Define

$$\tilde{d}_n(x) = d_n^*(x,P_n(\cdot,x)) = g(P_n(\cdot,x)) + \frac{1}{n} \sum_{i=1}^{n} \dot{g}(x_i,P_n(\cdot,x)); \tag{21}$$

$\tilde{d}_n$ should be a good estimator for g(P) since by (18) $g(P) \approx g(P_n) + \int \dot{g}(\cdot,P_n)dP \approx \tilde{d}_n(x)$ (21) is identical with the *'improved estimator sequence'* generalizing the improvement procedure based on the Newton-Raphson approximation to the solution of the maximum likelihood equation. This procedure was introduced by Pfanzagl (1982), 11.4.2 and shown in many examples to lead to asymptotically efficient estimators. In contrast to the usual motivation for (21) namely to improve P_n, we come to equation (21) by trying to make a MVUE in P^n (for $\tilde{g}$) to a global estimator. Extension of this idea to the construction for higher order approximations of g and to higher order improvements will be considered a forthcoming paper.

<u>Example</u>. Let $P_0 = \{P \in M^1(\mathbb{R}^1, \mathbb{B}^1), \int x^2 dP = m_2, \int x^4 dP < \infty\}$ and let $g(P) = \int x dP =: m_1(P)$. By Proposition 2 the MVUE for $g(Q^n) = m_1(Q)$ in P^n is

$$d_n^*(x_1,\ldots,x_n) = \frac{1}{n} \sum_{i=1}^{n} (x_i - c^*(x_i^2 - m_2)) \tag{22}$$

with $c^* = \frac{m_3(P)-m_2m_1(P)}{\int(x^2-m_2)^2dP}$, $m_i(P)$ denoting the i-th moment of P. The adaption step would now yield

$$\tilde{d}_n(x_1,\ldots,x_n) = \frac{1}{n}\sum_{i=1}^{n}(x_i - \frac{\hat{m}_3-m_2\hat{m}_1}{\hat{\sigma}_2^2}(x_i^2 - m_2)) \tag{23}$$

where $\hat{m}_i = \frac{1}{n}\sum_{k=1}^{n} x_k^i$, $\hat{\sigma}_2^2 = \frac{1}{n}\sum_{i=1}^{n}(x_i^2 - m_2)^2$

are the corresponding sample moment estimators of the moments of the distributions.

REFERENCES

Barankin, E.W. (1949). Locally best unbiased estimates. *Ann. Math. Statist.* 10, 477-501.

Basu, D. (1955). A note on the theory of unbiased estimation. *Ann. Math. Statist.* 26, 144-145.

Hoeffding, W. (1977). Some incomplete and boundedly complete families of distributions. *Ann. Statist.* 5, 278-291.

Lehmann, E.L. (1983). *Theory of Point Estimation.* Wiley, New York.

Pfanzagl, J. (1982). *Contributions to a General Asymptotic Statistical Theory.* Lecture Notes in Statistics 13, Springer.

Parzen, E. (1959). Statistical inference on time series by Hilbert space methods I. In: *Time Series Analysis Papers, ed. by E. Parzen, Holden Day.* 1967, pg. 251-382.

Portnoy, S. (1977). Asymptotic efficiency of minimum variance unbiased estimators. *Ann. Statist.* 5, 522-529.

Pukelsheim, F. (1981). On L_p-differentiable distributions, exponential families and the Cramer-Rao Inequality. Unpublished manuscript.

Sharma, D. (1973). Asymptotic equivalence for two estimators for an exponential family. *Ann. Statist.* 1, 973-980.

Smith, P.J. (1969). Structure of nonparametric tests of some multivariate hypotheses. Ph. D. thesis, Case Western Reserve University.

Stein, C. (1950). Unbiased estimates of minimum variance. *Ann. Math. Statist.* 21, 406-415.

Witting, H. (1985). *Mathematische Statistik I*. Teubner-Verlag.

Ludger Rüschendorf
Institut für Mathematische
Statistik der Universität Münster
Einsteinstraße 62
D-4400 Münster, BRD

Proc. of the 5th Pannonian Symp. on Math. Stat., Visegrád, Hungary 1985
W. Grossmann, J. Mogyoródi, I. Vincze, W. Wertz, eds

SHRINKAGE LEAST SQUARES ESTIMATION IN A GENERAL MULTIVARIATE LINEAR MODEL*

A.K. MD. E. SALEH, P.K. SEN
Ottawa Chapel Hill

For a general (possibly, non-normal) multivariate linear model, under the plausibility of a redundant subset of parameters, James-Stein rule based shrinkage least squares estimation theory is considered. Relative risk-efficiencies of the classical least squares estimators and their shrinkage as well as preliminary test versions are studied (in the light of their asymptotic distributional risks), and it is shown that none of the adaptive versions dominates over the other, though each of them fares well relative to the classical version.
AMS subject lassifications: 62C15, 62F12, 62J07

1. INTRODUCTION

For a p × n random matrix $\underset{\sim}{Y}_n$, consider a general multivariate linear model:

$$\underset{\sim}{Y}_n = \underset{\sim}{\beta}\underset{\sim}{A}_n + \underset{\sim}{E}_n;\ \underset{\sim}{E}_n = (\underset{\sim}{e}_1,\dots,\underset{\sim}{e}_n), \tag{1.1}$$

where the $\underset{\sim}{e}_i$ are independent and identically distributed random vectors (i.i.d.r.v.) with a distribution function F, defined on the Euclidean space R^p (for some $p \geq 1$), $\underset{\sim}{A}_n = (\underset{\sim}{a}_1,\dots,\underset{\sim}{a}_n)$ is a $q \times n$ matrix of known (regression) constants and $\underset{\sim}{\beta}$ is a $p \times q$ matrix of unknown (regression) parameters. It is assumed that $\underset{\sim}{E}_n$ has null mean and F admits a positive definite (p.d.), unknown dispersion matrix $\underset{\sim}{\Sigma}$. For suitable nonnegative integers q_1, q_2, consider the partition:

*This work was supported by a Grant from NSERC (Canada) A3088 and partially by the U. S. National Institutes of Health, Contract No. NIH-NHLBI-71-2243-L.

$$\underset{\sim}{\beta} = (\underset{p\times q_1}{\underset{\sim}{\beta}_1}, \underset{p\times q_2}{\underset{\sim}{\beta}_2}) \text{ and } A'_n = (\underset{n\times q_1}{A'_{1n}}, \underset{n\times q_2}{A'_{2n}});\ q=q_1+q_2 , \tag{1.2}$$

so that (1.1) may be written as

$$\underset{\sim}{Y}_n = \underset{\sim}{\beta}_1\underset{\sim}{A}_{1n} + \underset{\sim}{\beta}_2\underset{\sim}{A}_{2n} + \underset{\sim}{E}_n \quad . \tag{1.3}$$

We are primarily interested in the estimation of $\underset{\sim}{\beta}_1$ when $\underset{\sim}{\beta}_2$, although unknown, is suspected to be close to a pivot $\overset{o}{\underset{\sim}{\beta}}_2$ (say); without any loss of generality, we may take $\overset{o}{\underset{\sim}{\beta}}_2$ as $\underset{\sim}{0}$. For the model (1.1), let $\tilde{\underset{\sim}{\beta}}_n = (\tilde{\underset{\sim}{\beta}}_{1n}, \tilde{\underset{\sim}{\beta}}_{2n})$ be the (*unrestrained*) *least squares estimator* (ULSE) of $\underset{\sim}{\beta}$. Similarly, for the model (1.3) with $\underset{\sim}{\beta}_2 = \underset{\sim}{0}$, let $\hat{\underset{\sim}{\beta}}_{1n}$ be the (*restrained*) *RLSE* of $\underset{\sim}{\beta}_1$. Generally, under usual quadratic loss functions, when $\underset{\sim}{\beta}_2 = \underset{\sim}{0}$ (or is close to this pivot), $\hat{\underset{\sim}{\beta}}_{1n}$ has smaller risk than $\tilde{\underset{\sim}{\beta}}_{1n}$.

But, as $\underset{\sim}{\beta}_2$ moves away from $\underset{\sim}{0}$, $\hat{\underset{\sim}{\beta}}_{1n}$ may become considerably biased and inefficient too, while the performance of $\tilde{\underset{\sim}{\beta}}_{1n}$ remains unaffected by any departure of $\underset{\sim}{\beta}_2$ from $\underset{\sim}{0}$. In this context, a *preliminary test* (PT) for the null hypothesis $H_0:\underset{\sim}{\beta}_2=\underset{\sim}{0}$ may be incorporated in deciding on the choice between the restrained and the unrestrained LSE's, and such an adaptive estimator is termed a *preliminary test estimator* (PTE). The theory of PTLSE (in the general multivariate case) has been systematically studied in Saleh and Sen (1984, 1985a-d). Whereas the PTLSE is a convex combination of $\tilde{\underset{\sim}{\beta}}_{1n}$ and $\hat{\underset{\sim}{\beta}}_{1n}$ with (stochastic) binary coefficient (based on the PT), a smoother version (incorporating the PT statistic) is a natural contender, and we intend to concentrate here mainly on this *shrinkage LSE* (SLSE).

Incorporating the James-Stein (1961) rule, for simple non-normal as well as nonparametric models, shrinkage estimation theory has recently been considered by Sen (1984), Sen and Saleh (1985) and Saleh and Sen (1985b), among others. Identifying a shrinking neighbourhood of the null hypothesis, the asymptotic dominance of the shrinkage estimator over its classical counterpart has been established under quite general regularity conditions. Our main interest is to concentrate on the SLSE in this general multivariate linear model and to compare the situations with RLSE and PTLSE. Along with the preliminary notions, the proposed SLSE are presented in Section 2. The concept of *asymptotic distributional risk* (ADR) is laid down in Section 3, and in the light of this, the ADR of the different versions of LSE are studied there. The concluding section is devoted to the study of the asymptotic dominance of the estimators where the ADR results play the fundamental role.

2. THE PROPOSED SLSE

Note that for the model (1.1), the LSE of β is given by

$$\tilde{\beta}_n = (\tilde{\beta}_{1n}, \tilde{\beta}_{2n}) = (Y_n A_n')(A_n A_n')^{-1} = (Y_n A_n') C_n^{-1}; \tag{2.1}$$

$$C_n = A_n A_n' .$$

In this context, we assume that C_n is of full rank q (<n), and partition C_n as

$$C_n = \begin{pmatrix} C_{n,11} & C_{n,12} \\ C_{n,21} & C_{n,22} \end{pmatrix} = \begin{pmatrix} A_{n1}A_{n1}' & A_{n1}A_{n2}' \\ A_{n2}A_{n1}' & A_{n2}A_{n2}' \end{pmatrix}, \tag{2.2}$$

where $C_{n,jj}$ is of rank q_j, for j=1,2, and $q=q_1 + q_2$. Then, the *restricted LSE* of β_1 is given by

$$\hat{\beta}_{1n} = (Y_n A_{n1}') C_{n,11}^{-1} . \tag{2.3}$$

Note that an unbiased estimator of Σ (irrespective of H_0 being true of not) is

$$S_e^* = (n - q)^{-1} (Y_n - \tilde{\beta}_n A_n)(Y_n - \tilde{\beta}_n A_n)' . \tag{2.4}$$

Note also that for F being a multi-normal d.f., these LSE agree with the corresponding maximum likelihood estimators (MLE) and $(n - q)S_e^*$ (= S_e, say) has the Wishart distribution with n - q degrees of freedom (DF) and associated matrix Σ. In this case, for testing the null hypothesis H_0: $\underset{\sim}{\beta}_2 = \underset{\sim}{0}$ (against $\underset{\sim}{\beta}_2 \neq \underset{\sim}{0}$), the (normal theory) likelihood ratio test in based on the test statistic

$$L_{n1} = \{|S_e|/|S_e + S_H|\}^{n/2} , \tag{2.5}$$

where

$$\underset{\sim}{S}_H = \underset{\sim}{\tilde{\beta}}_{2n}\underset{\sim}{C}_{n,22.1}\underset{\sim}{\tilde{\beta}}'_{2n}$$

(2.6)

and

$$\underset{\sim}{C}_{n,22.1} = \underset{\sim}{C}_{n,22} - \underset{\sim}{C}^{-1}_{n,2}\ \underset{\sim}{C}_{n,11}\underset{\sim}{C}_{n,12}$$

For normal F, the exact distribution of L_{n1} (and $\mathcal{L}_n = -2 \log L_{n1}$), under H_0, as well as some asymptotic expansions are considered in Anderson (1958). For F not necessarily (multi-)normal, but possessing finite moments upto the second order (at least), the asymptotic distribution theory of $\mathcal{L}_n$ (under H_0 as well as local alternatives) has been worked out in Sen and Puri (1970). It follows that under H_0, $\mathcal{L}_n$ has closely the central chi square d.f. with pq_2 DF, so that the critical value of $\mathcal{L}_n$ (at the level significance α: $0 < \alpha < 1$) may be approximated by $\chi^2_{pq_2,\alpha}$, the 100α% point of the chi square d.f. with pq_2DF.

Note that by (2.1) - (2.3), $\underset{\sim}{\tilde{\beta}}_{1n} = \underset{\sim}{\tilde{\beta}}_{1n} + \underset{\sim}{\tilde{\beta}}_{2n}\underset{\sim}{C}_{n,21}\underset{\sim}{C}'^{-1}_{n,11}$, so that, the PTLSE, considered by Saleh and Sen (1984), is given by

$$\underset{\sim}{\tilde{\beta}}^{PT}_{1n} = \underset{\sim}{\tilde{\beta}}_{1n} I(C_n \leq \chi^2_{pq_2,\alpha}) + \underset{\sim}{\tilde{\beta}}_{1n} I(C_n > \chi^2_{pq_2,\alpha})$$

(2.7)

$$= \underset{\sim}{\tilde{\beta}}_{1n} + \underset{\sim}{\tilde{\beta}}_{2n}\underset{\sim}{C}_{n,21}\underset{\sim}{C}^{-1}_{n,}\ \ I(C_n \leq \chi^2_{pq_2,\alpha})$$

where I(A) stands for the indicator function of the set A. Various properties of $\underset{\sim}{\tilde{\beta}}_n$, $\underset{\sim}{\hat{\beta}}_{1n}$ and $\underset{\sim}{\hat{\beta}}^{PT}_{1n}$ have been studied by Saleh and Sen (1984, 1985a). In passing, we may remark that these estimators are defined for all possible values of p, q_1 and q_2 (i.e., without requiring that pq_2 is ≥ 3).

To introduce the SLSE, first, we consider a variant form of the James and Stein (1961) type estimator of $\underset{\sim}{\beta}_2$, where we tacitly assume that $pq_2 \geq 3$. Define

$$\hat{\underset{\sim}{\beta}}{}_{2n}^{JS} = \{1 - \tau C_n^{-1}\}\tilde{\underset{\sim}{\beta}}_{2n} , \tag{2.8}$$

where τ is a suitable positive number to be defined more specifically later on. Note that (2.8) attempts to shrink the LSE of $\underset{\sim}{\beta}_2$ towards the pivot $(\underset{\sim}{0})$. We propose the following SLSE of $\underset{\sim}{\beta}_1$:

$$\hat{\underset{\sim}{\beta}}{}_{1n}^{S} = \hat{\underset{\sim}{\beta}}_{1n} - \hat{\underset{\sim}{\beta}}{}_{2n}^{JS}\underset{\sim}{C}_{n,21}\underset{\sim}{C}_{n,11}^{-1} = \underset{\sim}{\beta}_{1n} + \underset{\sim}{C}_n^{-1}\tilde{\underset{\sim}{\beta}}_{2n}\underset{\sim}{C}_{n,21}\underset{\sim}{C}_{n,11}^{-1}. \tag{2.9}$$

The basic difference between the PTLSE in (2.7) and SLSE in (2.9) is that, in the first case, we have a zero-one valued indicator function, while, here, we have a smooth factor, both depending on the test statistics $\mathcal{L}_n$; however, both of these try to adjust for $\underset{\sim}{\beta}_2$ (being possibly different from $\underset{\sim}{0}$) on the basis of the test statistic $\mathcal{L}_n$. For large values of $\mathcal{L}_n$, both the PTLSE and SLSE are stochastically equivalent to the classical LSE $\tilde{\underset{\sim}{\beta}}_{1n}$, but for $\mathcal{L}_n$ not so large, they perform differently. Now, under H_0: $\underset{\sim}{\beta}_2 = \underset{\sim}{0}$, or under local alternatives, $\mathcal{L}_n$ is small (with a positive probability), while, for any fixed $\underset{\sim}{\beta}_2 \neq \underset{\sim}{0}$, as n becomes large, $\mathcal{L}_n$ increases indefinitely, in probability. Thus, for our subsequent analysis, we shall confine ourselves to local alternatives (containing H_0 as a special case), and under such a scheme, study the comparative performances of these estimators.

3. ASYMPTOTIC DISTRIBUTIONAL RISK OF THE ESTIMATORS

For an estimator $\underset{\sim}{\beta}_{1n}^{*}$ of $\underset{\sim}{\beta}_1$, we consider a *quadratic error loss function*

$$L(\underset{\sim}{\beta}_{1n}^{*}; \underset{\sim}{\beta}_1) = n(\overset{o}{\underset{\sim}{\beta}}{}_{1n}^{*} - \overset{o}{\underset{\sim}{\beta}}_1)' \overset{o}{\underset{\sim}{W}}(\overset{o}{\underset{\sim}{\beta}}{}_{1n}^{*} - \overset{o}{\underset{\sim}{\beta}}_1), \tag{3.1}$$

where $\overset{o}{\underset{\sim}{W}}$ is a $pq_1 \times pq_1$ matrix of full rank, and $\overset{o}{\underset{\sim}{\beta}}{}_{1n}^{*}$ and $\underset{\sim}{\beta}_1^{*}$ are the *rolled out* vector forms of $\underset{\sim}{\beta}_{1n}^{*}$ and $\underset{\sim}{\beta}_{.}$, respectively as given in Anderson (1984, page 290). The *risk* (= expected loss) for the estimator is therefore equal to

$$R(\underset{\sim}{\beta}^{*}_{1n};\ \underset{\sim}{W}^{o}) = EL(\underset{\sim}{\beta}^{*}_{1n};\ \underset{\sim}{\beta}_1) = n\mathrm{Tr}(\underset{\sim}{W}^{o} E[(\overset{o}{\underset{\sim}{\beta}^{*}_{1n}} - \overset{o}{\underset{\sim}{\beta}_1})(\overset{o}{\underset{\sim}{\beta}^{*}_{1n}} - \overset{o}{\underset{\sim}{\beta}_1})']). \quad (3.2)$$

Now, for the ULSE, RLSE and PTLSE, we may use (3.2) to compute the risk and compare them. This has been done in Saleh and Sen (1984, 1985a). The situation is somewhat different for the SLSE. The presence of $\mathcal{L}_n^{-1}$ in (2.9) raises the issue of the existence of second order moments of the estimator $\hat{\underset{\sim}{\beta}}^{S}_{1n}$. For normal F, $\mathcal{L}_n$ has a specific distribution which allows the computation of negative moments when pq_2 is ≥ 3. But, for an arbitrary F, the computation of the moments of $\mathcal{L}_n^{-1}$ may need stringent conditions which may be avoided by the alternative approach sketched below. From the results of Saleh and Sen (1984, 1985a), it follows that under quite general regularity conditions (for F admitting finite second order moments), $n^{\frac{1}{2}}(\tilde{\underset{\sim}{\beta}}_{1n} - \underset{\sim}{\beta}_1)$, $n^{\frac{1}{2}}(\hat{\underset{\sim}{\beta}}_{1n} - \underset{\sim}{\beta}_1)$, $n^{\frac{1}{2}}(\hat{\underset{\sim}{\beta}}^{PT}_{1n} - \beta_1)$ have all appropriate asymptotic distributions, and we shall see later on that the same is true for $n^{\frac{1}{2}}(\hat{\underset{\sim}{\beta}}^{S}_{1n} - \underset{\sim}{\beta}_1)$. As such, we may compute the risk of an estimator from its asymptotic distribution; such a mea-measure will be termed an *asymptotic distributional risk* (ADR). Under some additional regularity conditions, the ADR results agree with the asymptotic risk results, and, in general, these ADR results may be studied much more conveniently. As such, we shall mainly confine ourselves to the study of this ADR for the various estimators under consideration and pursue the *asymptotic dominance* of the SLSE and PTLSE on the others.

As has been mentioned before, for any fixed $\underset{\sim}{\beta}_2 \neq \underset{\sim}{0}$, the PTLSE and SLSE are both stochastically equivalent to $\underset{\sim}{\beta}_{1n}$. To avoid such limiting degeneracy and to obtain some meaningful results, we confine ourselves to a sequence $\{K_n\}$ of alternative hypotheses, where

$$K_n : \underset{\sim}{\beta}_2 = \underset{\sim}{\beta}_{2(n)} = n^{-\frac{1}{2}} \cdot \underset{\sim}{\gamma}_2;\ \underset{\sim}{\gamma}_2 = ((\gamma_{ij}))_{i=1,\dots,p;\,j=q_1+1,\dots,q}. \quad (3.3)$$

The null hypothesis relates to $\underset{\sim}{\gamma}_2 = 0$.

For a suitable estimator $\underset{\sim}{\beta}^*_{1n}$ of $\underset{\sim}{\beta}_1$, we denote by

$$G^*(\underset{\sim}{x}) = \lim_{n\to\infty} P\{n^{\frac{1}{2}}(\underset{\sim}{\beta}^{*o}_1 - \underset{\sim}{\beta}^o_1) \le \underset{\sim}{x} | K_n\}, \ \underset{\sim}{x} \in R^{pq_1}, \tag{3.4}$$

(and assume that G^* exists and is nondegenerate). Then, in (3.2), we may replace the actual mean product matrix by the mean product matrix for the d.f. G^* and define the ADR of $\underset{\sim}{\beta}^*_{1n}$ as

$$R(\underset{\sim}{\beta}^*_1; \underset{\sim}{W}^o) = \mathrm{Tr}\{\underset{\sim}{W}^o \int\ldots\int \underset{\sim}{x}\underset{\sim}{x}' dG^*(\underset{\sim}{x})\} = \mathrm{Tr}\{\underset{\sim}{W}^c \underset{\sim}{V}^*\}, \text{ say.} \tag{3.5}$$

Our primary goal is to study the ADR of all the ULSE, RLSE, PTLSE and SLSE, and to compare them in a meaningful way. In this context, we say that an estimator $\underset{\sim}{\beta}^{**}_{1n}$ *asymptotically dominates* over an alternative one $\underset{\sim}{\beta}^*_{1n}$, if

$$R(\underset{\sim}{\beta}^*_1; W^o) \ge R(\underset{\sim}{\beta}^{**}_1; W^o), \ (\underset{\sim}{\beta}_1, \underset{\sim}{\gamma}_2) \in R^{pq}, \tag{3.6}$$

with the strict inequality holding for some nonnull subset of R^{pq}. If for an estimator $\underset{\sim}{\beta}^{**}_{1n}$, (3.6) holds for every other estimator, then in the light of the proposed ADR, we say that $\underset{\sim}{\beta}^{**}_{1n}$ is an *asymptotically admissible estimator* of $\underset{\sim}{\beta}_1$.

For the study of the ADR results, we assume that the following regularity conditions hold:

$$\text{(i) } \max\{\underset{\sim}{a}_i' \underset{\sim}{C}_n^{-1} a_i : 1 \le i \le n\} \to 0 \quad \text{as } n \to \infty, \tag{3.7}$$

$$\text{(ii) } \lim_{n\to} n^{-1} \underset{\sim}{C}_n = \underset{\sim}{C} \quad \text{is p.d.,} \tag{3.8}$$

$$\text{and (iii) } \underset{\sim}{\Sigma} \text{ is p.d. with } \mathrm{Tr}(\underset{\sim}{\Sigma}) \in (0,\infty). \tag{3.9}$$

As in (2.2), we partition $\underset{\sim}{C}$ as $((\underset{\sim}{C}_{ij}))_{i,j=1,2}$ and also denote by $\underset{\sim}{C}_{ii.j} = \underset{\sim}{C}_{ii} - \underset{\sim}{C}_{ij}\underset{\sim}{C}_{jj}^{-1}\underset{\sim}{C}_{ji}$, for i,j = 1,2. Also, note that by (1.1) and (2.1),

$$n^{\frac{1}{2}}(\tilde{\underset{\sim}{\beta}}_n - \underset{\sim}{\beta}) = n^{\frac{1}{2}} \sum_{i=1}^{n} (\underset{\sim}{e}_i \underset{\sim}{a}_i')(n\underset{\sim}{C}_n^{-1}) , \tag{3.10}$$

where the $\underset{\sim}{e}_i$ are i.i.d.r.v. with finite second moments, so that by (3.7) through (3.9) and the classical (multivariate) central limit theorem,

$$n^{\frac{1}{2}}(\tilde{\underset{\sim}{\beta}}_n^o - \underset{\sim}{\beta}^o) \sim N_{pq}(\underset{\sim}{0}, \underset{\sim}{\Sigma} \otimes \underset{\sim}{C}^{-1}) , \tag{3.11}$$

where $\otimes$ stands for the Kronecker product of the two adjacent matrices. Note that (3.11) also holds under $\{K_n\}$, where for $\underset{\sim}{\beta}_2$, we need to substitute $n^{-\frac{1}{2}}\underset{\sim}{\gamma}_2$. Also, under (3.7) through (3.9) [viz., Sen and Puri (1970)], S_e^*, defined in (2.4), converges in probability to $\underset{\sim}{\Sigma}$, as $n \to \infty$, and

$$\mathcal{L}_n - \mathrm{Tr}(\underset{\sim}{S}_e^{*-1} \underset{\sim}{S}_H) \to 0, \text{ in probability, under } \{K_n\} \text{ as well as } H_0. \tag{3.12}$$

Therefore, by an appeal to Theorem 2.2 of Sen and Puri (1970), we conclude that

$$P\{\mathcal{L}_n \le x | K_n\} = H_{pq_2}(x;\delta_2);\ \delta = \mathrm{Tr}(\underset{\sim}{\Gamma}_2^o(\underset{\sim}{\Sigma}^{-1} \otimes \underset{\sim}{C}_{22.1})) \tag{3.13}$$

where $H_r(x;\ \delta)$ stands for the noncentral chi square d.f. with r DF and noncentrality parameter δ (≥ 0), and $\underset{\sim}{\Gamma}_2^o = ((\underset{\sim}{\gamma}_2^o\underset{\sim}{\gamma}_2^{o\prime}))$.

The following results follow directly from Saleh and Sen (1984, 1985a) under the regularity conditions (3.6) through (3.9) and under $\{K_n\}$:

$$R(\tilde{\underset{\sim}{\beta}}_{1n};\ \underset{\sim}{W}^o) = \mathrm{Tr}[\underset{\sim}{W}^o(\underset{\sim}{\Sigma} \otimes \underset{\sim}{C}_{11.2}^{-1})] \tag{3.14}$$

$$R(\hat{\underset{\sim}{\beta}}_{1n};\ \underset{\sim}{W}^o) = \mathrm{Tr}[\underset{\sim}{W}^o(\underset{\sim}{\Sigma} \otimes \underset{\sim}{C}_{11}^{-1})] + \underset{\sim}{\gamma}_2^{o\prime}W^o\underset{\sim}{\gamma}_2^o, \tag{3.15}$$

$$\begin{aligned} R(\hat{\underset{\sim}{\beta}}_{1n}^{PT};\ \underset{\sim}{W}^o) =& \mathrm{Tr}[\underset{\sim}{W}^o(\underset{\sim}{\Sigma} \otimes \underset{\sim}{C}_{11.2}^{-1})][1-H_{pq_2+2}(\chi_{pq_2,\alpha}^2;\ \delta_2)]+ \\ &+ \mathrm{Tr}[\underset{\sim}{W}^o(\underset{\sim}{\Sigma} \otimes \underset{\sim}{C}_{11}^{-1})]H_{pq_2+2}(\chi_{pq_2,\alpha}^2;\ \delta_2) \\ &+ (\underset{\sim}{\gamma}_2^{o\prime}W^o\underset{\sim}{\gamma}_2^o)[2H_{pq_2+2}(\chi_{pq_2,\alpha}^2;\ \delta_2) - \\ &- H_{pq_2+4}(\chi_{pq_2,\alpha}^2;\ \delta_2)] . \end{aligned} \tag{3.16}$$

Now, by virtue of (2.5), (3.6), (3.12) and the stochastic convergence of S_e^* to Σ, we conclude that under the assumed regularity conditions, under $\{K_n\}$,

$$\mathcal{L}_n - \mathrm{Tr}(\Sigma^{-1}(\tilde{\beta}_n C_n \tilde{\beta}_n')) \to 0, \text{ in probability, as } n\to\infty, \tag{3.17}$$

where

$$C_n^* = \begin{pmatrix} 0 & 0 \\ 0 & C_{n,22.1} \end{pmatrix} \tag{3.18}$$

so that by (2.9), (3.11) and (3.17), we obtain that under $\{K_n\}$,

$$\begin{aligned} n^{\frac{1}{2}}(\hat{\beta}_{1n}^S - \beta_1) &\overset{\mathcal{D}}{\to} UD_1 + \tau(UD_2+\gamma_2)C_{21}C_{11}^{-1}\ \mathrm{Tr}(\Sigma^{-1}[UD_2+\gamma_2)C_{22.1}(UD_2+\gamma_2)'])\}^{-} \\ &= UD_1 + \tau(UD_2^*+\gamma_2^*)\{\mathrm{Tr}(\Sigma^{-1}[(UD_2+\gamma_2)C_{22.1}(UD_2+\gamma_2)'])\}^{-1}, \end{aligned} \tag{3.19}$$

where

$$U(p \times q) \sim N_{pq}(0, \Sigma \otimes C), \tag{3.20}$$

$$D = (D_1 D_2) = C^{-1},\ D_2^* = D_2 C_{21} C_{11}^{-1} \text{ and } \gamma_2^* = \gamma_2 C_{21} C_{11}^{-1}. \tag{3.21}$$

For the computation of the ADR, we need to roll out the matrices in (3.19) in vector forms; this leads us to

$$\begin{aligned} n^{\frac{1}{2}}(\hat{\beta}_{1n}^S - \beta_1)^o &\overset{\mathcal{D}}{\to} (UD_1)^o + \\ &+ \tau(UD_2^*+\gamma_2^*)^o\{\mathrm{Tr}(\Sigma^{-1}[(UD_2+\gamma_2)C_{22.1}(UD_2+\gamma_2)'])\}^{-1} \end{aligned} \tag{3.22}$$

so that the ADR for the SLSE is given by

$$\begin{aligned} R(\hat{\beta}_1^S; W^o) &= E\{(UD_1)^{o\prime} W^o (UD_1)^o\} + \\ &+ 2\tau E\{(UD_2^*+\gamma_2^*)^{o\prime} W^o (UD_1)^o / \mathrm{Tr}(\Sigma^{-1}[(UD_2+\gamma_2)C_{22.1}(UD_2+\gamma_2)'])\} + \end{aligned}$$

$$+ \tau^2 E \left\{ \frac{(\underset{\sim}{U}\underset{\sim}{D}_2^* + \underset{\sim}{\gamma}_2^*)^{o\prime} \underset{\sim}{W}^o (\underset{\sim}{U}\underset{\sim}{D}_2^* + \underset{\sim}{\gamma}_2^*)^o}{[Tr(\underset{\sim}{\Sigma}^{-1}[(\underset{\sim}{U}\underset{\sim}{D}_2 + \underset{\sim}{\gamma}_2)\underset{\sim}{C}_{22.1}(\underset{\sim}{U}\underset{\sim}{D}_2 + \underset{\sim}{\gamma}_2)'])]^2} \right\} \tag{3.23}$$

It may be remarked that by (3.20) and (3.21),

$$\begin{aligned} E[(\underset{\sim}{U}\underset{\sim}{D}_1)^{o\prime} \underset{\sim}{W}^o (\underset{\sim}{U}\underset{\sim}{D}_1)^o] &= TR\{E[(\underset{\sim}{U}\underset{\sim}{D}_1)^o (\underset{\sim}{U}\underset{\sim}{D}_1)^{o\prime}]\underset{\sim}{W}^o\} \\ &= Tr\{(\underset{\sim}{\Sigma} \boxtimes \underset{\sim}{D}_1' \underset{\sim}{C}\underset{\sim}{D}_1)\underset{\sim}{W}^o\} = Tr\{(\underset{\sim}{\Sigma} \boxtimes \underset{\sim}{C}_{11.2}^{-1})\underset{\sim}{W}^o\} \end{aligned} \tag{3.24}$$

Further, we may partition $\underset{\sim}{U}$ as $(\underset{\sim}{U}_1, \underset{\sim}{U}_2)$ where $\underset{\sim}{U}_j$ is $q \times q_j$, $j=1,2$, and write

$$\underset{\sim}{D}_1 = \begin{pmatrix} \underset{\sim}{D}_{11.2} \\ \underset{\sim}{0} \end{pmatrix} + \underset{\sim}{D}_2 \underset{\sim}{D}_{22}^{-1} \underset{\sim}{D}_{21}, \text{ where } \underset{\sim}{D}_{22}^{-1} \underset{\sim}{D}_{21} = \underset{\sim}{C}_{21} \underset{\sim}{C}_{11}^{-1}, \tag{3.25}$$

and note that, conditional on $\underset{\sim}{U}\underset{\sim}{D}_2$ being held fixed, $\underset{\sim}{U}\underset{\sim}{D}_2^*$ is held fixed, while the $p \times q_1$ matrix $\underset{\sim}{U}_1 \underset{\sim}{D}_{11.2}$ has the pq -multinormal distribution with null mean matrix and a p.d. dispersion matrix. Hence, the *second term* on the right hand side of (3.23) may be written as

$$\begin{aligned} &-2\tau E\{(\underset{\sim}{U}\underset{\sim}{D}_2 + \underset{\sim}{\gamma}_2^*)^{o\prime} \underset{\sim}{W}^o (\underset{\sim}{U}\underset{\sim}{D}_2^*)^o / Tr(\underset{\sim}{\Sigma}^{-1}[(\underset{\sim}{U}\underset{\sim}{D}_2 + \underset{\sim}{\gamma}_2)\underset{\sim}{C}_{22.1}(\underset{\sim}{U}\underset{\sim}{D}_2 + \underset{\sim}{\gamma}_2)'])\} = \\ &= -2\tau E\{[((\underset{\sim}{U} + \underset{\sim}{\gamma}_o)\underset{\sim}{D}_2)^{o\prime} \underset{\sim}{W}^o (\underset{\sim}{U}\underset{\sim}{D}_2)^o] / Tr(\underset{\sim}{\Sigma}^{-1}[(\underset{\sim}{U} + \underset{\sim}{\gamma}_o)\underset{\sim}{D}_2 \underset{\sim}{C}_{22.1} \underset{\sim}{D}_2' (\underset{\sim}{U} + \underset{\sim}{\gamma}_o)'])\} \end{aligned} \tag{3.26}$$

where

$$\underset{\sim}{\gamma}_2 = \underset{\sim}{\gamma}_o \underset{\sim}{D}_2, \text{ so that } \underset{\sim}{\gamma}_2^* = \underset{\sim}{\gamma}_o \underset{\sim}{D}_2 \underset{\sim}{D}_{21} \underset{\sim}{C}_{11}^{-1} = \underset{\sim}{\gamma}_o \underset{\sim}{D}_2^* . \tag{3.27}$$

Note that if we define a pq-vector $\underset{\sim}{Z}$ by letting

$$\underset{\sim}{Z} = (\underset{\sim}{\Sigma} \boxtimes \underset{\sim}{C})^{-\frac{1}{2}} \{\underset{\sim}{U}^o + \underset{\sim}{\gamma}_o^o\} \text{ (so that } \underset{\sim}{Z} \sim N((\underset{\sim}{\Sigma} \boxtimes \underset{\sim}{C})^{-\frac{1}{2}} \underset{\sim}{\gamma}_o^o, \underset{\sim}{I}_p \boxtimes \underset{\sim}{I}_q)) \tag{3.28}$$

then, (3.26) may be written as

$$-2\tau E\{(Z'MZ)/(Z'M^{*}Z)-(Z'M\gamma^{o})/(Z'M^{*}Z)\};\ \gamma^{o}=(\Sigma\otimes C)^{-\frac{1}{2}}\gamma^{o}_{o}\ , \quad (3.29)$$

where both $\underset{\sim}{M}$ and $\underset{\sim}{M}^{*}$ are $pq\times pq$ positive-semi-definite matrices ; it is easy to verify that $\underset{\sim}{M}^{*}$ is an idempotent matrix of rank pq_2, so that $Tr(\underset{\sim}{M}^{*}) = pq_2$ [Note $\underset{\sim}{M}^{*} = I_p \times \underset{\sim}{C}^{\frac{1}{2}}\underset{\sim}{D}_2\underset{\sim}{C}_{22.1}\underset{\sim}{D}_2'\underset{\sim}{C}^{\frac{1}{2}}$]. Thus

$$\underset{\sim}{Z}'\underset{\sim}{M}^{*}\underset{\sim}{Z} \sim \chi^2_{pq_2}(\delta_2),\ \delta_2 = Tr[\Gamma_2^{o}(\underset{\sim}{\Sigma}^{-1} \otimes \underset{\sim}{C}_{22.1})] \quad (3.30)$$

and related distribution is defined in (3.13). Further

$$\underset{\sim}{M} = (\underset{\sim}{\Sigma}^{\frac{1}{2}} \otimes \underset{\sim}{C}^{\frac{1}{2}}\underset{\sim}{D}_2)\underset{\sim}{W}^{o}(\underset{\sim}{\Sigma}^{\frac{1}{2}} \otimes \underset{\sim}{C}^{\frac{1}{2}}\underset{\sim}{D}_2^{*})'. \quad (3.31)$$

The last term on the right hand side of (3.23) may also be written as

$$\tau^2 E\{(Z'MZ)/(Z'M^{*}Z)^2\} \quad (3.32)$$

Thus, by (3.23), (3.24), (3.32) and (3.33), we obtain

$$R(\underset{\sim}{\hat{\beta}}^{S}_{1n};\ \underset{\sim}{W}^{o}) = Tr[(\underset{\sim}{\Sigma} \otimes \underset{\sim}{C}^{-1}_{11.2})\underset{\sim}{W}^{o}]+\tau^2 E\{(Z'MZ)/(Z'M^{*}Z)^2\}-$$

$$- 2\tau E\{(\underset{\sim}{Z}'\underset{\sim}{M}\underset{\sim}{Z})/(\underset{\sim}{Z}'\underset{\sim}{M}^{*}\underset{\sim}{Z}) - (\underset{\sim}{Z}'\underset{\sim}{M}\underset{\sim}{\gamma}^{o})/(\underset{\sim}{Z}'\underset{\sim}{M}^{*}\underset{\sim}{Z})\} \quad (3.33)$$

In the next section, we shall simplify the above expression by appropriately choosing $\underset{\sim}{W}^{o}$ and determine the shrinkage constant τ. In passing we remark that if $\underset{\sim}{C}_{12}=\underset{\sim}{0}$, then $\underset{\sim}{C}_{11.2}=\underset{\sim}{C}_{11}$ and M is a null matrix. Also $\underset{\sim}{\gamma}_2^{*}=\underset{\sim}{0}$. As such, in this case, (3.14), (3.15), (3.16) and (3.33) are all the same, and, there is no point in the study of relative performance of the estimates. As such, in the sequel, it will be assumed that $\underset{\sim}{C}_{12} \neq \underset{\sim}{0}$.

4. ADR AND ADMISSIBILITY RESULTS

In order to compare the various estimates of $\underset{\sim}{\beta}_1$ in the light of ADR we choose

$$\underset{\sim}{W}^{o} = \underset{\sim}{\Sigma}^{-1} \otimes \underset{\sim}{C}_{11.2}\ . \quad (4.1)$$

This means we are considering Mahalanobis distance as the loss-function while τ is yet to be chosen. With this choice of $\underset{\sim}{W}^{o}$ we obtain

$$R(\tilde{\underset{\sim}{\beta}}_{1n};\ W^{o}) = pq_1\ , \tag{4.2}$$

$$R(\hat{\underset{\sim}{\beta}}_{1n};\ \underset{\sim}{W}^{o}) = p\mathrm{Tr}(\underset{\sim}{C}_{11}^{-1}\underset{\sim}{C}_{11.2}) + \underset{\sim}{\gamma}_2^{o\prime}(\underset{\sim}{\Sigma}^{-1} \otimes \underset{\sim}{C}_{11.2})\underset{\sim}{\gamma}_2^{o}, \tag{4.3}$$

$$\begin{aligned} R(\hat{\underset{\sim}{\beta}}{}^{PT}_{n};\ \underset{\sim}{W}^{o}) = {} & pq_1[1-H_{pq_2+2}(\chi^2_{pq_2,\alpha};\delta_2)] + \\ & + p\mathrm{Tr}(\underset{\sim}{C}_{11}^{-1}\underset{\sim}{C}_{11.2})\ H_{pq_2+2}(\chi^2_{pq_2,\alpha};\delta_2) + \\ & +\{\underset{\sim}{\gamma}_2^{o\prime}(\underset{\sim}{\Sigma}^{-1} \otimes \underset{\sim}{C}_{11.2})\underset{\sim}{\gamma}_2^{o}\}\ [2H_{pq_2+2}(\chi_{pq_2,\alpha};\delta_2) - \\ & - H_{pq_2+4}(\chi_{pq_2,\alpha};\ \delta_2)] \end{aligned} \tag{4.4}$$

Finally,

$$\begin{aligned} R(\hat{\underset{\sim}{\beta}}{}^{S}_{n};\underset{\sim}{W}^{o}) = {} & pq_1+\tau^2\{\mathrm{Tr}(\underset{\sim}{M})E(\chi^{-4}_{pq_2}(\delta_2))+\underset{\sim}{\gamma}^{o\prime}\underset{\sim}{M}\underset{\sim}{\gamma}^{o})E(\chi^{-4}_{pq_2+4}(\delta_2))\} \\ & -2\tau\{\mathrm{Tr}(\underset{\sim}{M})E(\chi^{-2}_{pq_2+2}(\delta_2))-2(\underset{\sim}{\gamma}^{o\prime}\underset{\sim}{M}\underset{\sim}{\gamma}^{o})E(\chi^{-4}_{pq_2+4}(\delta_2))\} \end{aligned} \tag{4.5}$$

where $\chi^2_{pq_2+r}(\delta_2)$ has the d.f. $H_{pq_2+r}(x;\delta_2)$. Note that by (3.31) and using $\underset{\sim}{W}^{o} = \underset{\sim}{\Sigma}^{-1} \otimes C_{11.2}$ we have

$$\underset{\sim}{M} = \underset{\sim}{I}_p \otimes \underset{\sim}{C}^{\frac{1}{2}}\underset{\sim}{D}_2^{*}\underset{\sim}{C}_{11.2}\underset{\sim}{D}_2^{*\prime}C^{\frac{1}{2}}\ . \tag{4.6}$$

Hence,

$$\begin{aligned} \mathrm{Tr}(\underset{\sim}{M}) & = p\mathrm{Tr}[(\underset{\sim}{I} - \underset{\sim}{C}_{22}^{-1}\underset{\sim}{C}_{22.1})\underset{\sim}{D}_2'\underset{\sim}{C}\underset{\sim}{D}_2\underset{\sim}{C}_{22.1}] = \\ & = p\mathrm{Tr}(\underset{\sim}{C}_{22}^{-1}\underset{\sim}{C}_{21}\underset{\sim}{C}_{11}^{-1}\underset{\sim}{C}_{12})\ . \end{aligned} \tag{4.7}$$

Let $q^* = \mathrm{Tr}(C_{22}^{-1}C_{21}C_{11}^{-1}C_{12})$ and that $0 \le q^* \le \min(q_1,q_2)$. (4.8)

Thus, $0 \le \mathrm{Tr}(M) \le p^*$, $p^* = \min(q_1,q_2)$. Now consider

$$\delta_2 = \underset{\sim}{\gamma}_2^{o\prime} \underset{\sim}{W}^o \gamma_2^o = \gamma_2^{o\prime}(\underset{\sim}{\Sigma}^{-1} \otimes \underset{\sim}{C}_{22.1})\underset{\sim}{\gamma}_2^o = \underset{\sim}{\gamma}^{o\prime}[\underset{\sim}{I}_p \otimes \underset{\sim}{C}^{\frac{1}{2}} \underset{\sim}{D}_2' \underset{\sim}{D}_{22.1} \underset{\sim}{D}_2 \underset{\sim}{C}^{\frac{1}{2}}]\underset{\sim}{\gamma}^o =$$

$$= \underset{\sim}{\gamma}^{o\prime} \underset{\sim}{M} \underset{\sim}{\gamma}^o \quad . \tag{4.9}$$

Therefore,

$\underset{\sim}{\gamma}^{o\prime} \underset{\sim}{M} \underset{\sim}{\gamma}^o / \delta_2 = (\underset{\sim}{\gamma}^{o\prime} \underset{\sim}{M} \underset{\sim}{\gamma}^o)/(\underset{\sim}{\gamma}^{o\prime} \underset{\sim}{M}^* \underset{\sim}{M}^* \underset{\sim}{\gamma}^o) \le \mathrm{Ch}_1(\underset{\sim}{M})$ by Courant

theorem on $\underset{\sim}{M}^* \underset{\sim}{\gamma}^o = h_1$, say (4.10)

where $\mathrm{Ch}_1(\underset{\sim}{M})$ is the largest characteristic root of $\underset{\sim}{M}$.

Then, we rewrite the risk of $\hat{\underset{\sim}{\beta}}_{1n}^S$ as

$$R(\hat{\underset{\sim}{\beta}}_n^S; \underset{\sim}{W}^o) = pq_1 - \tau\{2E(\chi_{pq_2+2}^{-2}(\delta_2)) - \tau E(\chi_{pq_2+2}^{-4}(\delta_2))\}\mathrm{Tr}(\underset{\sim}{M}) +$$

$$+ \tau(\tau + 4)(\gamma^{o\prime} \underset{\sim}{M} \underset{\sim}{\gamma}^o) E(\chi_{pq_2+4}^{-4}(\delta_2)) =$$

$$= pq_1 - \tau \mathrm{Tr}(\underset{\sim}{M})[\{2E(\chi_{pq_2+2}^{-2}(\delta_2)) - \tau E(\chi_{pq_2+2}^{-4}(\delta_2))\} -$$

$$- (\tau + 4)\{(\gamma^{o\prime} M \gamma^o)/\mathrm{Tr}(M)\} E(\chi_{pq_2+4}^{-4}(\delta_2)) \tag{4.11}$$

Now, consider the difference $R(\tilde{\underset{\sim}{\beta}}_{1n}; W^o) - R(\tilde{\underset{\sim}{\beta}}_{1n}^S; W^o)$ to compare $\hat{\underset{\sim}{\beta}}_{1n}^S$ and $\tilde{\underset{\sim}{\beta}}_{1n}$:

$$R(\tilde{\underset{\sim}{\beta}}_{1n}; \underset{\sim}{W}^o) - R(\hat{\underset{\sim}{\beta}}_{1n}^S; \underset{\sim}{W}^o) = \tau \mathrm{Tr}(\underset{\sim}{M})[\{2E(\chi_{pq_2+2}^{-2}(\delta_2)) - \tau E(\chi_{pq_2+2}^{-4}))\}$$

$$- (\tau+4)\{(\gamma^{o\prime} \underset{\sim}{M} \underset{\sim}{\gamma}^o)/\delta_2\}\{\delta_2/\mathrm{Tr}(\underset{\sim}{M})\} E(\chi_{pq_2+4}^{-4}(\delta_2)). \tag{4.12}$$

The right-hand-side is bounded from below by

$$pq^* \tau [\{2E(\chi_{pq_2+2}^{-2}(\delta_2)) - \tau E(\chi_{pq_2+2}^{-4}(\delta_2)) - (\tau+4)\frac{h_1}{pq^*} \delta_2 E(\chi_{pq_2+4}^{-4}(\delta_2)) \tag{4.13}$$

Using the relation

$$E(\chi_{pq_2+2}(\delta_2)) - (pq_2 - 2)E(\chi^{-4}_{pq_2+2}(\delta_2)) = \delta_2 E(\chi^{-2}_{pq_2+4}(\delta_2)) \tag{4.14}$$

(4.13) reduces to

$$pq^*[(2-g)E(\chi_{pq_2+2}(\delta_2)) - \{(pq_2-2)g - \tau\}E(\chi_{pq_2+4}(\delta_2))] \tag{4.15}$$

where $g = (\tau + 4)\frac{h_1}{pq^*}$. Note that $h_1 \leq 1$ and $pq^* \leq p \min(q_1, q_2)$ are non-negative numbers. Thus $\frac{h_1}{pq^*} < \frac{1}{pq^*}$. Also, in Stein-Rule estimation, one generally takes $\tau = pq_2 - 2$. If $1 \leq g \leq 2$ and $\tau = pq_2 - 2$, (4.15) is positive. If $0 \leq g \leq 1$, then $2 - g \geq 1$ and $(pq_2 - 2)g - \tau > -\tau$. Thus, (4.15) is bounded from below by

$$\begin{aligned} & pq^*\{E(\chi^{-2}_{pq_2+2}(\delta_2)) - (pq_2 - 2)E(\chi^{-4}_{pq_2+4}(\delta_2))\} \\ & = pq^*[\{E(\chi^{-2}_{pq_2+2}(\delta_2)) - (pq_2 - 2)E(\chi^{-4}_{pq_2+2}(\delta_2))\} + \\ & + E(\chi^{-4}_{pq_2+2}(\delta_2)) - E(\chi^{-4}_{pq_2+4}(\delta_2))\}] = \\ & = pq^*\{\delta_2 E(\chi^{-4}_{pq_2+4}(\delta_2)) + 4E(\chi^{-6}_{pq_2+4}(\delta_2))\} \geq 0. \end{aligned} \tag{4.16}$$

Thus, for $\tau = pq_2 - 2$ and $0 \leq g \leq 2$ implies that

$$R(\tilde{\underset{\sim}{\beta}}_{1n}; \underset{\sim}{W}^o) \geq R(\hat{\underset{\sim}{\beta}}^S_{1n}; \underset{\sim}{W}^o) \quad \underset{\sim}{\gamma}_2 \; . \tag{4.17}$$

This needs

$$h_1/pq^* \leq \frac{2}{\tau + 4} = \frac{1}{1 + \frac{1}{2}pq_2}, \text{ for } \tau = pq_2 - 2 \; . \tag{4.18}$$

Though both h_1 and pq^* are pre-specified numbers 4.18 may not hold in general. In such a case, we may note that $pq^* \geq 1$ (for $pq^* = 0$, there is no point in shrinking). Thus if

$h_1/pq^* \le (1 + \frac{1}{2}pq_2)^{-1}$, we need to consider a small value of $\tau > 0$ for which $g \le 2$ i.e.

$$\tau \le 2pq^*/h_1 - 4 - 2\,(pq^*/h_1 - 2) \tag{4.19}$$

which requires that $pq^*/h_1 > 2$. So that the domain of $\tilde{\underset{\sim}{\beta}}{}^S_{1n}$ rests upon initial design condition ($pq^*/h_1 > 2$). In particular if $pq^* = pq_2$ (i.e. $\underset{\sim}{C}_{22.1}$ is null), then $h_1 = 1$, then for all $pq_2 > 2$, $pq^*/h_1 > 2$. On the other hand, for $pq^* < pq_2$, we need h_1 to be small so that $pq^*/h > 2$.

Now, consider the comparison of $\tilde{\underset{\sim}{\beta}}_{1n}$ and $\tilde{\underset{\sim}{\beta}}{}^{PT}_{1n}$ relative to $\hat{\underset{\sim}{\beta}}{}^S_{1n}$. We have then

$$\begin{aligned} R(\hat{\underset{\sim}{\beta}}{}^S_{1n}; \underset{\sim}{W}^o) - R(\hat{\underset{\sim}{\beta}}_{1n}; \underset{\sim}{W}^o) &= \{pq_1 - p\mathrm{Tr}(\underset{\sim}{C}{}^{-1}_{11}\underset{\sim}{C}_{11.2})\} - \\ &- \tau\{\mathrm{Tr}(\underset{\sim}{M})E(\chi^{-2}_{pq_2+2}(\delta_2)) - (\tau + 4)(\underset{\sim}{\gamma}^{o\prime}\underset{\sim}{M}\underset{\sim}{\gamma}^o)\; E(\chi^{-4}_{pq_2+4}(\delta_2)) - \\ &- \tau\mathrm{Tr}(M)E(\chi^{-4}_{pq_2+2}(\delta_2)) - (\gamma^{o\prime}M^*\gamma^o) \end{aligned} \tag{4.20}$$

Under H_0: $\underset{\sim}{\gamma}_2 = 0$, (4.20) reduces

$$pq_1 - p\mathrm{Tr}(C^{-1}_{11}C_{11.2}) - \tau\mathrm{Tr}(\underset{\sim}{M})\{2(pq_2)^{-1} - \tau(pq_2)^{-1}(pq_2-2)^{-1}\}. \tag{4.21}$$

Note that

$$\begin{aligned} pq_1 - p\mathrm{Tr}(\underset{\sim}{C}{}^{-1}_{11}\underset{\sim}{C}_{11.2}) &= pq_1 - pq_1 + \mathrm{Tr}(\underset{\sim}{C}{}^{-1}_{11}\underset{\sim}{C}_{12}\underset{\sim}{C}{}^{-1}_{22}\underset{\sim}{C}_{21}) \\ &= \mathrm{Tr}(\underset{\sim}{C}{}^{-1}_{22}\underset{\sim}{C}_{21}\underset{\sim}{C}{}^{-1}_{11}\underset{\sim}{C}_{12}) = \mathrm{Tr}(\underset{\sim}{M}) = q^*. \end{aligned} \tag{4.22}$$

Thus, using (4.22), (4.21) reduces to

$$\mathrm{Tr}(\underset{\sim}{M})\{1 - \tau(pq_2)^{-1}[2 - \tau(pq_2 - 2)^{-1}]\}\;. \tag{4.23}$$

Hence, under H_0

$$R(\hat{\underset{\sim}{\beta}}{}^S_{1n}; W^o) \ge R(\hat{\underset{\sim}{\beta}}_{1n}; W^o) \tag{4.24}$$

whenever

$$\tau(pq_2)^{-1}[2 - \tau(pq_2 - 2)^{-1}] \le 1. \tag{4.25}$$

This is satisfied for $0 \le \tau \le (pq_2 - 2)$.

Similarly, under H_0: $\underset{\sim}{\gamma}_2 = \underset{\sim}{0}$, we may write

$$R(\hat{\underset{\sim}{\beta}}{}^{S}_{1n}; \underset{\sim}{W}^{o}) = R(\hat{\underset{\sim}{\beta}}{}^{PT}_{1n}; \underset{\sim}{W}^{o}) + \mathrm{Tr}(\underset{\sim}{M})\{H_{pq_2+2}(\chi^2_{pq_2,\alpha}; 0) - \tau(pq_2)^{-1}[2 - \tau(pq_2 - 2)^{-1}]\}. \tag{4.26}$$

Now, $H_{pq_2+2}(\chi^2_{pq_2,\alpha}; 0) \doteq 1 - \alpha - 2h_{pq_2+2}(\chi^2_{pq_2,\alpha}; 0)$ (4.27)

where $h_{pq_2+2}(\chi^2_{pq_2,\alpha}; 0)$ is the pdf of central chi-square distribution with $pq_2 + 2$ D.F. evaluated at $\chi^2_{pq_2,\alpha}$. Thus

$$R(\hat{\underset{\sim}{\beta}}{}^{S}_{1n}; \underset{\sim}{W}^{o}) \ge R(\hat{\underset{\sim}{\beta}}{}^{PT}_{1n}; \underset{\sim}{W}^{o}) \tag{4.28}$$

whenever,

$$2h_{pq_2+2}(\chi_{pq_2,\alpha}; 0) \le 1 - 2\tau\ pq_2)^{-1}[1 - \tfrac{\tau}{2}(pq_2 - 2)^{-1}] - \alpha. \tag{4.29}$$

For $0 \le \tau \le (pq_2 - 2)$, (4.28) holds.

Further, a comparison of $\hat{\underset{\sim}{\beta}}_{1n}$ and $\hat{\underset{\sim}{\beta}}{}^{PT}_{1n}$ under H_0 shows that

$$R(\hat{\underset{\sim}{\beta}}{}^{PT}_{1n}; \underset{\sim}{W}^{o}) = R(\hat{\underset{\sim}{\beta}}_{1n}; \underset{\sim}{W}^{o}) + \mathrm{Tr}(\underset{\sim}{M})\{1 - H_{pq_2+2}(\chi^2_{pq_2,\alpha}; 0)\} \tag{4.30}$$

where the second term is always positive. Thus, we may order the dominance of the estimators under H_0 as follows:

$$\hat{\underset{\sim}{\beta}}_{1n} > \hat{\underset{\sim}{\beta}}{}^{PT}_{1n} > \hat{\underset{\sim}{\beta}}{}^{S}_{1n} > \tilde{\underset{\sim}{\beta}}_{1n} \quad . \tag{4.31}$$

The picture changes as $\underset{\sim}{\gamma}_2$ moves away from $\underset{\sim}{0}$. First, we note that $\tilde{\underset{\sim}{\beta}}_{1n}$ has a constant risk equal to pq_1 and the risk of $\hat{\underset{\sim}{\beta}}_{1n}$ depends on γ_2, consequently, the risk of $\hat{\underset{\sim}{\beta}}_{1n}$ becomes unbounded as γ_2 moves furthest away from 0. As for the risk of $\hat{\beta}_{1n}^{PT}$, we note that as $||\gamma_2||$ goes to ∞, $H_{pq_2+p}(\chi^2_{pq_2,\alpha};\ \delta_2)$, and $(\gamma^{o\prime} M\gamma^o)H_{pq_2+\ell}(\chi^2_{pq_2,\alpha};\ \delta_2)$ $(\ell=2,4)$ approach zero and the risk of $\hat{\beta}_{1n}^{PT}$ approaches pq_1 (= the risk of $\tilde{\beta}_{1n}$). Also as γ_2 moves away from 0, the risk function of $\hat{\beta}_{1n}^{PT}$ increases monotonically to a maximum crossing the risk function of $\hat{\beta}_{1n}$, then monotonically decreases to the value pq_1. Hence, the risk function of $\hat{\beta}_{1n}^{PT}$ never blows up. Further (4.29) specifies the range of α $(0 < \alpha < 1)$ for any given $pq_2 \geq 3$ for which $\hat{\beta}_{1n}^{PT}$ dominates $\hat{\beta}_{1n}^{S}$ under H_0; it also tells us that increasing the value of α (the significance level of the PT on $\beta_2 = 0$) for any pq_2 (≥ 3) may lead to the opposite inequality (4.29), for which $\hat{\beta}_{1n}^{S}$ will dominate over $\hat{\beta}_{1n}^{PT}$. For γ_2 away from 0, $\hat{\beta}_{1n}^{S}$ may dominate over $\hat{\beta}_{1n}^{PT}$, but for small α $(0 < \alpha < 1)$ and γ_2 close to 0, the opposite conclusion is generally true. Thus, none of $\hat{\beta}_{1n}^{PT}$ and $\hat{\beta}_{1n}^{S}$ is inadmissible (asymptotically under $\{K_n\}$). Similar remarks have been made by Sclove, Morris and Radhakrishnan (1972) in their study of means of p-variate normal distribution. Actually $\hat{\beta}_{1n}^{S}$ and $\hat{\beta}_{1n}^{PT}$ share a common property that as $\delta_2 \to \infty$, their ADR converges to pq_1. But, $R(\hat{\beta}_{1n}^{S};\ W^o)$ is always below this limit, while $R(\hat{\beta}_{1n}^{PT};\ W^o)$ exceeds this limit for some intermediate value of δ_2.

Thus, from the point of robust-efficiency both PTLSE and SLSE may be advocated learning more towards SLSE where the size of δ_2 is generally unknown and unlikely to be small.

Finally, we note that the shrinkage factor used in (2.8) though works out well for the particular choice of $W^o = \Sigma^{-1} \otimes C_{11.2}$ may not be the most ideal one for arbitrary choice of W^o. For the multivariate normal mean estimation problem with unknown dispersion matrix, a more general shrinkage factor (involving the matrix W, the sample covariance matrix S_n and the smallest characteristic root of WS_n has been considered by Berger et al (1977). In view of the asymptotic multinormality results of LSE(s) a variant of shrinkage may be used and results on ADR of section 3 and 4 may be readily modified and asymptotic dominance of SLSE over ULSE, RLSE follows in the same line; the lack of dominance of SLSE over PTLSE also follows on the same line as in section 4. However, we have made a very reasonable choice of $\underset{\sim}{W}^{o}$ and in the light of that choice, the proposed SLSE is justified.

REFERENCES

1. Anderson, T. W. (1984). *An Introduction to Multivariate Statistical Analysis*. John Wiley, New York.
2. Berger, J. O., Bock, M. E., Brown, L. D., Casella, G. and Gleser, L. (1977). Minimax estimation of a normal mean vector for arbitrary quadratic loss and unknown covariance matrix. *Ann. Statist. 5*, 763-771.
3. James, W., and Stein, C. (1961). Estimation with quadratic loss. *Proc. 4th Berkeley Symp. Math. Statist. Probability* (ed. J. Neyman) *1*, 293-325.
4. Judge, G. G. and Bock, M. E. (1978). *The Statistical Implications of Pretest and Stein-rule Estimators in Economic*. North Holland Pub., Amsterdam.
5. Saleh, A. K. Md. E. and Sen, P. K. (1984). Least squares and rank order preliminary test estimation in general multivariate linear models. *Indian Statist. Inst. Golden Jub. Confer.: Statist. Appl. New Direct.* (ed. J. K. Ghosh et al), 237-253.
6. Saleh, A. K. Md. E. and Sen, P. K. (1985a). Nonparametric Preliminary Test Inference. In *Handbook of Statistics, Vol. 4: Nonparametric Methods* (ed. P. R. Krishnaiah and P. K. Sen), North Holland Publ., Amsterdam, pp. 275-297.
7. Saleh, A. K. Md. E. and Sen, P. K. (1985b). Nonparametric shrinkage estimators of location in a multivariate simple regression model. *Probability and Statistical Decision Theory: Proceedings of the 4th Pannonian Symposium on Mathematical Statistics*. Bad Tatzmanndorf, Austria, Sept. 4-10, 1983. Vol. A. Edited by F. Konecny, J. Mogyoródi and W. Wertz. Published by Akadémiai Kiadó, Budapest, 1985. pp. 311-324.

8. Saleh, A. K. Md. E., and Sen, P. K. (1985c). On shrinkage M-estimators of location parameters. *Commun. Statist. Theor. Meth. A14*, 2313-2329.

9. Saleh, A. K. Md. E., and Sen, P. K. (1986). Shrinkage least squares estimation in a parallelism problem. *Commun. Statist. Theor. Math. 15*. (in press).

10. Saleh, A. K. Md. E., and Sen, P. K. (1986). Relative performance of Stein-Rule and preliminary test estimators in linear models: Least-squares Theory. Submitted to *Communications in Statistics: Theory and Methods*.

11. Saleh, A. K. Md. E., and Sen, P. K. (1986). On shrinkage R-estimation in multiple regression model. *Commun. Statist. Theor. Meth. 15* (in press).

12. Sen. P. K. (1984). A James-Stein detour of U-statistics. *Commun. Statist. Theor. Meth. A13*, 2725-2747.

13. Sen, P. K. and Puri, M. L. (1970). Asymptotic theory of likelihood ratio and rank order tests in some multivariate linear models. *Ann. Math. Statist. 41*, 87-100.

14. Sen, P.K. and Saleh, A. K. Md. E. (1985). On some shrinkage estimators of multivariate locations. *Ann. Statist. 13*, 272-281.

15. Sen, P. K., and Saleh, A. K. Md. E. (1985a). On preliminary test and shrinkage M-estimation in linear models. Submitted to *Annals of Statistics*.

16. Sen, P. K. and Saleh, A. K. Md. E. (1985b). Nonparametric and robust shrinkage estimation in some multivariate linear models. *Proc. Symp. Recent Advances in Multivariate Analysis*, Indian Statistical Institute, Calcutta.

A. K. Md. Ehsanes Saleh
Carleton University
Colonel by Drive
OTTAWA-CANADA
K185B6

Pranah Kumer Sen
Department of Biostatistics
The University of North-Carolina
at Chapel Hill,
Chapel Hill N.C. 27514
USA

Proc. of the 5th Pannonian Symp. on Math. Stat., Visegrád, Hungary 1985
W. Grossmann, J. Mogyoródi, I. Vincze, W. Wertz, eds

ON THE ESTIMATION OF CROSS DISPERSION FUNCTION

L. SZEIDL

Budapest

This paper deals with estimation of the cross dispersion function, obtained from finite realization of stationary dependent ergodic processes. The present paper contains some results with respect to the absolute mean deviation of the estimate. The results without changes can be transferred to the case of discrete-time processes and used for the estimation of the auto dispersion function.
AMS subject classification: 60G10

1. INTRODUCTION

Rajbman's dispersional method, based on the regression and dispersion function of random processes plays an important role and can be used efficiently to nonparametric models for the identification of continuous nonlinear dynamic systems (see Rajbman, Chadeev (1980), Rajbman (1981)).

Let Y_t, X_t, $-\infty<t<\infty$, be real valued stochastic processes on the probability space $(\Omega,\mathcal{A},P)$. We assume that the following condition holds.

C1. The vector process $\{(Y_t, X_t, -\infty<t<\infty\}$ is a strictly stationary continuous in mean square ergodic process, $EY_t=0$ (for simplicity, without loss of generality), $EY_t^2 = EY_0^2 < \infty$.

Let $\Theta_{YX}(u)$, $-\infty<u<\infty$, denote the cross dispersion function of Y_t referring to X_t by

$$\Theta_{YX}(u) = E(E^2(Y_{t+u}|X_t)),$$

where $E(Y_{t+u}|X_t)$ means the conditional expectation of Y_{t+u}

referring to X_t. In an earlier paper Szeidl, Várlaki (1982) a simple estimate of cross dispersion function was given by the following method.

Let us divide the value set of process X_t into a finite N ($N \geq 2$ integer) number of intervals as follows

$$-\infty < s_1^N < s_2^N < \ldots < s_{N-1}^N < \infty, \quad S^N = \{s_1^N, \ldots, s_{N-1}^N\},$$

and denote the interval sequence by

$$\Delta_1^N = (-\infty, s_1^N), \; \Delta_2^N = [s_1^N, s_2^N), \; \ldots, \; \Delta_N^N = [s_{N-1}, \infty),$$

$$\Delta^N = \{\Delta_1^N, \Delta_2^N, \ldots, \Delta_N^N\}.$$

Then the estimate of the cross dispersion function $\Theta_{YX}(u)$ is for $0 \leq u < T$

$$\Theta_{YX}(u,\Delta^N,T) = \sum_{i=1}^{N} \left[\frac{1}{T-u} \int_0^{T-u} Y_{t+u}\, I(X_t \in \Delta_i^N)\, dt\right]^2 \left[\frac{1}{T}\int_0^T I(X_t \in \Delta_i)\, dt\right]^{-1} \quad (1)$$

where T is the time of observation, $I(X_t \in \Delta_i)$ denotes the indicator function of the event $\{X_t \in \Delta_i^N\}$. In the formula (1) and further on in this work we determine the fraction $^0/o$ as zero.

Now we formulate theorem 1, which in a more simple form can be found in Szeidl, Várlaki (1983).

Theorem 1. Suppose the condition C1. We assume that the sequence S^N, $N=2,3,\ldots$ satisfies the assuptions $S^N \subset S^{N+1}$ for all $N \geq 2$ and $S^\infty = \bigcup_N S^N$ is a dense set in the real axis. Then for arbitrary fixed $0 \leq u$

$$\lim_{N\to\infty} \lim_{T\to\infty} \Theta_{YX}(u,\Delta,T) = \Theta_{YX}(u) \qquad \text{w.p. 1.}$$

In practical applications of the estimate of cross dispersion function the most important problem is to determine the error of estimate. Let us see now, how to assess the absolute mean deviation of statistic $\Theta_{YX}(u,\Delta^N,T)$ from $\Theta_{YX}(u)$.

Making use of the triangle inequality we have

$$E|\Theta_{YX}(u) - \Theta_{YX}(u,\Delta^N,T)| \leqq |\Theta_{YX}(u) - \Theta_{YX}(u,\Delta^N)| +$$

$$+ E|\Theta_{YX}(u,\Delta^N) - \Theta_{YX}(u,\Delta^N,T)| = \qquad (2)$$

$$= \vartheta_1(u,\Delta^N) + \vartheta_2(u,\Delta^N,T),$$

where

$$\Theta_{YX}(u,\Delta^N) = \sum_{i=1}^{N} (EY_u I(X_o \in \Delta_i))^2 \, (EI(X_o \in \Delta_i))^{-1}.$$

It should be noted that the assessment of the first part in the right-hand side of (2) has pure analytical technique, but the second case has statistical method which is typical in the theory of time series analysis.

Let us begin with the examination of the quantity $\nu_1(u,\Delta^N)$. Since the number N will be fixed for the examination, therefore for simplicity we shall use the notation S, S_i, Δ, Δ_i instead of S^N, S_i^N, Δ^N, Δ_i^N.

Suppose that there exists the joint probability density function $f_{s,t}(r,z) = f_{Y_s, X_t}(r,z)$ of the random variables Y_s and X_t. Denote by $f_Y(r)$ and $f_X(z)$ the density function of R.V.'s Y_s and X_t, which in consequence of ergodicity are not dependent on s and t. Similarly, we can write for each real numbers t, u, r, $z \in \mathbb{R}_1$

$$f_{Y_{t+u}X_t}(r,z) = f_{u,o}(r,z).$$

Denote

$$\Phi(u,z) = \frac{\int_{-\infty}^{\infty} r \, f_{u,o}(r,z) \, dr}{f_X(z)}, \qquad z \in \mathbb{R}_1,$$

and

$$\varphi(u,\Delta_i) = \sup_{z',z'' \in \Delta_i} |\Phi(u,z') - \Phi(u,z'')|.$$

Here we note that $E(Y_u|X_o=z) = \Phi(u,z)$ P a.s. Denote $p_i=P(X_o \in \Delta_i)$. Now we give theorem 2.

Theorem 2. If the condition C1 holds and there exists the joint density function $f_{Y_s X_t}(r,z)$, then

$$\vartheta_1(u,\Delta) \leq 2E\{Y_0^2(I(X_0<S_1) + I(X_0>S_{N-1}))\} + +2\sum_{i=2}^{N-1} E|Y_u| I(X_0 \in \Delta_i) \cdot \varphi(u,\Delta_i)$$

Remark 1. We can establish an analogical result in the case, when there exists only the density function of R.V. X_t. There is no fundamental difference between the two cases, but in the second one the demonstration will be more difficult.

From the theorem 2. immediately follow the Corollary 1, 2 and 3.

Corollary 1. It is true, that

$$\vartheta_1(u,\Delta) \leq 2E|Y_0| \cdot \max_{1\leq i\leq N} \varphi(u,\Delta_i)$$

Corollary 2. If for some $\delta>0$ $(E|Y_0|^{2+\delta})^{1/2+\delta} \leq c < \infty$ holds, then

$$\vartheta(u,\Delta) \leq 2.C.((p_1)^{\delta/(2+\delta)} + (p_2)^{\delta/(2+\delta)}) + 2.C. \sum_{i=2}^{N-1} (p_i)^{\frac{1+\delta}{2+\delta}} \cdot \varphi(u,\Delta_i)$$

Corollary 3. If w.p. $1|Y_0| \leq C_1 < \infty$ holds, then

$$\vartheta_1(u,\Delta) \leq 2C_1(C_1(p_1+p_N) + \sum_{i=2}^{N-1} p_i \varphi(u,\Delta_i)).$$

Supposing that the function $\phi(u,z)$ satisfies in the variable z the Lipschitz$_1$ condition with the constant c_2, then

$$\vartheta_1(u,\Delta) \leq 2C_1(C_1(p_1+p_N) + C_2 \sum_{i=2}^{N-1} p_i \cdot |\Delta_i|) \leq$$

$$\leq 2C_1(C_1(p_1+p_N) + C_2 \cdot (1-p_1-p_N) \max_{2\leq i\leq N-1} |\Delta_i|),$$

where $|\Delta_i| = S_i - S_{i-1}$, $2 \leq i \leq N-1$.

Let us formulate condition C2 for theorems 3 and 4: C2. Process (Y_t, X_t) satisfies strong mixing condition with mixing function $\alpha_{YX}(s)$ and for $\gamma>0$

$$A_{YX}(\gamma) = \int_0^{\infty} (\alpha_{YX}(s))^{\gamma} ds<\infty.$$

For the distance $\vartheta_2(u,\Delta,T)$ we can obtain the next results.

Theorem 3. If the conditions C1, C2 hold for some $\delta>0$ with $\gamma = \frac{\delta}{2+\delta}$ and $(E|Y_0|^{2+\delta})^{1/(2+\delta)} \leqq C_3<\infty$, then

$$\vartheta_2(u,\Delta,T) \leqq C_4 \frac{1}{(T-u)^{\delta/(2+\delta)}} \sum_{i=1}^{N} \frac{1}{p_i}(E|Y_u I(X_0\in\Delta_i)|^{2+\delta})^{1/(2+\delta)}, \quad \text{if } 0<\delta<2, \tag{3}$$

$$\vartheta_2(u,\Delta,T) \leqq C_5 \frac{1}{(T-u)^{1/2}} \sum_{i=1}^{N} (p_i)^{-\frac{6+\delta}{4+2\delta}} (E|Y_u I(X_0\in\Delta_i)|^{2+\delta})^{1/(2+\delta)} \quad \text{if } 2 \leqq \delta \tag{4}$$

where $C_4 = \frac{\sqrt{2}}{(T-u)^{2/(2+\delta)}}(u\cdot C_3^2 + (4+6C_3^{2+\delta})\ A_{YX}(\frac{\delta}{2+\delta}))^{1/2} +$

$$+ (\frac{T}{T-u})^{2/(2+\delta)} \cdot C_3 \cdot (2A_{YX}(1))^{\delta/(2+\delta)},$$

$$C_5=\sqrt{2}(u\cdot C_3^2 + (4+6\cdot C_3^{2+\delta})\ A_{YX}\frac{\delta}{2+\delta}))^{1/2} + (\frac{2T}{T-u})^{1/2}.C_3(A_{YX}(1))^{1/2}.$$

Theorem 4.
If the conditions C1, C2 hold with $\gamma=1$ and further w.p. $1|Y_0| \leqq C<\infty$, then

$$\nu_2(u,\Delta,T) \leqq C^2(A_{YX}(1))^{1/2}((\frac{T}{T-u})^{1/2}+2\sqrt{2}) \frac{1}{(T-u)^{1/2}} \sum_{i=2}^{N} (p_i)^{-1/2} \tag{5}$$

Note that in the summation (3), (4) and (5) and further on we take indices i for which $p_i \neq 0$.

Remark 2. In the case of activ identification the distribution function of the input process X_t is given. In this situation the estimate of the cross dispersion function has the form

$$\hat{\Theta}_{YX}(u,\Delta,T) = \sum_{i=1}^{N} \frac{1}{p_i} \left(\frac{1}{T-u} \int_0^{T-u} Y_{t+u}\, I(X_t \quad \Delta_i)\, dt\right)^2.$$

It is easy to see that in this case (see the proof of theorem 3.)

$$\Theta_2(u,\Delta,T) \leqq \zeta_1(u,\Delta,T).$$

2. PROOF OF THEOREMS

Proof of Theorem 1. First we note that from the ergodicity of process (Y_t, X_t) follows

$$\lim_{T\to\infty} \Theta_{YX}(u,\Delta^N,T) = \lim_{T\to\infty} \sum_{i=1}^{N} \frac{\left(\frac{1}{T-u}\int_0^{T-u} Y_{t+u} I(X_t \in \Delta_i)\,dt\right)^2}{\frac{1}{T}\int_0^T I(X_t \in \Delta_i)\,dt} =$$

$$= \Theta_{YX}(u,\Delta^N) \quad P \text{ a.s.}$$

Now we prove the relation $\lim_{N\to\infty} \Theta_{YX}(u,\Delta^N) = \Theta_{YX}(u)$.

For every integer $N(N\geqq 2)$ let $\mathfrak{F}^0$ and $\mathfrak{F}^0_N$ denote the Borel σ-algebras of events generated by the R.V. X_0 and by the events $\{(X\in\Delta_i^N),\ 1\leqq i\leqq N\}$. It is obvious that

$$E(Y_u \mid \mathfrak{F}^0_N) = \sum_{i=1}^{N} \frac{E\, Y_u I(X_0 \in \Delta_i^N)}{EI(X_0 \in \Delta_i^N)} I(X_0 \in \Delta_i^N) \qquad P \text{ a.s.},$$

consequently

$$E(E^2(Y_u|\mathfrak{F}^0_N)) = \Theta_{YX}(u,\Delta^N).$$

Since the relation $\mathfrak{F}^0_2 \subset \mathfrak{F}^0_3 \subset \ldots$ holds, $\mathfrak{F}^0 = \sigma(\bigcup_N \mathfrak{F}^0_N)$ is true and $E(E^2(Y_u|\mathfrak{F}^0)) \leqq E\, Y_u^2 < \infty$, then from the dominated convergence theorem for conditional expectation follows (see theorem 44.19. in Parhasarathy, 1983)

$$\lim_{N\to\infty} E(Y_u|\mathfrak{F}_N^0) = E(Y_u|\mathfrak{F}^0) \text{ P a.s and in } \mathcal{L}_2 \text{ metric.}$$

Thus the theorem is proved.

Proof of Theorem 2. Let us notice, that

$$\vartheta_1(u,\Delta) = \left| \int_{-\infty}^{\infty} \Phi^2(u,z) f_X(z)dz - \sum_{i=1}^{N} \frac{\left[\int_{\Delta_i} \Phi(u,u)\ f_X(z)dz\right]}{\int_{\Delta_i} f_X(z)dz} \right| \leqq$$

$$\leqq \sum_{i=1}^{N} \left| \int_{\Delta_i} \left(\Phi^2(u,z) - \frac{\left[\int_{\Delta_i} \Phi(u,w) f_X(w)dw\right]^2}{\left(\int_{\Delta_i} f_X(w)dw\right)} \right) f_X(z)dz \right| \quad (6)$$

Since for arbitrary interval Δ'

$$\inf_{z\in\Delta'} \Phi(u,z) \int_{\Delta'} f_X(z)dz \leqq \int_{\Delta'} \Phi(u,z) f_X(z)dz \leqq \sup_{z\in\Delta'} \Phi(u,z) \int_{\Delta'} f_X(z)dz,$$

thus we have

$$\left| \int_{\Delta'} \left(\Phi^2(u,z) - \frac{\left(\int_{\Delta'} \Phi(u,w)\ f_X(w)dw\right)^2}{\left(\int_{\Delta'} f_X(w)dw\right)^2} \right) f_X(z)\ dz \right| \leqq$$

$$\leqq \int_{\Delta'} \left| \Phi(u,z) - \frac{\int_{\Delta'} \Phi(u,w) f_X(w)dw}{\int_{\Delta'} f_X(w)dw} \right| \left| \Phi(u,z) + \frac{\int_{\Delta'} \Phi(u,w) f_X(w)dw}{\int_{\Delta'} f_X(w)dw} \right| f_X(z)dz \leqq$$

$$\leqq \varphi(u,\Delta') \cdot \int_{\Delta'} \left(|\Phi(u,z)| + \frac{\int_{\Delta'} |\Phi(u,w)| f_X(w)\ dw}{\int_{\Delta'} f_X(w)\ dw} \right) f_X(z)\ dz \leqq$$

$$\leqq \varphi(u,\Delta')\cdot 2\int_{\Delta'}\int_{-\infty}^{\infty} |r| f_{u,o}(r,z)\,drdz = 2E(|Y_u| I(X_0 \quad \Delta'))\cdot \varphi(u,\Delta') \tag{7}$$

On the other hand

$$\left(\int_{\Delta'} \Phi(u,z) f_X(z)\,dz\right)^2 \leqq \int_{\Delta'} \Phi^2(u,z) f_X(z)\,dz \cdot \int_{\Delta'} f_X(z)\,dz,$$

$$\int_{\Delta'} \Phi^2(u,z) f_X(z)\,dz = \int_{\Delta'} \frac{\left[\int_{-\infty}^{\infty} r f_{u,o}(r,z)\,dr\right]^2}{f_X(z)}\,dz \leqq \int_{\Delta'}\int_{-\infty}^{\infty} r^2 f_{u,o}(r,z)\,drdz = EY_u^2 I(X_0 \in \Delta') \tag{8}$$

Combining (6), (7) and (8) the theorem is proved.

<u>Proof of Theorem 3.</u>

Let us denote

$$\zeta_1(u,\Delta,T) = \sum_{i=1}^{N} \frac{1}{p_i} E\left| (EY_u I(X \in \Delta_i))^2 - \left(\frac{1}{T-u}\int_0^{T-u} Y_{t+u} I(X_t \in \Delta_i)\,dt\right)^2 \right|.$$

and

$$\zeta_2(u,\Delta,T) = \sum_{i=1}^{N} \frac{1}{p_i} E\left\{ \frac{\left(\frac{1}{T-u}\int_0^{T-u} Y_{t+u} I(X_t \in \Delta_i)\,dt\right)^2}{\frac{1}{T}\int_0^T I(X_t \in \Delta_i)\,dt} \left|\frac{1}{T}\int_0^T I(X_t \in \Delta_i)\,dt - p_i\right| \right\}$$

It is obvious that

$$\vartheta_2(u,\Delta,T) \leqq \zeta_1(u,\Delta,T) + \zeta_2(u,\Delta,T). \tag{9}$$

Using the notation

$$w_i(t,u) = Y_{t+u} I(X_t \in \Delta_i), \quad 1 \leqq i \leqq N,$$

we obtain

$$\zeta_1(u,\Delta,T) = \sum_{i=1}^{N} \frac{1}{p_i} E\left| \frac{1}{T-u} \int_0^{T-u} (w_i(t,u)-Ew_i(t,u))dt \frac{1}{T-u} \int_0^{T-u} (w_i(t,u)+\right.$$

$$\left. +Ew_i(t,u))\ dt \right| \leq \sum_{i=1}^{N} \frac{1}{p_i} \left\{ E\left(\int_0^{T-u} (w_i(t,u)-Ew_i(t,u))dt \right)^2 \right\}^{1/2} \cdot$$

$$\cdot \left\{ E\left(\frac{1}{T-u} \int_0^{T-u} (w_i(t,u) + Ew_i\ (t,u))\ dt \right)^2 \right\}^{1/2} \tag{10}$$

It is not difficult to see that using Hölder's inequality

$$E\left(\frac{1}{T-u} \int_0^{T-u} (w_i(t,u)+Ew_i(t,u))dt \right)^2 \leq 4E(Y_u I(X_0 \in \Delta_i))^2 \leq$$

$$\leq 4\left(E|Y_u I(X_0 \in \Delta_i)|^{2+\delta}\right)^{2/(2+\delta)} (p_i)^{\delta/(2+\delta)}, \tag{11}$$

and

$$E\left(\frac{1}{T-u} \int_0^{T-u} (w_i(t,u)-Ew_i(t,u))dt \right)^2 = \frac{1}{(T-u)^2} \iint_0^{T-u} \mathrm{Cov}(w_i(t,u),$$

$$,w_i(s,u))dtds \leq \frac{2}{T-u} \int_{\infty}^{\infty} |\mathrm{Cov}(w_i(0,u))|ds \tag{12}$$

From the assumption of the theorem 17.2.2. in Ibragimov and Linnik (1971) follows that for every $s \geq u$ the following inequality holds

$$|\mathrm{Cov}(w_i,(s,u),w_i(o,u))| \leq (4+6\cdot E|Y_u I(X_0 \in \Delta_i)^{2+\delta})(\alpha_{YX}(s-u))^{\delta/(2+\delta)} \tag{13}$$

It is clear that for every s holds

$$|\mathrm{Cov}(w_i(s,u),w_i(0,u)| \leq E(w_i(0,u)^2 \leq (E|Y_u I(X_0 \in \Delta_i)|^{2+\delta})^{2/(2+\delta)} \cdot (p_i)^{\delta/(2+\delta)}$$

$$\tag{14}$$

Using relations (13) and (14) we have

$$\int_0^\infty |\mathrm{Cov}(w_i(s,u),w_i(0,u))|ds \leqq u(E|Y_0|^{2+\delta})^{2/(2+\delta)} + (4+6E|Y_0|^{2+\delta})A_{YX}\left(\frac{\delta}{2+\delta}\right) \quad (15)$$

Combining (10) - (12) and (15) we obtain

$$\zeta_1(u,\Delta,T) \leqq \{u.\left[E|Y_0|^{2+\delta}\right]^{2/(2+\delta)} + (4+6E|Y_0|^{2+\delta})A_{YX}\left(\frac{\delta}{2+\delta}\right)\}^{1/2}.$$

$$\cdot\sqrt{\frac{2}{T-u}}\sum_{i=1}^{N}(p_i)^{-\frac{4+\delta}{4+2\delta}}\left[E|Y_u I(X_0\in\Delta_i)|^{2+\delta}\right]^{1/(2+\delta)} \quad (16)$$

Now we estimate the value $\zeta_2(u,\Delta,T)$. Since

$$\left(\frac{1}{T-u}\int_0^{T-u} Y_{t+u} I(X_t\in\Delta_i)dt\right)^2 \leqq \frac{1}{T-u}\int_0^{T-u} Y^2_{t+u} I(X_t\in\Delta_i)dt \cdot \frac{1}{T-u}\int_0^{T-u} I(X_t\in\Delta_i)dt,$$

therefore

$$\zeta_2(u,\Delta,T) \leqq \sum_{i=1}^{N}\frac{1}{p_i}\frac{T}{T-u}E\{\frac{1}{T-u}\int_0^{T-u} Y^2_{t+u} I(X_t\in\Delta_i)dt\}\left|\frac{1}{T}\int_0^T (I(X_t\in\Delta_i)-p_i)dt\right| \quad (17)$$

If $0<\delta<2$, then

$$\left|\frac{1}{T}\int_0^T (I(X_t\in\Delta_i)-p_i)dt\right|^{\frac{2+\delta}{\delta}} \leqq \left|\frac{1}{T}\int_0^T (I(X_t\in\Delta_i) - p_i)\, dt\right|^2.$$

Since

$$E\left(\frac{1}{T}\int_0^T (I(X_t\in\Delta_i)-p_i)dt\right)^2 \leqq \frac{2}{T}\int_0^\infty |\mathrm{Cov}(I(X_s\in\Delta_i),\ I(X_0\in\Delta_i))|ds =$$

$$= \frac{2}{T}\int_0^\infty |I(X_s\in\Delta_i, X_0\in\Delta_i) - P(X_s\in\Delta_i)P(X_0\in\Delta_i)|ds \leqq$$

$$\leqq \frac{2}{T}\int_0^\infty \alpha_{YX}(s)\, ds = \frac{2}{T}A_{YX}(1), \quad (18)$$

thus we can obtain

$$\zeta_2(u,\Delta,T) \leqq \sum_{i=1}^{N} \frac{1}{p_i}\frac{T}{T-u}\left\{E\left(\frac{1}{T-u}\int_0^{T-u} Y^2_{t+u} I(X_0 \in \Delta_i)dt\right)^{(2+\delta)/2}\right\}^{2/(2+\delta)}$$

$$\left\{E\left(\frac{1}{T}\int_0^{T}(I(X_t \in \Delta_i) - p_i)dt\right)^{\frac{2+\delta}{\delta}}\right\}^{\delta/(2+\delta)}$$

$$\leqq \sum_{i=1}^{N} \frac{1}{T-u}\left[E|Y_u I(X_0 \in \Delta_i)^{2+\delta}\right]^{\frac{2}{2+\delta}}\left\{E\left(\frac{1}{T}\int_0^{T}(I(X_t \in \Delta_i)-p_i)dt\right)^2\right\}^{\delta/(2+\delta)} \tag{19}$$

$$\leqq \frac{T}{T-u}\left[E|Y_0|^{2+\delta}\right]^{1/(2+\delta)}\left[2A_{YX}^{(1)}\right]\cdot \frac{1}{T^{\delta/2+\delta}}\sum_{i=1}^{N}\frac{1}{p_i}\left[E|Y_u I(X_0 \in \Delta_i)|^{2+\delta}\right]^{1/(2+\delta)}$$

In the case $\delta \geqq 2$, it is easy to conclude using (17) that

$$\zeta_2(u,\Delta,T) \leqq \sum_{i=1}^{N}\frac{1}{p_i}\frac{T}{T-u}\left\{E\left(\frac{1}{T-u}\int_0^{T-u} Y^2_{t+u} I(X_t \in \Delta_i)dt\right)^2\right\}^{1/2} \cdot$$

$$\cdot\left\{E\left(\frac{1}{T}\int_0^{T}(I(X_t \in \Delta_i) - p_i\ dt\right)^2\right\}^{1/2} \leqq \tag{20}$$

$$\leqq \sum_{i=1}^{N}\frac{1}{p_i}\frac{T}{T-u}\left[E|Y_u I(X_0 \in \Delta_i)|^{2/\delta}\right]^{2/(2/(2+\delta)}\cdot(p_i)^{\frac{\delta-2}{2(2+\delta)}}$$

$$\sqrt{\frac{2}{T}}(A_{YX}(1))^{1/2}$$

Statements (16) and (19) imply (3) and by (16) and (20) we have (4). So our proof is complete.

Proof of Theorem 4. By (11) we get

$$E\left(\frac{1}{T-u}\int_0^{T-u}(w_i(t,u) + Ew_i(t,u))dt\right)^2 \leqq 4\cdot C^2\cdot p_i$$

Using the theorem 17.2.1. in Ibragimov and Linnik (1971) and (12) we can see that

$$E\left(\frac{1}{T-u}\int_0^{T-u}(w_i(t,u) - Ew_i(t,u))dt\right)^2 \leqq \frac{8\cdot C^2}{T-u} A_{YX}(1),$$

consequently by (10)

$$\zeta_1(u,\Delta,T) \leqq 4\sqrt{2}\cdot C^2\cdot (A_{YX}(1))^{1/2} \frac{1}{\sqrt{T-u}} \sum_{i=1}^{N} \frac{1}{\sqrt{p_i}} .$$

Since by (17) and (18)

$$\zeta_2(u,\Delta,T) \leqq \sum_{i=1}^{N} \frac{1}{p_i} C^2 \frac{T}{T-u} \left\{E\left[\frac{1}{T-u}\int_0^{T-u} I(X_t\in\Delta_i)dt\right]^2\right\}^{1/2} .$$

$$\cdot\left\{E\left[\frac{1}{T}\int_0^{T}(I(X_t\in\Delta_i) - p_i)dt\right]^2\right\}^{1/2} \leqq$$

$$\leqq C^2\left(\frac{T}{T-u} A_{YX}(1)\right)^{1/2}\cdot \frac{1}{\sqrt{T-u}} \sum_{i=1}^{N} \frac{2}{\sqrt{p_i}}$$

Thus, the theorem is proved.

REFERENCES

Rajbman, N.S. and Chadeev, V.M. (1980) Identification of industrial process, North-Holland, Amsterdam, New-York, Oxford.

Rajbman, N.S. (ed.) (1981) Dispersional identification (in Russian), Nauka, Moscow.

Szeidl, L. and Várlaki, P. (1982) A simple estimation method for dispersional identification of nonlinear dynamic systems, Discussion papers of IFAC Symposium on Theory and Application of Digital Control, New Delhi, vol. 2., 3-4.

Szeidl, L. and Várlaki, P. (1983) New results for dispersional identification of nonlinear systems, Transaction of the Ninth Prague Conference, 217-223.

Parthasarathy, K.R. (1983) Introduction to probability and measure (in Russian), MIR, Moscow.

Ibragimov, I. A. and Linnik, Yu.V. (1971) Independent and stationary sequences of random variables, Wolters-Noordhoff Publishing, Groningen.

L. Szeidl
Eötvös Loránd University
Computing Centre
Budapest, Bogdánfy u. 10/b.
H-1117, Hungary

Proc. of the 5th Pannonian Symp.
on Math. Stat., Visegrád, Hungary 1985
W. Grossmann, J. Mogyoródi, I. Vincze, W. Wertz, eds

CONDITIONS FOR STATIONARITY OF QUILO MODELS WITH GAUSSIAN RESIDUALS

G. TERDIK

Debrecen

The aim of this paper is to construct the transfer function system of QUILO models with Gaussian residuals and to give conditions for stationarity by the help of discrete Wiener models.
AMS subject classification: 60G10

1. INTRODUCTION

In a recent book T. Subba Rao and Gabr (1984) have shown that bilinear models can be used to investigate stationary non-Gaussian time series. Models which are Quadratic in Input and Linear in Output (QUILO) have appeared in controll theory, C. Bruni (1974). By the help of multiple Wiener-Ito integral we construct the transfer function system of the QUILO model. The explicit forms of transfer functions make it possible to give necessary conditions and sufficient conditions for stationarity. The cases considered by Pristley (1978), Hannan (1982), Subba Rao and Gabr (1984) and Terdik (1985) are special cases of the QUILO model.

2. SPECTRAL REPRESENTATION OF A DISCRETE WIENER MODEL

The discrete Wiener model with output X_t and white noise input $w_t \in N(0,\sigma^2)$ is defined by the following formula:

$$X_t = \sum_{n=0}^{\infty} \sum_{\underline{k}^{(n)} \in Z_1^n} \sum_{\underline{s}^{(n)} \in Z_+^n} g_{\underline{k}^{(n)}}(\underline{s}^{(n)}) A_{\underline{k}^{(n)}}(W_{t-s}, \dots, W_{t-s_n}), \tag{1}$$

where $Z_+^n = \{\underline{s}^{(n)} \in Z_1^n | 0 \le s_n < \dots < s_1\}$,

$Z_1 = \{1,2,\dots\}$ and $A_{\underline{k}^{(n)}}$ is the Q-polynomial system, Schetzen (1980), i.e.

$$A_{\underline{k}^{(n)}}(W_{t-s}, \dots, W_{t-s_n}) = \prod_{j=1}^{n} H_{k_j}(W_{t-s_j}) \quad ,$$

H_{k_j} denotes the k_j-th Hermite polynomial with leading coefficient 1 and parameter σ^2. We suppose that the output process X_t has finite second moment

$$\sum_{n=0}^{\infty} \sum_{\underline{k}^{(n)} \in Z_1^n} \sum_{\underline{s}^{(n)} \in Z_+^n} \underline{k}^{(n)}! \, \sigma^{2|\underline{k}^{(n)}|} g^2_{\underline{k}^{(n)}}(\underline{s}^{(n)}) < \infty \,. \tag{2}$$

Where $\underline{k}^{(n)}! = \prod_{j=1}^{n} k_j!$ and $|\underline{k}^{(n)}| = \sum_{j=1}^{n} k_j$. In this case it is easy to see that X_t is stationary in second order. The spectral measure F of W_t is the modified Lebesque measure $\sigma^2/2\pi \, d\omega$ on $D = [-\pi,\pi]$. Therefore from the Itô's formula, Major (1981), it follows that if $\underline{k}^{(n)} \in Z_1^n$ and $K_\ell = \sum_{j=1}^{\ell} k_j$, $K_0 = 0$, then

$$A_{\underline{k}^{(n)}}(W_{t-s_1}, \dots, W_{t-s_n}) = \int_{D^{|\underline{k}^{(n)}|}} \exp(it \sum_{j=1}^{K_n} \omega_j - i \sum_{\ell=1}^{n} s_\ell \sum_{j=K_{\ell-1}}^{K_\ell} \omega_j) W(d\underline{\omega}^{(K_n)}),$$

where $W(d\omega)$ denotes the random spectral measure of the series W_t.

Let the Fourier transform of the r-th degree kernels $g_{\underline{k}^{(n)}}(\underline{s}^{(n)})$, $|\underline{k}^{(n)}| = r$ be

$$\hat{g}_r(\underline{\omega}^{(r)}) = \sum_{n=1}^{r} \sum_{\substack{\underline{k}^{(n)} \in Z_1^n \\ |\underline{k}^{(n)}| = r}} \sum_{\underline{s}^{(n)} \in Z_+^n} g_{\underline{k}^{(n)}}(\underline{s}^{(n)}) e^{-i \sum_{\ell=1}^{n} \sum_{j=K_{\ell-1}}^{K_\ell} \omega_j} \tag{3}$$

and its symmetrization be $\tilde{g}_r = \text{sym}\ g_r$. We define $\tilde{g}_r$ as Priestly (1978) and also $\hat{g}_r$ an r-th degree transfer function of the Wiener model. As the multiple Wiener-Itô integral both $\hat{g}_r$ and $\tilde{g}_r$ are equal, we can write the output X_t in the form

$$X_t = \sum_{r=0}^{\infty} \int_{D^r} \tilde{g}_r(\underline{\omega}^{(r)}) e^{it \sum_{j=1}^{r} \omega_j} W(d\underline{\omega}^{(r)}) \quad , \tag{4}$$

i.e. the output X_t is sub-ordinated to the input $\{W_t\}$ with transfer functions $(\tilde{g}_o, \tilde{g}_1, \ldots, \tilde{g}_r, \ldots)$. The Wiener model has such a special type of transfer functions that the coefficients with non-negative indices of its inverse Fourier transform are only different from zero. This type of sub-ordinated series X_t will be referred to later as *regular*, i.e. when X_t is measurable with respect to the σ-algebra $\mathcal{B}_t = \mathcal{B}\ (W_s,\ s \le t)$.

3. TRANSFER FUNCTIONS OF THE QUILO MODEL

The QUILO $(P,Q,R,S,T_1,T_2,Z_1,Z_2,Z_3)$ model is defined by the equation

$$\sum_{k=0}^{P} a_k X_{t-k} = \sum_{k=0}^{Q} b_k W_{t-k} + \sum_{\ell=1}^{R} \sum_{k=0}^{S} C_{\ell,\ell+k} X_{t-\ell-k} W_{t-\ell} +$$

$$+ \sum_{\ell=1}^{T_1} \sum_{k=0}^{T_2} d_{\ell,\ell+k} W_{t-\ell-k} W_{t-\ell} + \tag{5}$$

$$+ \sum_{\ell=1}^{Z_1} \sum_{k=0}^{Z_2} \sum_{j=0}^{Z_3} \Psi_{\ell,\ell+k,\ \ell+k+j} X_{t-\ell-k-j} W_{t-\ell-k} W_{t-\ell}$$

where $a_o = b_o = 1$ and W_t are i.i.d. $\mathcal{N}(0,\sigma^2)$ random variables. We are giving the transfer functions of (5) when X_t is regularly stationary. Using the diagram formula for multiple Wiener-Itô integral we get a.,

$$W_{t-\ell-k}W_{t-\ell}=\int_{D^2} e^{i[(t-\ell-k)\omega_1+(t-\ell)\omega_2]}\, W(d\omega_1,d\omega_2)+\delta_0^k\sigma^2 ,$$

b.,

$$X_{t-\ell-k}W_{t-\ell}=\int_D e^{i(t-\ell)\omega}W(d\omega)\sum_{r=0}^{\infty}\int_{D^r}\hat{g}_r(\underline{\omega}^{(r)})e^{i(t-\ell-k)\sum_{j=1}^{r}\omega_j}\, W(d\underline{\omega}^{(r)}) =$$

$$=\sum_{r=0}^{\infty}\int_{D^{r+1}}\tilde{g}_r(\underline{\omega}^{(r)})e^{i[(t-\ell-k)\sum_{j=1}^{r}\omega_j+(t-\ell)\omega_{r+1}]}\, W(d\underline{\omega}^{(r+1)})+\delta_1^r\delta_\ell^0\sigma^2 ,$$

c.,

$$X_{t-k-\ell-j}\, W_{t-\ell-k}\, W_{t-k} =$$

$$=\int_D e^{i(t-\ell-k)\omega}W(d\omega)\int_D e^{i(t-k)\omega}W(d\omega)\sum_{r=0}^{\infty}\int_{D^r}\tilde{g}_r(\underline{\omega}^{(r)})e^{i(t-k-\ell-j)\sum_{n=1}^{r}\omega_n}\, W(d\underline{\omega}^{(r)})=$$

$$=\sum_{r=0}^{\infty}\int_{D^{r+2}}\tilde{g}_r(\underline{\omega}^{(r)})e^{i[(t-k-\ell-j)\sum_{n=1}^{r}\omega_n+(t-\ell-k)\omega_{r+1}+(t-k)\omega_{r+2}]}\, W(d\underline{\omega}^{(r+2)}) +$$

$$+\delta_r^1\delta_j^0\sigma^2\int_D e^{i(t-k)\omega}W(d\omega)+\delta_\ell^0\sigma^2\sum_{r=0}^{\infty}\int_{D^r}\tilde{g}_r(\underline{\omega}^{(r)})e^{i(t-k-j)\sum_{n=1}^{r}\omega_n}\, W(d\underline{\omega}^{(r)}) +$$

$$+\delta_r^1\delta_\ell^0\delta_j^0\sigma^2\int_D e^{i(t-k)\omega}\, W(d\omega) + \delta_r^0\delta_\ell^0\sigma^2\tilde{g}_0 ,$$

where δ_j^k is the Kronecker symbol. Now from (5) we get the following equations for the transfer functions

$$\sum_{k=0}^{P} a_k \tilde{g}_o = \sigma^2 \sum_{\ell=1}^{R} C_{\ell,\ell} + \sigma^2 \tilde{g}_o \sum_{\ell=1}^{Z_1} \sum_{j=0}^{Z_2} f_{\ell,\ell,\ell+j} \quad ,$$

$$\tilde{g}_1(\omega) \sum_{k=0}^{P} a_k e^{-ik\omega} = \sum_{k=0}^{Q} b_k e^{-ik\omega} + \tilde{g}_o \sum_{\ell=1}^{R} e^{-i\ell\omega} \sum_{k=0}^{S} C_{\ell,\ell+k} +$$

$$+ \tilde{g}_1(\omega)\sigma^2 \sum_{\ell=1}^{Z_1} \sum_{j=0}^{Z_3} f_{\ell,\ell,\ell+j}\, e^{-i(\ell+j)\omega} +$$

$$+ \sigma^2 \sum_{\ell=1}^{Z_1} \sum_{k=0}^{Z_2} f_{\ell,\ell+k,\ \ell+k}\, e^{-ik\omega} +$$

$$+ \sigma^2 \sum_{\ell=1}^{Z_1} f_{\ell,\ell,\ell}\, e^{-i\ell\omega} \ ,$$

$$\hat{g}_2(\omega_1,\omega_2) \sum_{k=0}^{P} a_k e^{-ik(\omega_1+\omega_2)} = \sum_{\ell=1}^{T_1} \sum_{k=0}^{T_2} d_{\ell,\ell+k} e^{-i[(\ell+k)\omega_1+\ell\omega_2]} +$$

$$+ \tilde{g}_1(\omega_1) \sum_{\ell=1}^{R} \sum_{k=0}^{S} C_{\ell,\ell+k} e^{-i[(\ell+k)\omega_1+\ell\omega_2]} +$$

$$+ \tilde{g}_o \sum_{\ell=1}^{Z_1} \sum_{k=0}^{Z_2} \sum_{j=0}^{Z_3} f_{\ell,\ell+k,\ell+k+j} e^{-i[(\ell+k)\omega_1+\ell\omega_2]} +$$

$$+ \sigma^2 \hat{g}_2(\omega_1,\omega_2) \sum_{\ell=1}^{Z_1} \sum_{j=0}^{Z_2} f_{\ell,\ell,\ell+j} e^{-i(\ell+j)(\omega_1+\omega_2)} \ ,$$

$$\hat{g}_r(\underline{\omega}^{(r)}) \sum_{k=0}^{P} a_k e^{-ik\sum_{n=1}^{r}\omega_n} =$$

$$= \hat{g}_{r-1}(\underline{\omega}^{(r-1)}) \sum_{\ell=1}^{R}\sum_{k=0}^{S} C_{\ell,\ell+k}\, e^{-i[(\ell+k)\sum_{n=1}^{r-1}\omega_n+\ell\omega_r]} +$$

$$+ \hat{g}_{r-2}(\underline{\omega}^{(r-2)}) \sum_{\ell=1}^{Z_1}\sum_{k=0}^{Z_2}\sum_{j=0}^{Z_3} f_{\ell,\ell+k,\ell+k+j}\, e^{-i[(\ell+k+j)\sum_{n=1}^{r-2}\omega_n+(\ell+k)\omega_{r-1}+\ell\omega_r]} +$$

$$+ \sigma^2\hat{g}_r(\underline{\omega}^{(r)}) \sum_{\ell=1}^{Z_1}\sum_{j=0}^{Z_3} f_{\ell,\ell,\ell+j}\, e^{-i(\ell+j)\sum_{n=1}^{r}\omega_n} .$$

We proved the following theorem

Theorem 1. If the QUILO model has a regular stationary solution X_t and all the roots of the polynomial

$$\sum_{k=0}^{P} a_k\, y^k + \sigma^2 \sum_{\ell=1}^{Z_1}\sum_{j=0}^{Z_2} f_{\ell,\ell,\ell+j}\, y^{\ell+j} \tag{6}$$

associated with the model (5) are outside of the unit circle, then the transfer functions of X_t are given by the symmetrization of the following functions:

$$\hat{g}_o = \sigma^2\left(\sum_{\ell=1}^{R} C_{\ell,\ell} + \sum_{\ell=1}^{T_1} d_{\ell,\ell}\right)\left(\sum_{k=0}^{P} a_k - \sigma^2\sum_{\ell=1}^{Z_1}\sum_{j=0}^{Z_3} f_{\ell,\ell,\ell+j}\right)^{-1} ,$$

$$\hat{g}_1(\omega)=\left(\sum_{k=0}^{Q} b_k e^{-ik\omega} + \hat{g}_o \sum_{\ell=1}^{R} e^{-i\ell\omega}\sum_{k=0}^{S} C_{\ell,\ell+k} + \sigma^2\sum_{\ell=1}^{Z_1}\sum_{k=0}^{Z_2} f_{\ell,\ell+k,\ell+k}\, e^{-ik\omega} +\right.$$

$$\left. + \sigma^2\sum_{\ell=1}^{Z_1} f_{\ell,\ell,\ell} e^{-i\ell\omega}\right)\left(\sum_{k=0}^{P} a_k e^{-ik\omega} - \sigma^2\sum_{\ell=1}^{Z_1}\sum_{j=0}^{Z_3} f_{\ell,\ell,\ell+j} e^{-i(\ell+j)\omega}\right)^{-1} ,$$

$$\hat{g}_2(\omega_1,\omega_2) = \left\{\sum_{\ell=1}^{T_1}\sum_{k=0}^{T_2} d_{\ell,\ell+k}\, e^{-i[(\ell+k)\omega_1+\ell\omega_2]} + \right.$$

$$+ \hat{g}_1(\omega_1)\sum_{\ell=1}^{R}\sum_{k=0}^{S} c_{\ell,\ell+k}\, e^{-i[(\ell+k)\omega_1+\ell\omega_2]} +$$

$$\left. + \hat{g}_0 \sum_{\ell=1}^{Z_1}\sum_{k=0}^{Z_2}\sum_{j=0}^{Z_3} f_{\ell,\ell+k,\ell+k+j}\, e^{-i[(\ell+k)\omega_1+\ell\omega_2]}\right\} \times$$

$$\times \left\{\sum_{k=0}^{P} a_k e^{-ik(\omega_1+\omega_2)} - \sigma^2 \sum_{\ell=1}^{Z_1}\sum_{j=0}^{Z_3} f_{\ell,\ell,\ell+j} e^{-i(\ell+j)(\omega_1+\omega_2)}\right\}^{-1} ,$$

$$\hat{g}_r(\underline{\omega}^{(r)}) = \left(\hat{g}_{r-1}(\underline{\omega}^{(r-1)})\sum_{\ell=1}^{R}\sum_{k=0}^{S} c_{\ell,\ell+k}\, e^{-i[(\ell+k)\sum_{n=1}^{r-1}\omega_n+\ell\omega_r]} + \right.$$

$$\left. + \hat{g}_{r-2}(\underline{\omega}^{(r-2)})\sum_{\ell=1}^{Z_1}\sum_{k=0}^{Z_2}\sum_{j=0}^{Z_3} f_{\ell,\ell+k,\ell+k+j} e^{-i[(\ell+k+j)\sum_{n=1}^{r-2}\omega_n+(\ell+k)\omega_{r-1}+\ell\omega_r]}\right\} \times$$

$$\times \left\{\sum_{k=0}^{P} a_k e^{-ik\sum_{n=1}^{r}\omega_n} - \sigma^2 \sum_{\ell=1}^{Z_1}\sum_{j=0}^{Z_3} f_{\ell,\ell,\ell+j}\, e^{-i(\ell+j)\sum_{n=1}^{r}\omega_n}\right\}^{-1} .$$

4. CONDITIONS FOR STATIONARITY

We are now in the position to give conditions for stationarity of QUILO models. Let the Wiener model X_t with Gaussian white noise input $W_t \in (0,\sigma^2)$ be given in the form

$$X_t = \sum_{r=0}^{\infty} \int_{D^r} \tilde{g}_r(\underline{\omega}^{(r)})\, e^{it\sum_{n=1}^{r}\omega_n} W(d\underline{\omega}^{(r)}) .$$

The necessary and sufficient condition for regular stationarity of X_t is that

$$\sum_{r=0}^{\infty} r!\,||\tilde{g}_r(\underline{\omega}^{(r)})||_F^2 < \infty \tag{7}$$

The condition (7) is equivalent to the condition

$$\sum_{n=0}^{\infty} \sum_{\underline{k}^{(n)}\in Z_1^n} \sum_{\underline{s}^{(n)}\in Z_+^n} \sigma^{2|\underline{k}^{(n)}|}\, \underline{k}^{(n)}!\, g^2_{\underline{k}^{(n)}}(\underline{s}^{(n)})<\infty \ . \tag{8}$$

To check conditions (7) and (8) for concrete QUILO models even in the simplest cases are very hard because of the symmetrization of the transfer functions. Therefore, we use the transfer functions $\hat{g}_r(\underline{\omega}^{(r)})$ and consider the pure QUILO model i.e. R=S=0.

Theorem 2. I. The necessary condition that the QUILO ($P,Q,0,0,T_1,T_2,Z_1,Z_2,Z_3$) model (5), where $f_{\ell,\ell+k,\ell+k+j}=0$, if $k+j>0$, be regular stationary is that

a., all the roots of the polynomial (6) be outside of the unit circle and

b.,

$$\frac{\sigma^2}{2\pi}\int_D \left| \frac{\sum_{\ell=1}^{Z_1} f_{\ell,\ell,\ell}\, e^{-i\ell\omega}}{\sum_{m=0}^{P} a_m e^{-im\omega} - \sigma^2 \sum_{\ell=1}^{Z_1} f_{\ell,\ell,\ell} e^{-ie\omega}} \right|^2 d\omega = \varepsilon < 1 \ . \tag{9}$$

II. The sufficient condition that the QUILO ($P,Q,0,0,T_1,T_2,Z_1,Z_2,Z_3$) model (5) be regular stationary is that fulfill I.a., and that c., $\sqrt[3]{6}\,\varepsilon < 1$, where ε is given by (9).

Proof. Put

$$\alpha(\omega) = \sum_{m=0}^{P} a_m e^{-im\omega} - \sigma^2 \sum_{\ell=1}^{Z_1} \sum_{j=0}^{Z_3} f_{\ell,\ell,\ell+j}\, e^{-i(\ell+j)\omega}, \tag{10}$$

$$\gamma(\underline{\omega}^{(r)}) = \sum_{\ell=1}^{Z_1} \sum_{k=0}^{Z_2} \sum_{j=0}^{Z_3} f_{\ell,\ell+k,\ell+k+j}\, e^{-i[(\ell-k+j)\sum_{n=1}^{r-2}\omega_n + (\ell+k)\omega_{r-1} + \ell\omega_r}.$$

By the Theorem 1 the transfer functions of the model are given by

$$\hat{g}_r(\underline{\omega}^{(r)}) = \hat{g}_{r-2}(\underline{\omega}^{(r-2)})\ \gamma(\underline{\omega}^{(r)})\alpha^{-1}\left(\sum_{n=1}^{r}\omega_n\right),$$

i.e.

$$\hat{g}_{2r}(\underline{\omega}^{(2r)}) = \prod_{m=2}^{r} \gamma(\underline{\omega}^{(2m)}) \cdot \alpha^{-1}\left(\sum_{n=1}^{2m}\omega_n\right)\hat{g}_2(\omega_1,\omega_2),$$

and

$$\hat{g}_{2r+1}(\underline{\omega}^{(2r+1)}) = \hat{g}_1(\omega_1) \sum_{m=1}^{r} \gamma(\underline{\omega}^{(2m+1)})\alpha^{-1}\left(\sum_{n=1}^{2m+1}\omega_n\right),$$

respectivelly. From this follows that

$$\|\hat{g}_{2r+1}(\underline{\omega}^{(2r+1)})\|_F = \frac{\sigma^{2(r2+1)}}{(2\pi)^{2r+1}} \left(\int_D \left|\frac{\gamma(\omega)}{\alpha(\omega)}\right|^2 d\omega\right)^{2r} \int_D |\hat{g}_1(\omega)|^2 d\omega.$$

Taking into consideration that $\underline{k}^{(n)}! \le (3!)^{r/3}$ if $|\underline{k}^{(n)}| = r$ and $1 \le k_j \le 3$, $j=1,2,\ldots,$ we proved the Theorem 2.

The same reasoning leads to the following result.

Theorem 3. Under the condition I.a. of Theorem 2 and

$$\sqrt[3]{6}\ \frac{\sigma^2}{2\pi} \int_D \left| \frac{\gamma(\underline{\omega}^{(3)})}{\alpha\left(\sum_{j=1}^{3} \omega_j\right)} \right|^2 d\omega_1 < 1 \quad ,$$

for all $(\omega_2, \omega_3) \in D^2$, the QUILO $(P,Q,0,0,T_1,T_2,Z_1,Z_2,Z_3)$ model has a regular stationary solution.

Example: The necessary condition that the model

$$X_t + aX_{t-1} = \sum_{k=0}^{Q} b_k W_{t-k} + \sum_{k=1}^{T_1} \sum_{\ell=0}^{T_2} C_{k,k+\ell} W_{t-k} W_{t-k-\ell} +$$

$$+ f\, X_{t-1} W_{t-1}^2$$

be regular stationary is that

$$\frac{\sigma^2}{2\pi} f^2 + (\sigma^2 f - a)^2 < 1 \quad .$$

The sufficient condition is that

$$\sqrt[3]{6}\left(\frac{\sigma^2}{2\pi} f^2 + (\sigma^2 f - a)^2\right) < 1 \quad .$$

REFERENCES

Arató, M. (1982), Linear Stochastic Systems with Constant Coefficients, Lecture Notes in Control and Inf. 45, Springer Verlag.

Brillinger, D. R. (1970), The identification of polynomial systems by means of higher oerder spectra. J. Sound Vib. 12, 301-314.

Bruni, C. Di Pillo, G. Koch, G. (1974), An appealing class of "Nearly linear" systems in theory and application, IEEE Transaction on Aut. and Contr. vol. AC-19, No 4, pp. 334.

Dobrushin, R. L. (1979), Gaussian and their subordinated self-similar random generalized fields. Ann. Prob. 7, 1-28.

Granger, C.W.F. and Anderson, A.P. (1978), An Introduction to Bilinear Time Series Models. Vandenhoek and Rjprecht Gottingen.

Hannan, E.F. (1982), A note on bilinear time series models. Stochastic Processes and their Applications 12, 221-224.

Helson, H. and Lowdenslager, D. (1958), Prediction theory and Fourier series in several variables. Acta Mathematica 99, 165-202.

Major, P. (1981), Multiple Wiener-Ito Integrals. Lecture Notes in Maths., 849, New York: Springer-Verlag.

Priestley, M.B. (1978), Nonlinear models in time series analysis. The Statistician 27, 159-176.

Quinn, B.G. (1982), Stationarity and invertibility of simple bilinear models. Stochastic Processes and their Applications 12, 225-230.

Ruberti, A., Isidori, A. and d'Allessandro, P. (1972), Theory of Bilinear Dynamical Systems, New York, Springer-Verlag.

Rudin, W. (1967), Fourier Analysis on Groups. New York: Wiley.

Schetczen, M. (1980), The Volterra and Wiener Theories of Nonlinear Systems. New York: Wiley.

Subba Rao, T. (1981), On the theory of bilinear time series models. F.R. Statistic. Soc. B 43, 244-255.

Subba Rao, T. and Gabr, M.M. (1984), An Introduction to Bispectral Analysis and Bilinear Time Series Models. Lecture Notes in Statistics 24, New York: Springer-Verlag.

Terdik, G. (1985), Transfer functions an conditions for stationarity of bilinear models with Gaussian residuals Proc. R. Soc. Lond. A 400, 315-330.

Terdik György
Department of Mathematics
Kossuth L. University,
H-4010. Debrecen, Pf. 12.
Hungary

Proc. of the 5th Pannonian Symp. on Math. Stat., Visegrád, Hungary 1985
W. Grossmann, J. Mogyoródi, I. Vincze, W. Wertz, eds

UNIFORM ALMOST SURE CONVERGENCE OF WEIGHTED SUMS AND APPLICATIONS IN REGRESSION MODELS

R. THRUM

Berlin (GDR)

It is shown that for i.i.d. r.v!s $X_1,\dots,X_n,\dots$ with expectation zero and finite p-th absolute moment $(p\geq 2)$ the weightes sums $\sum_{i=1}^{n} a_{n,i}X_i \,/\, n^{1/p}$ tend to zero a.s. provided that $\sum_{1\leq i\leq m} a^2_{n,i} \equiv 1$. Moreover this convergence is uniform with respect ot $\{a_{n,i}\}$. The method of proof is based on two "elementary" tools. An inequality for moments of quadratic forms is used and a special Toeplitz summation procedure is defined which allows the application of a strong law of large numbers.

AMS subject classification: 60F62

I. MAIN RESULT

Theorem 1. Let $\{X_i/i\geq 1\}$ be a sequence of independent identically distributed random variables (i.i.d. r.v.'s) with $E|X_1|^p < \infty$ for some $p \geq 2$ and $EX_1 = 0$. Suppose that $\{a_{n,i}, i=1,\dots,n, n\geq 1\}$ is a sequence of nonrandom coefficients normalized by $\sum_{i=1}^{n} a^2_{n,i} = 1$ for all $n \geq 1$. Then $U_n = \sum_{1\leq i\leq n} a_{n,i}\, X_i \,/\, n^{1/p}$ tends uniformly almost surely (a.s.) to zero in the sense that for arbitrary $\varepsilon > 0$

$$\sup P(\sup_{n\geq N}|U_n| > \varepsilon) \xrightarrow[N\to\infty]{} 0$$

where the first supreum is taken over all normalized sequences $\{a_{n,i}\}$.

Remarks.

1. Without taking the supremum over all normalized sequences

$\{a_{n,i}\}$ and for $p = 2$ this theorem and related statements were proved by Y.S. Chow (1966). While the proof of Chow uses truncation techniques our method is "elementary" and seems to be applicable to other problems, too. Prof. H. Läuter brought ot my attention the usefulness of a uniform version of the theorem in order to show strong consistency of least-square estimators in nonlinear regression where the coefficients $a_{n,i}$ usually depend on unknown parameters. This stimulated a uniform version. Here we outline the proof with main emphasis on a more explicit formulation of a certain stochastic Toeplitz lemma (Lemma 3).

2. Consider the extreme cases $a_{n,k} \equiv 1/n^{1/2}$ and $a_{n,n} = 1$, $a_{n,k} = 0$ for $k < n$. In the first case one get with $U_n = \sum_{1 \leq i \leq n} X_i / n^{1/p + 1/2}$ a classical strong law of large numbers. The second case with $U_n = X_n / n^{1/p}$ shows that the theorem cannot be improved. Indeed, for i.i.d. r.v.'s $X_n/n^{1/p}$ tends to zero a.s. if and only if $E|X_1|^p < \infty$.

II. APPLICATIONS

1. Least-square estimators in linear regression

Here and in the following E denotes expectation an T transpose. Consider the least-square estimator $\hat{\theta}_n = (F_n^T F_n)^{-1} F_n^T Y_n$ where it is assumed that the regression matrices

$$F_n = \begin{pmatrix} f_1^T \\ \vdots \\ f_n^T \end{pmatrix}$$

have full rank $r < n$ and Y_n follows a linear model $Y_n - F_n\theta = \varepsilon_n = (X_1, \ldots, X_n)^T$ with i.i.d. error variables X_i. Define for $0 \neq e \in R^r$, $e^T(\hat{\theta}_n - \theta) =$

$$(e^T (F_n^T F_n)^{-1} e \cdot n^{2/p})^{1/2} \left(\sum_{1 \leq k \leq n} a_{n,k} X_k / n^{1/p} \right) .$$

Then $a_{n,k} = e^T (F_n^T F_n)^{-1} f_k \, (e^T (F_n^T F_n)^{-1} e)^{1/2}$ and $\sum_{K=1}^{N} a_{n,k}^2 = 1$. If $EX_1 = 0$, $E|X_1|^p < \infty$ and $\lambda_{max}(F_n^T F_n)^{-1} = 0(n^{2/p})$ then theorem 1 implies the strong consistency of $\hat{\theta}_n$. However, the best possible result for this problem is due to T.L. Lai, H.Robbins and C.Z. Wei (1979) who showed even under weaker assumptions on the error variables X_i but exploiting the special structure of the $a_{n,i}$ in this model that $\lambda_{max}(F_n^T F_n)^{-1} = o(1)$ is a necessary and sufficient condition for $\hat{\theta}_n \xrightarrow[N\to\infty]{} \theta$ a.s.

2. Strong consistency of the sample variance

If $P_n = \sum_{i=1}^{r} p_n(i) p_n(i)^T$ denotes the orthogonal projector on the range of an arbitrary regression matrix $F_n = \begin{pmatrix} f_{n,1}^T \\ \vdots \\ f_{n,n}^T \end{pmatrix}$ then the sample variance $s_n^2 = \|Y_n - P_n Y_n\|^2 / n$ is a strongly consistent estimator for $E\,X_1^2 = \sigma^2$ (see e.g. [4]). Under the previous assumptions the following refinement is true:

Corollary: For all $\varepsilon > 0$ $P(\sup_{n\geq N} n^{(p-2)/p} |(s_n^2 - \sigma^2)| > \varepsilon)$ tends to zero, uniformly with respect to $\{F_n\}$, if $N \to \infty$.

Proof:

$$n^{(p-2)/p}(s_n^2 - \delta^2) = \left[n^{(p-2)/p} \left(\sum_{i=1}^{n} (X_i^2 - \delta^2) / n \right) \right] - \sum_{j=1}^{r} U_n^2(j)$$

where $U_n(j) = \varepsilon_n^T p_n(j) / n^{1/p} = \sum_{1\leq i\leq n} a_{n,i}(j) X_i / n^{1/p}$

and $\sum_{1\leq i\leq n} a_{n,i}^2(j) = \|p_n(j)\|^2 = 1$. To every $U_n(j)$ $j=1,\ldots,r$ Theorem 1 can be applied while the classical strong law of large numbers (Lemma 2) insures $\sum_{1\leq i\leq n} (X_i^2 - \sigma^2) / n^{2/p} \xrightarrow[n\to\infty]{} 0$ a.s.

3. Nonlinear regression

A standard method in nonlinear regression ($Y_i = f_i(\theta) + X_i$, $i = 1,\ldots,n$) is the minimization of the sum of squares $S_n(\theta) = \sum_{1\leq i\leq n} (Y_i - f_i(\theta))^2$ with respect to θ in order to obtain an estimator $\hat{\theta}_n$. If θ_o is the parameter of the "true" distribution then

$$S_n(\theta) = S_n(\theta_o) + d_n^2(\theta,\theta_o) + 2\, d_n(\theta,\theta_o)\, U_n(\theta)$$

where

$$d_n^2(\theta,\theta_o) = \sum_{i=1}^{n} (f_i(\theta) - f_i(\theta_o))^2 \quad \text{and}$$

$$U_n(\theta) = \sum_{i=1}^{n} ((f_i(\theta_o) - f_i(\theta))/ d_n(\theta,\theta_o))X_i \; / \; n^{1/p} \quad .$$

From this representation and the uniform version of the theorem in the case p=2 H. Läuter proved a theorem concerning the strong consistency of $\hat{\theta}_n$. Using his idea the conditions on the desing in terms of $d_n(\theta,\theta_o)$ can be weakened if $p > 2$.

III. AUXILARY RESULTS

<u>Lemma 1:</u> Suppose that $\{X_n / n \geq 1\}$ is a sequence of independent r.v.'s with $E|X_i|^p < \infty$ for some $p \geq 2$, $EX_i = 0$ for $i \geq 1$. Let $\{a_{i,j} \mid i \geq 1;\ j \geq 1\}$ be a double array of real numbers with $a_{i,j} = a_{j,i}$ for $i \neq j$ and $a_{i,i} = 0$ for $i \geq 1$. Then for $Z_n = \sum_{i,j \leq n} a_{i,j} X_i X_j$ there exists a constant C(p) not depending on n such that

$$E|Z_n|^p \leq c(p) \; (E\, Z_n^2)^{p/2} \left(\max \{E\, X_i^{\;p} \,/ (E\, X_i^2)^{p/2} \mid i=1,\ldots,n;\ E\, X_i^2=0\} \right)^2 .$$

The proof makes use of Burkholders inequality for martingale differences and an inequality of P. Whittle for linear forms and is omitted here.

<u>Lemma 2:</u> If $\{X_i \mid i \geq 1\}$ is a sequence of i.i.d. r.v.'s with $E|X_1|^p < \infty$ for $p \in (0,2)$ then $\sum_{i=1}^{n} (X_i - a)/n^{1/p} \xrightarrow[n\to\infty]{} 0$ a.s. where $a = EX_1$ if $p \in [1,2)$ and arbitrary finite otherwise.

This is the law of large numbers given in Neveu's book [2] (Theorem IV.7.1.)

<u>Lemma 3:</u> Let $\{c_{n,k} \mid n \geq 1;\ k \geq 1\}$ be a double array of real numbers for which

1. $\sum_{i=1}^{\infty} |c_{n,i}| \leq c$ for $n \geq 1$,

2. $\lim_{n\to\infty} c_{n,i} = 0$ for $i \geq 1$.

Further let $S_1,\dots,S_n,\dots$ be a sequences of r.v.'s for which $S_n \xrightarrow[n\to\infty]{} 0$ a.s.. Then to given $\varepsilon > 0$ and $\delta > 0$ there exists a natural number $N_0 = N(\varepsilon, \delta, c)$ such that $P(\sup_{n\geq N} |\sum_{i=1}^{\infty} c_{n,i} S_i| > \varepsilon) \leq \delta$ for $N \geq N_0$. In particular $\sum_{i=1}^{\infty} c_{n,i} S_i \xrightarrow[n\to\infty]{} 0$ a.s.

Remark: $\sum_{i=1}^{\infty} c_{n,i} S_i$ is well defined because of

$$|\sum_{i=1}^{\infty} c_{n,i} S_i| \leq \sum_{i=1}^{\infty} |c_{n,i}||S_i| \leq (\sup_{i\geq 1} |S_i|) < \infty \qquad \text{a.s.}$$

IV. PROOFS

1. Proof of Theorem 1. The square of U_n can be written as $U_n^2 = V_n + W_n$ where $V_n = \sum_{i=1}^{n} a_{n,i}^2 \; X_i^2 / n^{2/p}$ and $W_n = \sum_{i\neq j} a_{n,i} \, a_{n,j} X_i X_j \, / \, n^{2/p}$. Because of $n^{4/p} \, E \, W_n^2 = 2\left(\sum_{i\neq j} (a_{n,i} \, a_{n,j})^2\right) (E \, X_1^2)^2 \leq 2 (E \, X_1^2)^2$ (see e.g. [3]) and from Lemma 1 it follows that $E|W_n|^p \leq B / n^2$ where B does not depend on n. Therefore $P(\sup_{n\geq N} |W_n| > \;) \leq B/\varepsilon^p \, (\sum_{i\geq N} i^{-2}) \xrightarrow[N\to\infty]{} 0$.

Secondly, observe that $0 \leq V_n \leq \tilde{V}_n + E \, X_1^2 / n^{2/p}$ and $|\tilde{V}_n| \leq |\tilde{U}_n|$ where $\tilde{V}_n = \sum_{i=1}^{n} a_{n,i}^2 \; \tilde{X}_i / n^{2/p}$; $\tilde{U}_n = \sum_{i=1}^{n} \tilde{a}_{n,i} \; \tilde{X}_i / n^{2/p}$; $\tilde{a}_{n,i} = a_{n,i}^2 \, (\sum_{i=1}^{n} a_{n,i}^4)^{-1/2}$; $\tilde{X}_i = X_i^2 - E \, X_1^2$. Now let us consider two cases: If $p \in [2,4)$ then $\tilde{p} = p/2 \in [1,2)$ and by rearranging terms $\tilde{V}_n = \sum_{i=1}^{\infty} c_{n,i} S_i$ where $S_i = \sum_{j=1}^{i} \tilde{X}_j / i^{1/\tilde{p}}$ and $c_{n,i} = (i/n)^{1/\tilde{p}} \, (a_{n,i}^2 - a_{n,i+1}^2)$ if $1 \leq i \leq n-1$, $c_{n,n} = a_{n,n}^2$, $c_{n,j} = 0$ if $j > n$. Because of $E|\tilde{X}_i|^{\tilde{p}} < \infty$, $E\tilde{X}_i = 0$, $\sum_{i=1}^{\infty} |c_{n,i}| \leq 2$ and $|c_{n,i}| \leq 2(i/n)^{1/\tilde{p}}$ for $i < n$ Lemma 2 and Lemma 3 can be applied and yield

$$\sup_{\{a_{n,i}\}} P(\sup_{n \geq N} |\tilde{V}_n| > \varepsilon) \quad \sup_{\{c_{n,i}\}} P(\sup_{n \geq N} |\tilde{V}_n| > \varepsilon) \xrightarrow[N\to\infty]{} 0$$

for all $\varepsilon > 0$. If otherwise $\tilde{p} = (p/2) \geq 2$ the same procedure can be repeated to $\tilde{U}_n$ which has the same structure as U_n. After a finite number of steps the desired result follows.

2. Proof of Lemma 3. To given $\varepsilon > 0$, $\delta > 0$ choose $M_o = M(\varepsilon, \delta, c)$ such that $P(c \sup_{k > M_o} |S_k| > \varepsilon) \leq \delta/2$. This is possible because of $s_n \xrightarrow[n\to\infty]{} 0$ a.s. By assumption 2. $d_k(i) = \sup_{n \geq k} |c_{n,i}|$ nonincreasing sequences with $d_k(i) \xrightarrow[k\to\infty]{} 0$ for $i = 1, \ldots, M_o$. Therefore one can find an $N(M_o, \varepsilon, \delta)$ such that

$P(d_N(i)|S_i| > \varepsilon | 2M_o) \leq \delta/2M_o$ for all $N \geq N(M_o, \varepsilon, \delta)$ and and $1 \leq i \leq M_o$. Hence for $N \; N(M_o, \varepsilon, \delta)$ it hold

$$P(\sup_{n \geq N} |\sum_{i=1}^{\infty} c_{n,i} \; S_i| > \varepsilon \quad \leq$$

$$P(\sup_{n \geq N} |\sum_{i=1}^{M_o} c_{n,i} \; S_i| > \varepsilon/2) + P(\sup_{n \geq N} |\sum_{i > M_o} c_{n,i} \; S_i| > \varepsilon/2) \leq$$

$$\sum_{i=1}^{M_o} P(d_N(i)|S_i| > \varepsilon/2M_o) + P(c \sup_{i > M} |S_i| > \varepsilon/2) \leq$$

$$M_o(\delta/2M_o) + \delta/2 = \delta \quad .$$

ACKNOWLEDGEMENT

I am grateful to Prof. H. Läuter for usefut discussions.

V. REFERENCES

1. Chow, Y.S. (1966), Some Convergence Theorems for Independent Random Variables. Ann. Math. Statist. 37, 1482-1492.
2. Neveu, J. (1964), Bases Matematique du Calcul de Probabilités. Masson et cie, Paris.
3. Kleffe, J., Thrum, R. (1983), Inequalities for Moments of

Quadratic Forms with Applications to a.s. Convergence.
4. Schmidt, W. (1976), Strong Consistency of Variance Estimation and Asymptotic Theory for Tests of the Linear Hypothesis in Multivariate Linear Models, Math. Operetionsforsch. Statist. 7, 701-705.
5. Lai,T.L., Robbins,H. and Wei,C.Z. (1979), Strong Consistency of Least Squares Estimates in Multiple Regression II. J.Multivariate Anal. 9,343-361.

R.Thrum,
Humboldt-University, Sektion Mathematik,
DDR 1086 Berlin
Unter den Linden 6, Postfach 1297.

Proc. of the 5th Pannonian Symp. on Math. Stat., Visegrád, Hungary 1985
W. Grossmann, J. Mogyoródi, I. Vincze, W. Wertz, eds

STRUCTURE OF THE STATE SPACE IN STOCHASTIC KINETICS

J. TÓTH

Budapest

The continuous time discrete state stochastic model of complex chemical reactions is defined: it is a Markovian jump process. A precondition of the uniqueness of the stationary distribution of the model is investigated. A necessary condition and a sufficient one are given in terms of quantities with chemical meaning called stoichiometric coefficients.
AMS subject classification: 60J27, 80A30.

1. INTRODUCTION

Let us consider a vessel containing a finite number of chemical components or species and let us suppose that among these components elementary reactions take place. As a result of these elementary reactions the quantities of the chemical components change. The usual stochastic model for this phenomenon (known for 50 years) is a continuous time discrete state Markovian jump process.

A branch of the literature on this model strives to state general results on all the reactions or on great classes of reactions simultaneously. The most important results in this spirit are due to T. G. Kurtz (see e.g. Kurtz (1981)), for further results see Arnold (1980) and works of other but general character by Lemarchand (1980), Érdi et al. (1981) and Tóth et al. (1983). These authors almost exclusively dealt with similarities and differences between the stochastic and the usual deterministic model.

The present paper is devoted to the general problems of the stochastic model of complex chemical reactions in itself. The purpose is to investigate irreducibility of the stochastic model. We define the graph of possible transitions corresponding to this continuous time discrete state Markovian jump processes as usual, and we investigate the connectedness of this graph in terms of the defining quantities (mostly in terms of the parameters called stoichiometric coefficients).

2. DEFINITIONS

Definition 1. A (complex chemical) *reaction* is a quadruple $\langle\mathcal{M},\mathcal{R},(\alpha,\beta),w\rangle$ consisting of a *mechanism* $\langle\mathcal{M},\mathcal{R},(\alpha,\beta)\rangle$ and a *kinetics* w, where

(i) $\mathcal{M}$ and $\mathcal{R}$ are finite sets of M and R elements respectively ($M,R \in \mathbb{N}$);

(ii) α and β are MxR matrices of nonnegative integers;

(iii) for all $r\in\mathcal{R}$ w_r is a real valued function on $\mathbb{R}^M$ or on $\mathbb{R}^{\mathcal{M}}$; (this kind of identification will be made here and below), i.e. $w:\mathcal{R} \to (\mathbb{R}_0^+)^{(\mathbb{R}^M)}$.

A usual way of displaying the reaction $\langle\mathcal{M},\mathcal{R},(\alpha,\beta),w\rangle$ is the following:

$$\sum_{m=1}^{M} \alpha(m,r)X(m) \xrightarrow{w_r} \sum_{m=1}^{M} \beta(m,r)X(m) \quad (r=1,\ldots,R) \tag{1}$$

and here (1) is called the rth *elementary reaction*, X(m) is the mth *chemical component* or *species*, the nonnegative integers $\alpha(m,r)$ and $\beta(m,r)$ are the *stoichiometric coefficients* expressing the number of X(m) needed to and produced in the rth elementary reaction. The matrix $\gamma:=\beta-\alpha$ and its columns $\gamma(\cdot,r)$ - *the elementary reaction vectors* - will often be used below.

Definition 2. The stochastic model of the reaction $\langle\mathcal{M},\mathcal{R},(\alpha,\beta),w\rangle$ is a continuous time discrete state Markovian jump process with the state space $\mathbb{Z}^M$ and with the infinitesimal transition probabilities

$$a_{np} := \sum_{r:\gamma(\cdot,r)=p-n} w_r(n) \qquad (n,p \in \mathbb{Z}^M).$$

The kinetics w usually fulfils either

(C) $w_r(n) > 0 \iff n \geq \alpha(\cdot,r) \qquad (r \in \mathcal{R},\ n \in \mathbb{Z}^M)$

or

(P) $w_r(n) > 0 \iff \operatorname{sign} n \gtreqless \operatorname{sign} \alpha(\cdot,r) \ (r \in \mathcal{R},\ n \in \mathbb{Z}^M)$.

An example to fulfil (C) is

$$\bar{\bar{w}}_r(n) := k(r) \prod_{m=1}^{M} \prod_{j_m=0}^{\alpha(m,r)-1} (n_m - j_m) \quad (r \in \mathcal{R}, k(r) \in \mathbb{R}^+,\ n \in \mathbb{Z}^M).$$

while an example to fulfil (P) is

$$\bar{w}_r(n) := k(r) \prod_{m=1}^{M} n_m^{\alpha(m,r)} \quad (r \in \mathcal{R},\ k(r) \in \mathbb{R}^+,\ n \in \mathbb{Z}^M)$$

E.g. the Poisson process is the stochastic model of the reaction $0 \to X$ (here 0 stands for a right hand side where all the stoichiometric coefficients are zero), similarly the linear birth and death process is the stochastic model of the reaction $X \to 2X$, $X \to 0$ with appropriate kinetics.

Definition 3. Let the directed graph (or synonimously, the relation) Γ of possible transitions be defined as follows:

$$\Gamma := \{(n,p) \in \mathbb{Z}^M \times \mathbb{Z}^M;\ \exists r \in \mathcal{R} (p-n = \gamma(\cdot,r)) \wedge (w_r(n) > 0)\}.$$

This graph is generally used in the theory of discrete state Markov processes. For an arbitrary set $X \subset \mathbb{Z}^M$ let us denote by $X_{\rightarrow}$ the union of the vertices of X and those vertices of Γ that can be reached from the vertices of X by a directed route.

3. RESULTS

The following statement obviously holds.

Remark 1. In the case of kinetics fulfilling condition (C) one has $(\mathbb{N}_0^M)_{\rightarrow} \subset \mathbb{N}_0^M$.

The meaning of this remark is that these types of kinetics reflect that aspect of physical reality that starting from a vector of nonnegative numbers of the chemical components the state of the process will remain a nonnegative vector for all time. This result is the stochastic analogue of a theorem by Vol'pert related to the deterministic case (Vol'pert, 1972).

The above statement is *not* true in the case of kinetics fulfilling condition (P) as it can be seen on the example $2X \to 0$. Here Γ is:

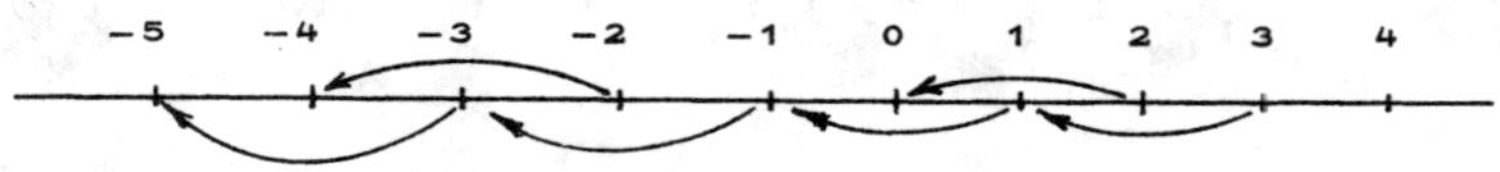

while in case (C) it is

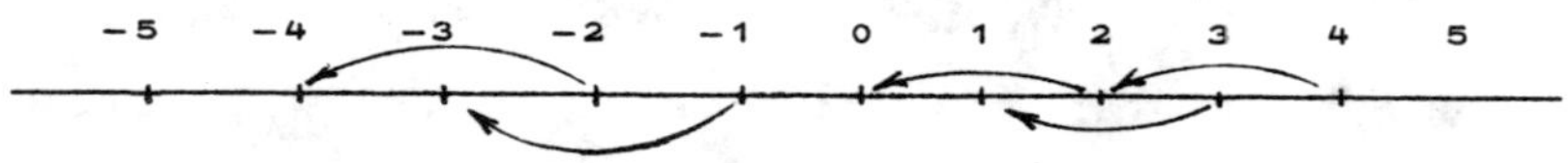

Further on we restrict the domain of the kinetics to $\mathbb{N}_0^M$. Γ will also be used in this naturally restricted sense.

A notion of central importance in chemical kinetics will be defined now.

Definition 4. The mechanism $\langle \mathcal{M}, \mathcal{R}, (\alpha, \beta) \rangle$ is said to be *reversible*, if for all $r \in \mathcal{R}$ there exists an $r' \in \mathcal{R}$ such that $\alpha(\cdot, r') = \beta(\cdot, r)$ and $\beta(\cdot, r') = \alpha(\cdot, r)$.

Being a natural assumption that $\alpha(\cdot, r) = \alpha(\cdot, r')$, $\beta(\cdot, r) = \beta(\cdot, r')$ implies $r = r'$ we take it for granted. Therefore uniqueness of r' in the definition is a consequence. Another easy consequence of the definition is the following statement.

Remark 2. Let us suppose that in the reaction $\langle \mathcal{M}, \mathcal{R}, (\alpha, \beta), w \rangle$ the underlying mechanism $\langle \mathcal{M}, \mathcal{R}, (\alpha, \beta) \rangle$ is reversible and the kinetics w satisfies (C), then Γ is symmetric.

The example $2X \to 0$ shows that the statement is generally false in the nonreversible case while the reaction $2X \rightleftharpoons X$ endowed with kinetics satisfying (P) shows that reversibility alone is not enough to imply the conclusion, as the corresponding Γ looks like

Finally, the reaction 0→X←2X endowed with a kinetics satisfying (C) has a symmetric Γ:

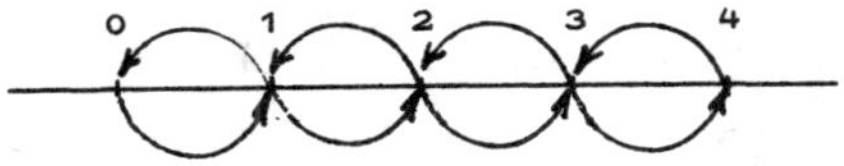

It is known that in the case of a finite state space connectedness of the graph of a Markov process or irreducibility is a necessary and sufficient condition for the stationary distribution to be unique. Irreducibility is obviously a necessary condition in the infinite state case too. In what follows, a necessary and a sufficient condition will be given for irreducibility.

It can be seen that neither the state space of the reaction 0→2X nor the state space of the reaction 0→X+Y is connected. These examples suggest necessary conditions.

Theorem 1. The stochastic model of a reaction endowed with arbitrary kinetics can only be irreducible if

(i) $M \leq R$ and the row vectors of γ are linearly independent;

and

(ii) the greatest common divisor of the elements of every row is 1.

Proof. In order to prove (i) let us suppose, on the contrary, that the row vectors of γ are linearly dependent. Then - as the number of linearly independent rows is equal to the number of independent columns - the number of the linearly independent columns generate a nontrivial subspace of $\mathbb{Z}^M$, let this be denoted by S. Then a trajectory starting from $x_o \in \mathbb{N}_o^M$ cannot leave the set $x_o + S$ that is, points outside this set cannot be reached from this part of the state space.

In order to prove (ii) we remark that irreducibility implies the existence of nonnegative integers $\nu^1, \ldots, \nu^R \in \mathbb{N}_o$ such that

$$p-n = \nu^1\gamma(\cdot,1) + \ldots + \nu^R\gamma(\cdot,R) \ . \tag{2}$$

(The meaning of the numbers ν^r is how many times the rth elementary reaction should take place in order to change the state from n to p.)

Solvability of equation (2) may be reformulated stating that for each $q \in \mathbb{Z}^M$ there exists a vector $\nu_q \in \mathbb{N}_o^R$ such that

$$\gamma\nu_q = q \ . \tag{3}$$

The existence of a solution to (3) for all $q \in \mathbb{Z}^M$, however, is equivalent to the existence of a solution to all of the equations

$$\gamma\nu_m = e_m \qquad (m=1,2,\ldots,M) \tag{4}$$

where the e_ms are the elements of the standard base.

Let us consider the mth scalar equation of (4):

$$\sum_{r=1}^{R} \gamma(m,r)\nu_m^r = 1 \ , \tag{5}$$

(here $\nu_m := (\nu_m^1, \ldots, \nu_m^R)^T$). It is well known from elementary number theory (see e.g. Niven and Zuckerman, 1970, Sec. 5.4) that (5) has a solution if and only if the greatest common divisor of the coefficients $\gamma(m,1), \ldots, \gamma(m,R)$ is 1.

It is obvious that the effective upper bound is the number R_o of linearly independent elementary reaction vectors: $R_o := \dim \operatorname{span}\{\gamma(\cdot,r);\ r \in R\}$, therefore in the reversible case at least $M \leq R/2$ can be stated.

Translating condition (ii) into the language of chemistry it is implied by the fact that in each elementary reaction at least one of the components changes by one.

Theorem 2. The stochastic model of a reversible chemical reaction endowed with kinetics satisfying condition (C) is irreducible, except possibly a set of points that is contained in a union of sets $\bar{\Gamma}_1,\ldots,\bar{\Gamma}_I$ for which there exists such a coordinate axis that the projection of $\bar{\Gamma}_i$ onto it is bounded, if

(i) the rows of γ are independent (thus $M\le R$);

(ii) (possibly after renumbering) there exists a decomposition $\gamma=(\gamma_1,\gamma_2)$ such that γ_1 is a nonsingular square matrix with an inverse γ_1^{-1} of integers.

Proof. Let us renumber (if necessary) the elementary reactions in such a way that the first M columns of γ form a nonsingular square matrix γ_1 with an inverse of integer elements and let us rewrite equation (4) in the form

$$[\gamma_1\ \gamma_2]\begin{bmatrix}\nu_1^m\\ \\ \nu_2^m\end{bmatrix} = e_m$$

or

$$\gamma_1\nu_1^m = e_m - \gamma_2\nu_2^m \qquad (m=1,2,\ldots,M).$$

On applying γ_1^{-1} to this last equality one gets

$$\nu_1^m = \gamma_1^{-1}(e_m - \gamma_2\nu_2^m)\ (m=1,2,\ldots,M).$$

This formula yields the R-M parameter solution to (4) and it shows that if ν_2^M is an arbitrarily chosen vector of R-M integers: $\nu_2\in\mathbb{Z}^{R-M}$, then both ν_1^m and $\nu^m:=(\nu_1^m\ \nu_2^m)$ are vectors of integers. The solution to (3) is then

$$\nu_q = \sum_{m=1}^{M} q^m\nu^m$$

if $q=(q^1,\ldots,q^M)$. The coordinates of ν_q are integers as well. These integers can be taken nonnegative because of reversibility. This implies the existence of a route from n to n+q if $\forall r$ $n \geq \alpha(\cdot,r)$.

Thus it may happen that p-n can be expressed as the linear combination of the column vectors of γ, but the probability of leaving n is still zero. This can only happen when there exists an $r \in \mathcal{R}$ such that $n-\alpha(\cdot,r) \not\geq 0$. But this set of unaccessible points is surely contained in the union of the sets

$$\{q \in \mathbb{N}_o^M;\ q < \alpha(\cdot,r)\} \qquad (r \in \mathcal{R})$$

of which all have a finite projection onto at least one of the coordinate axes.

The fact that the kinetics satisfies (C) was used in the last step of the proof through the symmetry of Γ.

Condition (ii) of the theorem is fulfilled if det $\gamma=1$.

REFERENCES

Arnold, L. (1980). On the consistency of the mathematical models of chemical reactions, *Dynamic of Synergetic Systems*, 107-118 (H. Haken, ed.), Springer-Verlag, Berlin-Heidelberg-New York.

Érdi, P., Tóth, J. and Hárs, Vera (1981). Some kinds of exotic phenomena in chemical systems, *Qualitative Theory of Differential Equations*, 205-229 (M. Farkas, ed.), János Bolyai Mathematical Society and North Holland Publ. Co., Amsterdam-London.

Kurtz, T.G. (1981). *Approximation of Population Processes*, SIAM Philadelphia PA.

Lemarchand, H. (1980). Asymptotic solution of the master equation near equilibrium transition: The stationary solutions, *Physica* 101A, 518.

Niven, I. and Zuckerman, H.S. (1970). *An Introduction to the Theory of Numbers* (Third ed.), J. Wiley and Sons Inc., New York-London-Sydney-Toronto.

Tóth, J., Érdi, P. and Török, T. L. (1983). Significance of the Poisson distribution in the stochastic models of complex chemical reactions, *Alk. Mat. Lapok 9*, 175-195. (In Hungarian).

Vol'pert, A. I. (1972). Differential equations on graphs, *Mat. sb.* 88, 578-588. (In Russian).

János Tóth
Computer and Automation Institute
Hungarian Academy of Sciences
Budapest, H-1014
Úri u. 49, Hungary

Proc. of the 5th Pannonian Symp. on Math. Stat., Visegrád, Hungary 1985
W. Grossmann, J. Mogyoródi, I. Vincze, W. Wertz, eds

PARAMETER DEPENDENT ASYMPTOTIC NORMALITY FOR VECTOR ELEMENTARY GAUSSIAN PROCESSES

S. VERES

Budapest

This paper gives an example of a new type of asymptotic normatlity, where an appropriate change of the parameters implies normality. The first theorem shows that an appropriate change of the drift matrix of a Gaussian Markov process results the asymptotic normality of the associated measures in a new sense, where the covariance of the normal vector variable depends on the parameter itself very much. The second result is more restrictive and essentially states that if the maximal real part of the eigenvalues tends to $-\infty$ then the measures generated by the processes are asymptotically normal.

AMS subject classification: 62M09

1. INTRODUCTION

We shall consider p-dimensional elementary Gaussian processes satisfying stochastic differential equations of the type

$$d\xi_t = A\xi_t dt + B^{1/2} dW_t, \quad 0<t<T, \quad \xi_o = 0, \tag{1.1}$$

where W_t, $t>o$, is a p-dimensional Wiener process, A and B are $p \times p$ matrices, A is stable in the Ljapunov sense and B is a symmetric positive definite matrix. For the properties of elementary Gaussian processes we refer to the works of Arató (1982) and Arató, Kolmogorov and Sinai (1962). It is known by a Baxter type theorem that the matrix B can be estimated consistently on the basis of observations on any finite time interval. (See e.g. Arató (1982) p. 126.) This implies that from a statistical point of view it is sufficient to consider instead of (1.1) the processes, which are solutions of equations

$$d\xi_t = A\xi_t dt + dW_t, \quad 0<t<T, \ \xi_o = 0 \tag{1.2}$$

The transformation

$$\begin{aligned} t' &= t\,T^{-1} & \xi'_{t'} &= T^{-1/2}\xi_t \\ W'_{t'} &= T^{-1/2}W_t & A' &= T\,A \end{aligned} \tag{1.3}$$

enables us to treat the special case where T=1 :

$$d\xi'_{t'} = A'\xi'_{t'}dt' + dW'_{t'} \quad , \ 0 < t' < 1, \ \xi'_o = 0 \tag{1.4}$$

Thus the maximum likelihood (ML) estimation problem of (1.1) can be reduced to the case of (1.4), where our observation is on the time interval $[0,1]$ and the distribution of the estimator depends only on the generating drift parameter matrix A.

In this setting of the problem our main result essentially states that if the maximum of the real parts of the eigenvalues of A tends to $-\infty$, then the ML estimates of A are approximately normal. However, interestingly, the covariance of the approximate normal distribution depends on the parameter matrix A itself. We note that our result does not directly follows from asymptotic local normality (see for the definitions Ibragimov and Hasminkii (1980) or Basawa and Prakasa Rao (1980)) but using the transformation (1.3) it implies asymptotic local normality with $T\to\infty$ and therefore it is its generalization. In the scalar case this problem was solved and also the exact distribution of the ML estimate was given by Arató (1962, 1982). In this paper we concentrate on the formulation of the new type of asymptotic normality and for the sake of simplicity it is assumed that the eigenvalues of A and $A+A^T$ have multiplicity one.

Now, we give some comments on our notations. For a given drift matrix A, μ_A denotes the probability measure induced by the solution of equation (2.1) on the measurable space $(C[0,1]^p;\mathcal{B}_1^p)$. The notation $\mathcal{L}(\zeta_n) \to N(0,C)$, $n\to\infty$, means the convergence in distribution of the variables ζ_n to the normal N(0,C) distribution. Convergence of matrices is to be understood componentwise. $A \otimes B$ denotes the Kronecker product of matrices A and B. If $A_1, A_2, \ldots, A_p$ denotes the row vectors of matrix A, then vec(A) is defined by vec(A)= $(A_1\ A_2 \ldots A_p)^T$. For $u\ R^{p^2}$ $\underline{u}$ is the p x p matrix satisfying vec($\underline{u}$). The norm is as usual $||A||^2 = tr(A^TA)$. If it is not otherwise stated the indeces i,j,k,ℓ,m,n run always through the numbers 1,2, ... p in the sums denoted by Σ.

2. FORMULATION OF THE ASYMPTOTIC NORMALITY

In this section we give a theorem on the parameter dependent asymptotic normality of processes satisfying the equation

$$d\xi_t = A\xi_t dt + dW_t, \quad 0<t<1, \ \xi_o = 0. \tag{2.1}$$

Theorem 1. Denote by $-\lambda_M$ the largest real part of the eigenvalues of A. Let Γ_2 (A) defined by

$$[\Gamma_2(A)]_{i,j} = \lambda_M^2 D^2 \left(\int_0^1 \xi_t^i(A)\xi_t^j(A)dt\right), \ i,j=1,2, \ldots, p.$$

Then there is a representation

$$\log \frac{d\mu_{A+\lambda_M^{1/2}\underline{u}}}{d\mu_A}(\xi(A)) = u^T \Delta(A,\xi(A)) - \frac{1}{2} u^T \Gamma(A,\xi(A))\, u, \tag{2.2}$$

where $\Delta(A,\xi(A))$ and $\Gamma(A,\xi(A))$ are p-dimensional vector and r-order matrix valued random variables, respectively. For each $v \in R^{p^2}$ there is a random variable $\eta(v)$ of distribution $N(0, v^T\Gamma(A)v)$ with $\Gamma(A) = E\Gamma(A,\xi(A))$ such that for any $\delta>0$

$$P(|\eta(v) - v^T\Delta(A,\xi(A))|>\delta) < \delta + \frac{p^6}{\delta^6} \sum_{ijk} v^2_{(k-1)p+i}\, v^2_{(k-1)p+j} \cdot [\Gamma_2(A)]_{ij} \tag{2.3}$$

Furthermore the equalities

$$\Gamma(A) = \Gamma_o(A) \otimes I_p, \quad [\Gamma_o(A)]_{ij} = E\lambda_M \int_o^1 \xi_t^i(A)\xi_t^j(A)dt, \ i,j=1, \ldots, p. \quad (2.4)$$

and

$$E||\Gamma(A) - \Gamma(A,\xi(A))||^2 = p \sum_{ij} [\Gamma_2(A)]_{ij} \quad (2.5)$$

are satisfied.

<u>Remarks</u>. Approximation (2.3) can be useful by the fact that the coefficients of $\Gamma_2(A)$ are usually very small, when reasonable conditions are imposed. In this sense (2.2)-(2.5) implies the approximate normality of ML estimates of A. Theorem 2 and its proof gives a special case and analysis of the decrease of $\Gamma_2(A)$.

<u>Proof</u>. By a vectorial version of Theorem 7.19 of Liptser and Shiryayev (1977) it follows for another matrix A′ that

$$\log \frac{d\mu_{A'}}{d\mu_A}(\xi(A)) = \int_o^1 \xi_t^T(A)(A'-A)dW_t - \frac{1}{2}\int_o^1 ||(A'-A)\xi_t(A)||^2 dt \quad (2.6)$$

Introduce the notation

$$\underline{u} = (\lambda_M)^{-1/2} \operatorname{vec}(A'-A)$$

and let $\underline{\xi}_t(A) = \Gamma_t(A) \otimes I_p$.

Thus (2.6) can be rewritten as

$$\log \frac{d\mu_{A+\lambda_M^{1/2}\underline{u}}}{d\mu_A}(\xi(A)) = u^T\lambda_M^{1/2} \int_o^1 \underline{\xi}_t(A)dW_t - \frac{1}{2}u^T\lambda_M \int_o^1 \underline{\xi}_t(A)\underline{\xi}_t^T(A)dt\, u \quad (2.7)$$

Now, let

$$\Delta(A,\xi(A)) = \lambda_M^{1/2} \int_o^1 \underline{\xi}_t(A)dW_t \quad (2.8)$$

and

$$\Gamma(A,\xi(A)) = \lambda_M \int_0^1 \underline{\xi}_t(A)\,\xi_t^T(A)\,dt \tag{2.9}$$

which results representation (2.2).

For an arbitrarily chosen $v \in R^{p^2}$ define $f_t = f_t(v,A)$ as

$$f_t = \begin{cases} \underline{\xi}_t(A), & \text{if } \; 0<t<1 \\ \underline{1}\;, & \text{if } \; 1 \; t \; 1+C_v,\; \underline{1} = (1,1 \;\ldots\; 1)^T, \end{cases}$$

where $C_v = \lambda_M v^T \Gamma(A) v/p$.
Introduce the Markov time

$$\tau = \inf \{t>0: \int_0^t f_s f_s^T ds = \lambda_M v^T \Gamma(A) v\}$$

and using it let

$$\eta(v) = \int_0^\tau f_t^T \, dW_t$$

Applying an argumentation similar to that of Kutoyants (1980) or Basawa and Prakasa Rao (1980) it follows for any $\delta>0$ that

$$P(|\eta(v)-v^T\Delta(A,\xi(A))|>\delta) < \delta + P(v^T(\int_0^1 \underline{\xi}_t \underline{\xi}_t^T dt - E\int_0^1 \underline{\xi}_t \underline{\xi}_t^T dt)v > \delta^3) <$$

$$<\delta + \sum_{ijk} P(|v_{(k-1)p+i}(\int_0^1 \xi_t^i \xi_t^j \, dt - E\int_0^1 \xi_t^i \xi_t^j dt) \; v_{(k-1)p+j}| > \frac{\delta^3}{p^3}) <$$

$$<\delta + \frac{p^6}{\delta^6} \sum_{ijk} v^2_{(k-1)p+i} [\Gamma_2(A)]_{ij} \; v^2_{(k-1)p+j},$$

where in the last step the Chebyshev inequality was used. The equalities (2.4) and (2.5) are obtained directly by the definitions and the proof is finished.

3. THE AFFECT OF THE ASYMPTOTIC BEHAVIOUR OF EIGENVALUES

Denote by $\lambda_1^n, \lambda_2^n, \ldots, \lambda_p^n$ the eigenvalues of A_n. Let $\lambda_M^n = \max_i \mathrm{Re}\lambda_i^n$. Then the following theorem holds.

Theorem 2. Assume that the following conditions are satisfied for a sequence of normal drift matrices $A_1, A_2, A_3, \ldots$.

(i) $\lambda_M^n \to -\infty$

(ii) $-\lambda_M^n (A_n + A_n^T)^{-1} \to \Gamma_\infty$ as $n\to\infty$.

Then

$$\log \frac{d\mu_{A_n+\sqrt{\sigma_m^n}u}}{d\mu_{A_n}}(\xi(A_n)) = u^T\Delta(A_n,\xi(A_n)) - \frac{1}{2}u^T(\Gamma_\infty \otimes I_p)u + \alpha(u,A_n,\xi(A_n)) \quad (3.1)$$

where

$$\mathcal{L}(\Delta(A_n,\xi(A_n))) \to N(0,\Gamma_\infty \otimes I_p) \quad (3.2)$$

as $n\to\infty$, and uniformly on each bounded set $B \subset R^{p^2}$

$$\mathcal{L}(u,A_n,\xi(A_n)) \overset{p}{\to} 0, \qquad u \in B \quad (3.3)$$

as $n\to\infty$.

Proof. For the proof we need some preliminary lemmas.

Lemma 1. Under conditions (i) and (ii)

$$\Gamma_o(A_n) = \lambda_M^n E \int_0^1 \xi_t(A_n)\xi_t^T(A_n)dt \to \Gamma_\infty \quad (3.4)$$

as $n\to\infty$.

Proof. Let Q_n be unitary matrices transforming $A_n+A_n^T$ to the diagonal form $\Sigma_n = Q_n(A_n+A_n^T)Q_n^*$. Then

$$\lambda_m^n E\int_0^1 \xi_t(A_n)\xi_t^T(A_n)dt = \lambda_m^n \int_0^1 E(\int_0^t e^{A_n(t-s)}dW_s \int_0^t dW_s^T e^{A_n^T(t-s)})\,dt =$$

$$= \lambda_M^n \int_0^1 (A_n+A_n^T)^{-1}(e^{(A_n+A_n^T)t} - I_p)dt =$$

$$= \lambda_M^n \left[(A_n+A_n^T)^{-2}(e^{(A_n+A_n^T)}-I_p) - (A_n+A_n^T)^{-1} \right] =$$

$$= Q_n(\lambda_M^n \Sigma_n^{-2})\ (e^{\Sigma_n} - I_p)\ Q_n^* - \lambda_m^n(A_n+ A_n^T)^{-1}$$

On the last right hand side the first term tends to 0 and the second converges by condition (iii), from which the statement of the lemma follows.

Lemma 2. Assume that ζ is an m-dimensional vector valued random variable and for each $x\in R^m$ there is a normally distributed random variable $\eta(x)$ with mean 0 and variance x^TGx such that

$$P(|\eta(x) - x^T\zeta|>\delta) < K(\delta), \quad \delta>0. \tag{3.5}$$

Then ζ is approximately normally distributed with variance G in the sense that its characteristic function $\varphi_\zeta(x)$ satisfies

$$|\varphi_\zeta(x) - \exp(-\tfrac{1}{2}x^TGx)| < \min_{\delta>0} (\delta+2K(\delta)), \quad x\in R^m. \tag{3.6}$$

Proof. The proof follows by the straightforward application for any $\delta>0$ of the following sequence of inequalities:

$$|\exp(-\tfrac{1}{2}x^T\Gamma x) - E\exp(ix_\zeta^T)| = |E\exp(i\eta(x)) - E\exp(ix_\zeta^T))| <$$

$$< \int_{|\eta(x)-x_\zeta^T|<\delta} |\exp(i\eta(x)) - \exp(ixp(ix_\zeta^T)| \ dP + \int_{|\eta(x)-x_\zeta^T|\geq\delta} 2dP$$

$$< \int_{|\eta(x)-x^T_\zeta|<\delta} \delta \; dP + 2\,P(|\eta(x)-x^T\zeta|>\delta) < \delta + 2\cdot K(\delta).$$

The solution of the the differential equation (2.1) is $\xi_t = \int_0^t \exp(A\,(t-s))dW_s$. For the sake of simpler notations we shall use A, Λ, T, Q ... etc, instead of A_n, Λ_n, T_n, Q_n ... etc. for a while. For $s<t$

$$E\xi_t\xi_s^T = E(\int_0^t e^{A(t-u)}dW_u \int_0^s dW_u^T e^{A^T(s-u)}) = e^{A(t-s)} \int_0^s e^{(A+A^T)(s-u)}du$$

$$= e^{A(t-s)}(A+A^T)^{-1}(e^{(A+A^T)s}-I) = Qe^{\Lambda(t-s)}\Sigma^{-1}(e^{\Sigma s} - I)Q^*,$$

where Q is the unitary matrix diagonalizing both A and A^T and therefore $\Sigma = Q(A+A^T)Q^* = \Lambda + \Lambda^*$ with $\Lambda = QAQ^* = \mathrm{Diag}(\lambda_1,\lambda_2, \ldots,\lambda_p)$, $\Sigma = \mathrm{Diag}(\sigma_1,\sigma_2, \ldots, \sigma_p)$, $\sigma_i = 2\,\mathrm{Re}\lambda_i$, $i=1, \ldots, p$. Then

$$[\Gamma_2(A)] = E\lambda_M^2(\int_0^1 \xi_t^i\xi_t^j\,dt - E\int_0^1 \xi_t^i\xi_t^j\,dt)^2 =$$

$$= 2\lambda_M^2 \int_0^1\int_0^t E\xi_t^i\xi_s^i E\xi_t^j\xi_s^j + E\xi_t^i\xi_t^j\,E\xi_t^j\xi_s^i\,ds\,dt =$$

$$= 2\lambda_M^2 \int_0^1\int_0^t [(\sum_k Q_{ik}e^{\lambda_k(t-s)}\sigma_k^{-1}(e^{\sigma_k s}-1)Q^*_{ki})(\sum_m Q_{jm}e^{\lambda_m(t-s)}\sigma_m^{-1}(e^{\sigma_m s}-1)Q^*_{mj}) +$$

$$+(\sum_k Q_{ik}e^{\lambda_k(t-s)}\sigma_k^{-1}(e^{\sigma_k s}-1)Q^*_{kj})(\sum_m Q_{jm}e^{\lambda_m(t-s)}\sigma_m^{-1}(e^{\sigma_m s}-1)Q^*_{mi})]dsdt$$

Introduce the notations $D_{ijk} = Q_{ik}Q^*_{kj}$, $i,j,k = 1,2, \ldots,p$. Rewriting (3.7) we have that

$$[\Gamma_2(A)]_{ij} < 8\sum_{km}(D_{iik}D_{jjm} + D_{ijk}D_{jim})\lambda_M^2\sigma_k^{-1}\sigma_m^{-1}\int_0^1\int_0^t e^{(\lambda_k+\lambda_m)(t-s)}dsdt$$

Here the last terms can be majorized in the following way:

$$|\lambda_M^2\sigma_k^{-1}\sigma_m^{-1}\int_0^1\int_0^t e^{(\lambda_k+\lambda_m)(t-s)}dsdt| = |\lambda_M^2\sigma_k^{-1}\sigma_m^{-1}\int_0^1 \frac{e^{(\lambda_k+\lambda_m)t}-1}{\lambda_k+\lambda_m}dt| =$$

$$= \left| \lambda_M^2 \sigma_k^{-1} \sigma_m^{-1} \left(\frac{e^{\lambda_k+\lambda_m}-1}{(\lambda_k+\lambda_m)^2} - \frac{1}{\lambda_k+\lambda_m} \right) \right| < \frac{-1}{4\lambda_M},$$

and the last inequality is satisfied if $\lambda_M \leqq -1$ holds.

Now we return to the use of the superscript n. Note that the sequence Q_{ij}^n is uniformly bounded. It is easily seen that

$$D_{ijk} \to 0, \quad [\Gamma_2(A_n)]_{ij} \to 0, \quad i,j,k = 1,2, \ldots, p$$

as $n\to\infty$. Putting λ_M^n for the normalizing term λ_M in Theorem 1 and defining $\Delta(A_n,\xi(A_n))$ and $\Gamma(A_n,\xi(A_n))$ as in (2.8) and (2.9) we have that for the Radon-Nikodym derivative

$$\log \frac{d\mu_{A_n+\sqrt{\lambda_m^n}\,\underline{u}}}{d\mu_{A_n}}(\xi(A_n)) = u^T \Delta(A_n,\xi(A_n)) - \frac{1}{2} u^T \Gamma(A_n,\xi(A_n))u$$

holds. By Theorem 1 for each $v \in R^{p^2}$ and $n \in N$ there is a variable $\eta^n(v)$ with mean 0 and variance $v^T \Gamma(A_n)v$ such that for any $\delta>0$

$$P(|\eta^n(v) - v^T \Delta(A_n,\xi(A_n))| > \delta + \delta^{-6} \cdot R_n,$$

where $R_n = R_n(A_n,v) = \frac{p^6}{\delta^6} \sum_{ijk} v^2_{(k-1)p+1} v^2_{(k-1)p+j} \quad (\Gamma_2(A_n))_{ij}$

Now, Lemma 2 can be applied for $\zeta_n = \Delta(A_n,\xi(A_n))$ and by some easy computations it follows that

$$|\varphi_{\zeta_n}(v) - \exp(-\tfrac{1}{2} v^T \Gamma(A_n)v)| < 3 \cdot 2^{1/7} R_n^{1/7} + 2^{-1/7} R_n^{6/7}. \qquad (3.9)$$

Using (3.8) it is obtained that for each $v \in R^{p^2}$

$$R_n(A_n,v) \to 0 \qquad \text{as } n\to\infty$$

and by the weak convergence criterion

$$\mathcal{L}(\zeta_n) \rightarrow N(0, \Gamma_\infty \otimes I_p)$$

is implied.

Finally, the stochastic convergence

$$\Gamma(A_n, \xi(A_n)) \overset{p}{\rightarrow} \Gamma_\infty \otimes I_p$$

easily follows by the application of Lemma 1 and (2.3), (2.4) completing the proof of the theorem.

4. ACKNOWLEDGEMENT

The author is indebted to Professor M. Arató for helpful discussions and to one of the referees for valuable remarks.

REFERENCES

Arató, M. (1982). Linear Stochastic Systems with Constant Coefficients. Lecture Notes in Control and Information Sci. 45, Springer Vlg.

Arató, M., Kolmogorov, A. N. and Sinai, J.G. (1962). On parameter estimation of complex stationary Gaussian Markov processes (in Russian). Doklady Akad. Nauk SSSR 146 747-750.

Basawa, I. V and Prakasa Rao, B.L.S. (1980). Statistical Inference for Stochastic Processes. Academic Press, London-New York.

Ibragimov, I.A. and Hasminskii, R. Z. (1978). Asymptotic Theory of EStimation (in Russian). Nauka, Moscow.

Kutoyants, U. A. (1980). Parameter Estimation of Stochastic Processes (in Russian). Akad. Nauk Arm. SSR, Yerevan.

Liptser, R.S. and Shiryayev, A. N. (1977). Statistics of Random Processes I. Springer Vlg., New York - Berlin.

S. Veres
Department of Probability Theory and Statistics
Eötvös Loránd University,
Budapest, Muzeum krt. 6-8.
H-1088

Proc. of the 5th Pannonian Symp. on Math. Stat., Visegrád, Hungary 1985
W. Grossmann, J. Mogyoródi, I. Vincze, W. Wertz, eds

ON THE OVERPARAMETRIZED MAXIMUM LIKELIHOOD ESTIMATION OF ARMA PROCESSES

S. VERES

Budapest

This paper deals with the asymptotic distribution of maximum likelihood estimators and log likelihood ratios of Gaussian ARMA processes in the case of overparametrization, where the Fisher information matrix of the estimated parameters is asymptotically singular. First the asymptotic distributions of log likelihood ratios are derived. Then examples and simulations are given for the distributions.
The second order overparametrization is treated without detailed proofs which are analogous to the first order case. Finally, a testing procedure is proposed for the orders of processes.
AMS subject classification: 62 M09

1. INTRODUCTION

This paper discusses the asymptotic distribution of maximum likelihood (ML) estimates of stationary Gaussian autoregressive moving average (ARMA) processes y_t, $t\in Z$, satisfying equations of the type

$$y_t + a_1 y_{t-1} + \dots + a_m y_{t-m} = b_o e_t + b_1 e_{t-1} + \dots + b_\ell e_{t-\ell} \tag{1.1}$$

where e_t, $t\in Z$, is a Gaussian white noise process, i.e. e_t, $t\in Z$ is of distribution $N(0,1)$ and $Ee_t e_s = \delta_s^t$. We shall consider the case of overparametrization, where the proper model orders m_o and ℓ_o are strictly overestimated by $m > m_o$ and $\ell > \ell_o$. In this situation the model is first order informationally singular with respect to the observed process (see Veres 1985), which means that the models under consideration can faithfully describe the observed process, however, the asymptotic Fisher information matrix is singular. Hannan (1982) has dealt with a similar problem for the overparametrization of the pure white

noise process, where he has shown the inconsistency of Akaike's criteria AIC caused by estimated poles of the transfer function being very closed to the unit circle. He shows that AIC is inconsistent with probability 1. The following results are essentially the extentions of his result on the computation of the asymptotic distribution of a log likelihood ratio to arbitrary orders. However, the parametrizations and the ways of proofs are different from those of Hannan (1982).
We give the asymptotic distribution of the log likelihood ratio statistic in an implicit way, which provides a statistical tool for deciding between model orders. This is a contribution to earlier results in this direction: Akaike' criteria (1974), F-tests, methods based on singularity of the information matrix (Van Overbeek and Ljung 1982, Söderström 1975, 1977), methods based on pole-zero cancellation (Söderström 1975), test of residuals (Bohlin 1978), tests based upon the Lagrangian multiplier procedure (Poskitt and Tremayne 1981).

In the case of continuous time ARMA processes the problem of order determination can be solved by different methods as testing the order of mean square differentiability of the process (see Arató, 1982.)

Using the backshift operator B on t equation (1.1) can be rewritten as

$$y_t = T(B)e_t, \qquad T(B) = Q(B)/P(B)$$

where $P(B) = 1 + a_1B + \ldots + a_mB^m$, $Q(B) = b_o + b_1B + \ldots + b_\ell B^\ell$.
The polinomials P(B) and Q(B) can be decomposed into prime factors in T(B):

$$T(B) = \frac{b(1-\mu_1B)\ldots(1-\mu_pB)\ (1-(\beta_{11}+\beta_{12})B)(1-(\beta_{11}-i\beta_{12})B)\,.}{(1-\lambda_1B)\ \ldots(1-\lambda_rB)\ (1-(\alpha_{11}+i\alpha_{12})B)(1-(\alpha_{11}-i\alpha_{12})B)\,.}$$

$$\frac{\ldots(1-(\beta_{q1}+i\beta_{q2})B)(1-(\beta_{q1}-i\beta_{q2})B)}{\ldots(1-(\alpha_{s1}+i\alpha_{s2})B)(1-(\alpha_{s1}-i\alpha_{s2})B)}.$$

Then the parameters in our transfer function can be formed in a parameter vector Θ , which is described by four structural parameters p, q, r, s,

$$\Theta = (b,\mu_1,\ldots,\mu_p,\beta_{11},\beta_{12},\ldots,\beta_{q1},\beta_{q2},\lambda_1,\ldots,\lambda_r,\alpha_{11},\alpha_{12},\ldots,\alpha_{s1},\alpha_{s2}).$$

For any $\delta>0$ define the sets $W_\delta = (-1+\delta, 1-\delta)$ and

$$D_\delta=\{(t,v)\in R^2:\ t^2+v^2 < (1-\delta)^2 \ \& \ b\neq 0\},\quad W_\delta^o=\{t\in W_\delta:\ t\neq\mu_j^o, t\neq\lambda_j^o, \forall j\},$$

$$D_\delta^o=\{(t,v)\in D_\delta:\ (t,v)\neq(\alpha_{j1}^o,\alpha_{j2}^o),\ (t,v)\neq(\beta_{j1}^o,\beta_{j2}^o),\ \forall j\}$$

For $p>0$, $q>0$, $r>0$, $s>0$ let $H_\delta = H_\delta(p,q,r,s)$ be the set of parameter vector Θ such that $\mu_i \in W_\delta (i=1,\ldots,p)$ $\lambda_j \in W_\delta$, $(j=1,\ldots,r)$ $(\beta_{i1},\beta_{i2})\in D_\delta$ $(i=1,\ldots,q)$ and $(\alpha_{i1},\alpha_{i2})\in D_\delta (i=1,\ldots,s)$ are satisfied. Let $\Theta'^o = (b^o,\mu_1^o,\ldots,\mu_p^o, \beta_{11}^o, \beta_{12}^o,\ldots,\beta_{q1}^o,\beta_{q2}^o,\lambda_1^o, \ldots,\lambda_r^o,\alpha_{11}^o,\alpha_{12}^o,\ldots,\alpha_{s1}^o,\alpha_{s2}^o) \in H_\delta$ be a once for all fixed parameter vector generating the observed process, where it is assumed that $\mu_i^o \neq u_j^o$, $\lambda_i^o \neq \lambda_j^o$, $(\beta_{i1}^o,\beta_{i2}^o) \neq (\beta_{j1}^c,\beta_{j2}^o)$ and $(\alpha_{i1}^o,\alpha_{i2}^o) \neq (\alpha_{j1}^o,\alpha_{j2}^o)$ if $i\neq j$.

Denote by $H_\delta^o = H_\delta^o(p+1,q,r+1,s)$ the set of parameter vectors which generate essentially the same ARMA processes (processes with the same finite dimensional distributions) as Θ^o.

The log likelihood function is defined as usual by

$$L^T(\underline{y}_T,\Theta) = \frac{1}{2}(\log\det\Gamma_T(\Theta), + \underline{y}_T^*\Gamma_T^{-1}(\Theta)\underline{y}_T)$$

where

$$\Gamma_T(\Theta) = \begin{bmatrix} c(o) & c(1) & c(2) & \ldots & c(T-1) \\ c(1) & c(o) & c(1) & \ldots & c(T-2) \\ c(2) & c(1) & c(o) & \ldots & c(T-3) \\ \vdots & & & \ddots & \vdots \\ c(T-1) & \ldots & & & c(o) \end{bmatrix}, \underline{y}_T = (y_1,\ldots,y_T)$$

and c(i), i=0,1,2,... are the autocovariances of the process with parameter Θ. A^* will be used for the transpose of a matrix A, and $\bar{a}$ is the conjugate of the complex number a.

2. ASYMPTOTIC DISTRIBUTION OF THE LOG LIKELIHOOD RATIOS.

In this section we state our main results concerning the two basic types of likelihood ratios.

<u>Theorem 1.</u> Assume that the observed stationary Gaussian ARMA process is generated by a parameter $\Theta^o \in H_\delta(p,q,r,s)$.

a/ The asymptotic distribution of the log likelihood ratio

$$L^T(\underline{y}_T, \Theta^o) - \inf_{\Theta \in H_\delta(p+1,q,r+1,s)} L^T(\underline{y}_T, \Theta) \tag{2.1}$$

is the distribution of

$$\frac{1}{2} \max_{t \in W_\delta^o} ||\eta_\infty(t)||^2 \tag{2.2}$$

where $\eta_\infty(t)$, $t \in W_\delta^o$, is a zero mean Gaussian vector process with covariance function

$$k_\infty(t,s) = \Gamma(t,t)^{-1/2}\Gamma(t,s)\Gamma(s,s)^{-1/2}, \quad t,s \in W_\delta^o$$

and

$$\Gamma(t,s) = \begin{bmatrix} I(\Theta^o) & J(\Theta^o,s) \\ J(\Theta^o,t)^* & (1-ts)^{-1} \end{bmatrix} \quad t,s \in W_\delta^o .$$

with coefficients given in (2.5) - (2.7).

b/ The asymptotic distribution of the log likelihood ratio

$$\inf_{\theta \in H_\delta(p,q,r,s)} L^T(\underline{y}_T,\theta) - \inf_{\theta \in H_\delta(p+1,q,r+1,s)} L^T(\underline{y}_T,\theta) \tag{2.3}$$

is the distribution of

$$\frac{1}{2} \max_{t\in W_\delta} \zeta_\infty(t)^2 ,$$

where $\zeta_\infty(t)$, $t\in W_\delta^o$, is a zero mean Gaussian process with covariance function

$$c_\infty(t,s)) = [1/(1-ts) - J(\Theta^o,t)^* I(\Theta^o)^{-1} J(\Theta^o,s)] \tag{2.4}$$

$$[(1/(1-t^2) - J(\Theta^o,t)^* I(\Theta^o)^{-1} J(\Theta^o,t))(1/(1-s^2) -$$

$$- J(\Theta^o,s)^* I(\Theta^o)^{-1} J(\Theta^o,s))^{-1/2}],\ t,s\in W_\delta^o$$

Here $I = I(\Theta^o)$ is the mxm Fisher information matrix for estimating Θ^o $(m=1+p+2q+r+2s)$. Its components are the following, where the indices i,j,k,ℓ run over all their possible values.

$$I_{bb} = 3/(b^c)^2,\ I_{b\mu_i} = I_{b\lambda_j} = 0, \tag{2.5}$$

$$I\nu_i\vartheta_j = \gamma\ /(1-\nu_i^o\vartheta_j^o) \quad (\nu=\lambda,\mu;\ \vartheta=\lambda,\mu),\quad \text{where } \gamma_\vartheta^\nu = 1,$$

if $\nu=\vartheta$ and $\gamma_\vartheta^\nu=-1$, if $\nu\neq\vartheta$.

For brevity define the functions $f_1(a,b,c) = 2(1-ab)/A$, $f_2(a,b,c) = 2ab/A$, $A=(1-ab)^2+(ac)^2$, $a, b, c \in R$.

$$I_{\nu_i\vartheta_{j\ell}} = \gamma_\vartheta^\lambda\ \gamma_\beta^\nu f_\ell(\nu_i^o, \vartheta_{j1}^o, \vartheta_{j2}^o) \qquad (\nu=\lambda,\mu;\ \vartheta=\alpha,\beta), \tag{2.6}$$

$$I_{\nu_{ik}\vartheta_{j\ell}} = \gamma_a^\nu \gamma_\beta^\vartheta (h_1 - (-1)^\ell h_2 - (-1)^k \bar{h}_2 + (-1)^{k+\ell}\bar{h}_1)),$$

where $h_1 = 1/(1-\nu_i\vartheta_j)$, $h_2 = 1/(1-\nu_i\vartheta_j)$ with $\nu_i = \nu_{i1}^o + i\nu_{i2}^o$, $\vartheta_j = \vartheta_{j1}^o + i\vartheta_{j2}^o$ $(\nu=\alpha,\beta\ :\ \vartheta=\alpha,\beta)$

$J = J(\Theta^o,t)$, $t\in W_\delta$ is a column vector with components

$$J_b = 0,\ J_{\nu_i} = \gamma_\lambda^\nu\ /(1-\nu_i^o t) \quad (\nu=\lambda,\mu), \tag{2.7}$$

$$J_{\vartheta_{j\ell}} = \gamma_\beta f_\ell(t, \vartheta_{j1}^o, \vartheta_{j2}^o) \qquad (\vartheta=\alpha,\beta).$$

3. EXAMPLES AND SIMULATIONS

Assume that the observed process is a first order stationary autoregressive process $y(t)$, $t\epsilon Z$, satisfying the equation

$$y_t - \lambda_o y_{t-1} = b_o e_t, \tag{3.1}$$

where e_t is a Gaussian white noise process. The maximum of the likelihood function is to be computed over $H_\delta(0,0,1,0)$ and $H_\delta(1,0,2,0)$. In the higher order model the transfer function is $T(B) = b(1-\mu B)/(1-\lambda_1 B)(1-\lambda_2 B)$, and the parameter to be estimated is $\Theta = (b,\mu,\lambda_1,\lambda_2)$. In this special case the function $\Gamma(t,s)$ takes the form

$$\Gamma(t,s) = \begin{bmatrix} 3/b_o^2 & 0 & 0 \\ 0 & \dfrac{1}{1-\lambda_o^2} & \dfrac{1}{1-\lambda_o s} \\ 0 & \dfrac{1}{1-\lambda_o t} & \dfrac{1}{1-ts} \end{bmatrix}$$

and

$$k_\infty(t,s) = \Gamma(t,t)^{-1/2}\Gamma(t,s)\Gamma(s,s)^{-1/2}, \qquad t,s\epsilon W_\delta^o.$$

Furthermore, by (2.4) we have that

$$c_\infty(t,s) = \left(\frac{1}{1-ts} - \frac{1-\lambda_o^2}{(1-\lambda_o t)(1-\lambda_o s)}\right)$$

$$\left(\left(\frac{1}{1-t^2} - \frac{1-\lambda_o^2}{(1-\lambda_o t)^2}\right)\left(\frac{1}{1-s^2} - \frac{1-\lambda_o^2}{(1-\lambda_o s)^2}\right)\right)^{-1/2}$$

$$t,s\epsilon W_\delta^o.$$

As an illustration we have computed the distribution of

$$\frac{1}{2} \max_{t\epsilon W_\delta^o} \zeta_\infty(t)^2 \tag{3.2}$$

for $\delta=0.1$ by simulation of a discrete time process with covariance structure $c_\infty(t,s)$, $t,s\in W_\delta$. By 1000 replications and using the discrete time points $t_i = -C.9 + k\cdot 0.05$ $(k=1,\ldots,36)$ the sample distribution of

$$\frac{1}{2}\max_{0<k<36} \zeta_\infty(t_k)^2 \tag{3.2}$$

was computed and it is described in fig. 1 for $\lambda_o=0.1$, $\lambda_o=0.9$. For verfying the goodness of the approximation of (3.1) by (3.2) we note that the increase of the number n of discrete time points $t_1,\ldots,t_n$ did not significantly modify the form ot the sample distribution function. For each value of λ_o almost the same asymptotic distribution functions were obtained, which shows the insensitivness of the asymptotic distribution of the LR with respect to the true parameter.

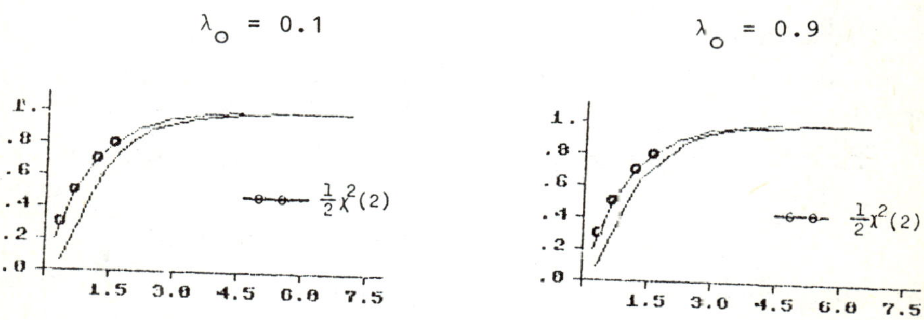

Figure 1. Asymptotic distribution of LR's of the first type.

4. ASYMPTOTIC DISTRIBUTION OF ESTIMATORS

In this section we state a theorem on the maximum likelihood estimates of parameters when estimating in the overparametrized structure. Assume that the observed process is generated by a parameter $\Theta^o\in H_\delta(p,q,r,s)$ and denote by V^o_{ij} $(i=1,\ldots,p+1,\ j=1,\ldots,r+1)$ the set of parameter vectors Θ in $H^o_\delta(p+1,q,r+1,s)$ such that for their components $\mu_i=\lambda_j\in W_\delta$, and the remaining components of Θ are some appropriate permutations of Θ^o.

Then $H_\delta^o(p+1,q,r+1,s) = \bigcup\{V_{ij}^o: 1<i<p+1, 1<j<r+1\}$ is the set of parameters corresponding to the observed process.

For $d>0$ introduce the sets $U_{ij}(d) = \{\Theta\epsilon H_\delta: \inf_{\Theta_1 \epsilon V_{ij}^o} ||\Theta-\Theta_1||^2<d^2\}$ i.e. $U_{ij}(d)$ is the neighbourhood of the set V_{ij}^o with radius d. Let $U_{ij}^T(d) = U_{ij}(dT^{-1/2})$. For any $d>0$ we shall consider the asymptotic distribution of $L^T(y_T,\Theta)$ on the sets $U_{ij}^T(d)$ $(1<i<p+1,\ 1<j<r+1)$, where the log likelihood function has similar distributions due to symmetry reasons.

Then Lemma 2 provides a sufficient characterization of the asymptotic distribution of

$$L^T(y_T;t,\Theta_t^o+T^{-1/2}u) - L^T(y_T;t,\Theta_t^o),\ \Theta_t^o+T^{-1/2}u\epsilon U_{ij}^T(d),\ t\epsilon W_\delta^o$$

as $T\to\infty$. From Lemma 2 the following theorem can be deduced for the asymptotic behaviour of maximum likelihood estimators computed over $U_{ij}^T(d)$.

Theorem 2. By reordening the components of the parameter vector denote by $(\hat{\Theta}^T \hat{\mu}_i^T,\ \hat{\lambda}_j^T)$, the maximum likelihood estimates computed over the sets $U_{ij}^T(d_T)$, where $d_T\to\infty$ and $T^{-1/2}d_T\to 0$. Let $\Delta_\infty(t)$, $t\epsilon W_\delta^o$, be a zero mean Gaussian process with covariance function $\Gamma(t,s)$, $t,s\epsilon W_\delta^o$. Then the asymptotic distribution of

$$(T^{1/2}(\hat{\Theta}^T - \Theta^o),\ \hat{\mu}_i^T,\ \hat{\lambda}_j^T) \qquad \text{as} \quad T\to\infty$$

is the same as the distribution of

$$(\Delta_\infty^*(\tilde{t})\Gamma(\tilde{t},\tilde{t})^{-1},\ \tilde{t})$$

where

$$\tilde{t} = \arg\max_{t\epsilon W_\delta^o} \Delta_\infty^*(t)\Gamma(t,t)^{-1}\Delta_\infty(t).$$

5. SECOND ORDER OVERPARAMETRIZATION

In Theorem 1 the asymptotic distributions of likelihood ratios were given for first order overparametrization. The above results can be extended in a similar way also for those cases where the log likelihood function is to be minimized over the domain $H_\delta^2 = H_\delta(p+2,q,r+2,s) \cup H_\delta(p,q+1,r,s+1)$, which might be called second order overparametrization. Introduce the notations $W_\delta^2 = W_\delta \times W_\delta$ and $V_\delta = W_\delta^2 \cup [D_\delta \times \{0\}]$.

<u>Theorem 3.</u> Assume that the observed stationary Gaussian ARMA process is generated by a parameter $\Theta^0 \in H_\delta(p,q,r,s)$.

a./ The asymptotic distribution of the log likelihood ratio

$$\inf_{\Theta \in H_\delta} L^T(y_T,\Theta^0) - \inf_{\Theta \in H_\delta^2} L^T(y_T,\Theta) \tag{5.1}$$

is the same as the distribution of

$$\frac{1}{2} \max_{w \in V_\delta} ||\eta_\infty(w)||^2,$$

where $\eta_\infty(w)$, $w \in V_\delta$, is a vector valued zero mean Gaussian random field with covariance function

$$K_\infty^2(w_1,w_2) = \Gamma_2(t_1,s_1,t_1,s_1)^{-1/2}\Gamma_2(t_1,s_1;t_2,s_2)\Gamma_2(t_2,s_2;t_2,s_2)^{-1/2}$$

where

$$\Gamma_2(t_1,s_1;t_2,s_2) = \begin{bmatrix} I(\Theta^0) & J(\Theta^0,t_2,s_2) \\ J(\Theta^0;t_1,s_1)^* & K(t_1,s_1;t_2,s_2) \end{bmatrix}$$

and the components of $I(\Theta^0)$, $J(\Theta^0;t,s)$, $K(t_1,s_1;t_2,s_2)$ are given in (2.5) - (2.7) and (5.3) - (5.5), respectively.

b./ The asymptotic distribution of the log likelihood

ratio

$$\inf_{\Theta \in H_\delta} L^T(y_T,\Theta) - \inf_{\Theta \in H_\delta^2} L^T(y_T,\Theta)$$

is the same as the distribution of

$$\frac{1}{2} \max_{w \in V_\delta} ||\xi_\infty(w)||^2, \tag{5.2}$$

where ξ_∞ (w), $w \in V_\delta$, is a two dimensional vector valued zero mean Gaussian random field with covariance function

$$C_\infty(w_1,w_2) = R(t_1,s_1;t_1,s_1)^{-1/2}\ R(t_1,s_1;t_2,s_2)\ R(t_2,s_2;t_2,s_2)^{-1/2}$$

with

$$R(t_i,s_i;t_j,s_j) = K(t_i,s_i;t_j,s_j) - J(\Theta^o,t_i,s_i)I(\Theta^o)^{-1}J(\Theta^o,t_j,s_j), \quad (i,j=1,2).$$

$J=J(\Theta^o,t,s)$ is a mx2-dimensional matrix ($m=1+p+2q+r+2s$) with the following coefficients.

$$J_{b1}=0,\ J_{b2}=0,\ J_{i\ell}=\gamma^\nu f_\ell(\nu_i^o,t,s)\ (\nu=\lambda,\ \mu;\ \ell=1,2) \tag{5.3}$$

$$J_{\nu_{jk}\ell} = \gamma_\alpha^\nu(h_1-(-1)^\ell h_2-(-1)^k h_2 +(-1)^{k+\ell}h_1)\ (\nu = \alpha,\beta\ ;\ k,\ell = 1,2.)$$

where $h_1=1/(1-hq)$ and $h_2 = 1/(1-hq)$ with $h=\nu_{j1}^o + i\nu_{j2}^o, q=t+is$.

If $w_1 = (t_1,s_1) \in W_\delta^2$ and $w_2 = (t_2,s_2) \in W_\delta^2$ then

$$K(t_1,s_1,t_2,s_2)_{k\ell} = 1/(1-t_k s_\ell)\quad (k,\ell = 1,2). \tag{5.4}$$

If $w_1=(t_1,s_1) \in W_\delta^2$ and $w_2= (t_2,s_2) \in W_\delta^2$ then

$$K(t_1,s_1,t_2,s_2)_{k\ell} = 1/(1-t_k s_\ell)\ (k,\ell = 1,2). \tag{5.4}$$

If $w_1 = (t_1,s_1) \in W_\delta^2$ and $w_2 = (t_2,s_2,0) \in D_\delta^o \times\{0\}$ then

$$K(t_1,s_1,t_2,s_2)_{k\ell} = f_\ell(\delta_1^k t_1 + \delta_2^k s_1,t_2,s_2)\ (k,\ell = 1,2).$$

If $w_1 =(t_1,s_1,0) \in D_\delta^o\times\{0\}$ and $w_2 = (t_2,s_2,0) \in D_\delta^o \times\{0\}$ then

$$K(t_1,s_1,t_2,s_2)_{k\ell} = g_1-(-1)^\ell g_2-(-1)^k g_2+(-1)^{k+\ell}g_1\ (k,\ell=1,2), \tag{5.5}$$

where $g_1=1/(1-q_1q_2)$, $g_2=1/(1-q_1q_2)$ with $q_1=t_1+is_1$, $q_2=t_2+is_2$.

6. APPLICATION TO HYPOTHESIS TESTING

Suppose that we have computed our maximum likelihood estimate $\hat{\Theta}_1$ for an ARMA(p,q) model, where p>0, q>0, and that the computed Fisher information matrix (Hesse matrix) is approximately singular which suggests us to reduce our model to ARMA(p-1, q-1). Compute the ML estimate $\hat{\Theta}_o$ for the model ARMA(p-1,q-1). Now, how to decide wether the estimated ARMA(p,q) model is significantly better then the estimated model of type ARMA(p-1,q-1) or not. Recently the Lagrangian multiplier procedure was proposed by Poskitt and Tramayne (1981) for such a decision, which proved to be more powerful then some earlier methods.

In the present context we propose another method of testing based on the asymptotic distribution of the log likelihood ratio (LR). Compute the asymptotic distribution of the LR for the parameter $\hat{\Theta}_o$ obtained as the ML estimate of the ARMA (p-1,q-1). Denote by q_α the α-quantile $(1/2<\alpha<1)$ of this distribution. Then a critical region

$$\{y_T: L^T(\underline{y}_T,\hat{\Theta}_o) - L^T(\underline{y}_T,\hat{\Theta}_1) > q_\alpha\} \tag{6.1}$$

can be given for a decision between the models ARMA(p,q) and ARMA(p-1,q-1), where the last one is the null hypothesis. Since the asymptotic distribution of the likelihood function is little sensitive on the true parameter, therefore the test given by (6.1) can be considered as an extension of the classical χ^2-test.

APPENDIX

Proof of Theorem 1.

Denote by V_{ij}^{o} $(i=1, \ldots, p+1;\ j=1, \ldots, r+1)$ the set of parameter vektors Θ in $H_{\delta}^{o}(p+1,q,r+1,s)$ such that for their components $\mu_i = \lambda_j \in W_{\delta}$, and the remaining components of Θ are some appropriate permutations of the components of Θ^{o}. Then $H_{\delta}^{o}(p+1,q,r+1,s) = \bigcup \{V_{ij}^{o}: 1<i<p+1,\ 1<j<r+1\}$. For $d>0$ introduce the sets $U_{ij}(d) = \{\Theta \in H_{\delta}: \inf_{\Theta_1 \in V_{ij}^{o}} ||\Theta - \Theta_1||^2 < d^2\}$ and

$$\Sigma(d) = \{\Theta \in H_{\delta}: \inf_{\Theta_o \in H_{\delta}^{o}} ||\Theta - \Theta_o||^2 < d^2\},$$

where $H_{\delta} = H_{\delta}(p+1,q,r+1,s)$.
For $d>0$, $T>0$ we shall use also the notations $\Sigma^T(d) = \Sigma(dT^{-1/2})$, $U_{ij}^T(d) = U_{ij}(dT^{-1/2})$.

For any $d>0$ we shall consider the asymptotic bahaviour of $L^T(y_T;\Theta)$ on the sets $\Sigma^T(d)$. Since $\Sigma^T(d) = \{U_{ij}^T(d): 1<i<p+1, 1<j<r+1\}$ and the behaviour of the likelihood function over the sets $U_{ij}^T(d)$ have evident symmetries, therefore it is sufficient to limit our consideration to a specific $U_{ij}^T(d)$, e.g. to $U_{11}^T(d)$.

Denote by v_{11} the vector which is parallel to the set V_{11}^{o}, and let $\frac{dL^T}{dv_{11}}(y_T;\Theta)$, $\Theta \in U_{11}(d)$, be the directional derivative of $L^T(y_T,\Theta)$ at the parameter Θ. The proof of Theorem 1 is essentially based on two propositions and on a theorem of Gikhman and Skorokhod. The first proposition states that the directional derivatives of L^T are stochastically uniformly bounded in a contracting neighbourhood of the true parameter set.

Lemma 1. For any $\varepsilon>0$ there is a $K>0$ and T_o such that

$$P\Big(\sup_{\Theta \in U_{11}(d)} \Big|\Big| \frac{\partial}{\partial v_{11}} L^T(y_T,\Theta)\Big|\Big|^2 < K\Big) > 1-\varepsilon,\quad T>T_o.$$

The second proposition refers to the asymptotic common distribution of partial likelihood ratios defined as follows.

For any $t\varepsilon W_\delta^o$ define the set $\tilde{V}(t) = \{\Theta=(b,\mu_1,\ldots,\alpha_{s2})\epsilon U_{11}(d): \mu_1=t\}$ and the partial likelihood estimators by

$$\hat{\Theta}^T(t) = \arg\min_{\Theta\epsilon\tilde{V}(t)} L^T(\underline{y}_T;\Theta) \qquad \text{(A.1)}$$

If $t\epsilon W_\delta$ is fixed then $\hat{\Theta}^T(t)$ and $2(L^T(y_T,\Theta^o) - L^T(y_T,\hat{\Theta}^T(t)))$ are asymptotically $N(0,\Gamma(t,t)^{-1})$ and $\chi^2(\nu)(\nu=2+p+2q-r+2s)$ distributed, respectively. In the following we shall describe the asymptotic behaviour of the likelihood function within the framework of local asymptotic normality (LAN) of probability measures defined by random processes (see e.g. Roussas 1972).

$L^T(y_T;\Theta)$ is a stochastic random field on $U_{11}^T(d)$. Introduce the section sets $\tilde{V}_d^T(t) = U_{11}^T(d)\ \tilde{V}(t)$. Let the random processes $\eta^T(t;y_T)$, $t\epsilon W_\delta^o$, and $\xi^T(t;y_T)$, $t\epsilon W_\delta^o$ be defined by

$$\eta_d^T(t,y_T) = L^T(y_T,\Theta^o) - \inf_{\Theta\epsilon\tilde{V}^T(t)} L^T(y_T,\Theta)$$

and

$$\xi_d^T(t,y_T) = \inf_{\Theta\epsilon H_\delta(p,q,r,s)} L^T(y_T;\Theta) - \inf_{\Theta\epsilon\tilde{V}_d^T(t)} L^T(y_T;\Theta).$$

Using Lemma 1 it is not difficult to see that the following stochastic limit relations are satisfied. For any $\varepsilon>0$

$$\lim_{h\to o}\ \sup_T P\Big(\sup_{\substack{|t-t|<h\\ t,t\epsilon W^o}} |\eta_d^T(t';y_T) - \eta_d^T(t'';y_T)| > \varepsilon\Big) = 0.$$

Let $V_d^T(t)$ be the set of parameter vectors obtained from the parameter vectors of $\tilde{V}_d^T(t)$ by omitting the component of μ_1, and conversely, denote by $L^T(y_T;\mu_1,\Theta_t)$, $\mu_1\epsilon W_\delta^o$, $\Theta_t\epsilon V_d^T(t)$ the likelihood function at a parameter vector $\Theta_t(\mu_1)\varepsilon H_\delta(p+1,q,r+1,s)$, which is composed in an appropriate way from μ_1 and Θ_t.
Let $\Theta_t^o \epsilon V_d^T(t)$ denote the parameter vectors corresponding to the true process, i.e. $\nu_1 = t$ and all the remaining components of Θ_t^o are those of Θ^o except μ_1 which is omitted.

The asymptotic Fisher information matrix $I(\Theta^o)$ for estimating Θ^o can be composed from the second order derivatives of

$$\frac{1}{4\pi i}\oint_c \frac{T_o(z)T_o(z^{-1})}{T(z)T(z^{-1})}\,\frac{dz}{z},\quad c=\{z\in C:|z|=1\},$$

where $T_o(z)$ is the transfer function associated with parameter Θ^o and the parameter variables are in $T(z)$. Let $\tilde{I}(\Theta^o,t)$, $t\in W_\delta^o$ be the asymptotic information matrix computed in a similar way for the estimation of Θ_t^o in $V_d^T(t)$. Then $J(\Theta^o,t)$ is defined by the splitting

$$\tilde{I}(\Theta^o,t) = \begin{bmatrix} I(\Theta^o) & J(\Theta^o,t) \\ J(\Theta^o,t)^* & \frac{1}{1-t^2}\end{bmatrix}$$

Lemma 2.

a./ For each fixed $t\in W_\delta^o$ the set of probability measures generated by processes with parameters $\Theta_t\in V_d^T(t)$ is locally asymptoticallay normal at Θ_t^o, i.e. on any bounded set $U\subset R^o$, $\nu=2+p+2q+r+2s$, a representation

$$L^T(y_T;t,\Theta_t^o+T^{-1/2}u) - L^T(y_T;t,\Theta_t^o) = u^*\Delta_T(y_T;t,\Theta_t^o) - \frac{1}{2}u^*\Gamma(t,t)u + \alpha_T(y_T;t,\Theta_t^o,u),\ u\in U, \tag{A.3}$$

holds, where

$$\Delta_T(\underline{y}_T;t,\Theta_t^o) \to N(0,\Gamma(t,t)) \text{ as } T\to\infty, \tag{A.4}$$

in distribution and

$$\alpha_T(\underline{y}_T;t,\Theta_t^o,u)\to 0 \tag{A.5}$$

in probability as $T\to\infty$ and both (A.4) and (A.5) hold uniformly for $u\in U$, $t\in W_\delta^o$. Here $\Delta_T(y_T;t,\Theta_t^o)$ can be chosen as

$$\Delta_T(y_T;t,\Theta_t^o) = T^{-1/2}\frac{\partial}{\partial\Theta_t}L^T(y_T;t,\Theta_t)\,|_{\Theta_t=\Theta_t^o} \tag{A.6}$$

b./ The variables given in (A.6) are such that the finite

dimensional distributions of the processes $\Delta_T(y_T;t,\Theta^o_t)$, $t\epsilon W^o_\delta$, converge to the corresponding finite dimensional distributions of a zero mean Gaussian process $\Delta_\infty(t)$, $t\epsilon W^o_\delta$, with covariance function $\Gamma(t,s)$, $t,s\epsilon W^o_\delta$.

<u>Proof</u>. Part a/ of Lemma 2 can be proved by standard techniques. For proving part b/ denote by $\tau(\nu_1,\Theta)\ \epsilon V^T_\delta(t)$ the vector composed from $\nu_1 \epsilon W^o_\delta$ and $\Theta\epsilon H_\delta(p,q,r,s)$ in an appropriate way. Let $t_1,t_2,\ldots,t_n\epsilon W^o_\delta$ be fixed numbers. Then

$$\Delta_T(y_T;t_i,\Theta^o_{t_i}) = \begin{pmatrix} \dfrac{\partial\tilde{L}}{\partial\mu_1} \\ \dfrac{\partial\tilde{L}}{\partial\Theta} \end{pmatrix}\Bigg|_{\substack{\mu_1=t_i \\ \Theta=\Theta^o_i}} = T^{-1}\,\frac{\partial^2 L^T;t_i,\tau(\mu_1,\Theta))}{\partial(\mu_1,\Theta)\,\partial(\mu_1,\Theta)}\Bigg|_{\substack{\mu_1=t_i \\ \Theta=\Theta^o_i}} \begin{pmatrix} \hat{\mu}^i_1-t_i \\ \hat{\Theta}_{t_i}-\Theta^o \end{pmatrix} + \tag{A.7}$$

$$\left\| \begin{matrix} \hat{\mu}_1-t_i \\ \hat{\Theta}_{t_i}-\Theta^o \end{matrix} \right\|^2 \quad (i=1,2,\ldots,n).$$

Consider now another estimation problem by fixing $t_1, t_2,\ldots,t_n$ and estimating the parameters $\underline{\kappa} = (\kappa_1,\kappa_2,\ldots,\kappa_n)$ and those of $T(B)$ in the transfer function

$$T_n(B) = \frac{(1-\kappa_1 B)\ \ldots\ (1-\kappa_n B)}{(1-t_1 B)\ \ldots\ (1-t_n B)}\,T(B).$$

In accordance with the above notation the likelihood function which is dependent on the parameter vectors $\underline{\kappa},\Theta\ \epsilon\ H_\delta(p,q,r,s)$ will be denoted by $\tilde{\tilde{L}}^T(\underline{y}_T;\underline{t},\bar{\tau}(\underline{\kappa}),\Theta))$. The gradient process of $\tilde{\tilde{L}}^T$

$$C^T_{\underline{t}} = T^{1/2}\,\frac{\partial\tilde{\tilde{L}}^T}{\partial(\kappa,\Theta)}\,(y_T;\underline{t},\bar{\tau}(\kappa,\Theta))\Bigg|_{\substack{\kappa=t \\ \Theta=\Theta}} = T^{1/2}\begin{pmatrix} \dfrac{\partial L^T}{\partial\kappa_1} \\ \cdot \\ \cdot \\ \cdot \\ \dfrac{\partial L^T}{\partial\Theta} \end{pmatrix}\Bigg|_{\substack{\kappa=t \\ \Theta=\Theta^o}} \tag{A.8}$$

is a martingale with respect to the sequence $\mathcal{F}_T=\sigma(y_1,y_2,\ldots,y_T)$ for which a martingale CLT can be applied. The asymptotic distribution of C_t^T is $N(0,\Gamma_n(t))$, where

$$\Gamma_n = \begin{bmatrix} \Gamma_n^{11}(\underline{t}) & \Gamma_n^{12}(\underline{t}) \\ \Gamma_n^{21}(\underline{t}) & \Gamma_n^{22}(\underline{t}) \end{bmatrix} \quad \left[\Gamma_n^{22}(\underline{t})\right]_{i,j} = \frac{1}{1-t_i t_j} \quad (i,j = 1,2,\ldots,n)$$

and

$$\Gamma_n^{21}(t)^* = \Gamma_n^{12}(\underline{t}) = (J(\Theta^o,t_1)\ \ldots J(\Theta^o,t_n)),\ \Gamma_n^{11}(\underline{t}) = I(\Theta^o).$$

Since some components of $C_{\underline{t}}^T$ and $\Delta_T(\underline{y}_T;t_i,\Theta^o_{t_i})$ are coincident:

$$\left.\frac{\partial \tilde{\tilde{L}}^T}{\partial \kappa_i}\right|_{\substack{\kappa=\underline{t}\\ \Theta=\Theta^o}} = \left.\frac{\partial \tilde{L}^T}{\partial \mu_1}\right|_{\substack{\mu_1=t_i\\ \Theta=\Theta^o}} \quad \text{and} \quad \left.\frac{\partial \tilde{\tilde{L}}^T}{\partial \Theta}\right|_{\substack{\kappa=\underline{t}\\ \Theta=\Theta^o}} = \left.\frac{\partial \tilde{L}^T}{\partial \Theta}\right|_{\substack{\mu_1=t_i\\ \Theta=\Theta^o}} \quad (i=1,2,\ldots,n)$$

therefore it follows that $\Delta_T(y_T;t_1,\Theta^o)$, $\Delta_T(y_T,t_2,\Theta^o),\ldots$ $\ldots,\Delta_T(y_T,t_n,\Theta^o)$ are asymptotically normally distributed with covariances

$$\lim_{T\to\infty} E(\Delta_T(y_T;t_i,\Theta^o)\Delta_T;t_j,\Theta^o)^*) = \begin{bmatrix} I(\Theta^o) & J(\Theta^o,t_j) \\ J(\Theta^o,t_i)^* & (1-t_i t_j)^{-1} \end{bmatrix} \quad (i,j=1,\ldots,n)$$

Comparing (A.7) with the Taylor expansion

$$L^T(\underline{y}_T;t_i,\tau(t_i,\Theta^o)) - L^T(\underline{y}_T;t_i,\tau(\hat{\mu}_1^i,\hat{\Theta}_{t_i})) = \tag{A.9}$$

$$(\hat{\mu}_1-t_i,\hat{\Theta}_{t_i}-\Theta^o)\ \frac{\partial^2 L^T(y_T,t_i,\tau(\hat{\mu}_1,\Theta))}{\partial(\mu_1,\Theta)\,\partial(\mu_1,\Theta)} \begin{pmatrix} \hat{\mu}_1-t_i \\ \hat{\Theta}_{t_i}-\Theta^o \end{pmatrix} + o\left\|\begin{matrix} \hat{\mu}_1-t_i \\ \hat{\Theta}_{t_i}-\Theta^o \end{matrix}\right\|^2$$

$$(i=1,\ldots,n).$$

it can easily be derived that the asymptotic common distribution of

$$\tilde{L}^T(\underline{y}_T;t_1,\Theta^o_{t_1})-\tilde{L}^T(\underline{y}_T;t_1,\tau(\hat{\mu}^1_1,\hat{\Theta}_{t_1})),\ldots,\ \tilde{L}^T(\underline{y}_T;t_n,\Theta^o_{t_n}) -$$

$$- \tilde{L}^T(\underline{y}_T;t_n,\tau\ (\hat{\mu}_1,\hat{\Theta}_{t_n})\quad \text{is the same as that of}$$

$$||\eta_\infty(t_1))||^2,\quad ||\eta_\infty(t_2)||^2,\ldots,||\eta_\infty(t_n)||^2.$$

Using an argumentation similar to the classical one (see Wilks (1963)) it can be proved that the asymptotic common distribution of $L^T(\underline{y}_T;t_1,\tau(\hat{\mu}^i_1,\hat{\Theta}_{t_i}))-L^T(\underline{y}_t;\hat{\Theta})$ ..., $L^T(\underline{y}_T;t_n,\tau(\hat{\mu}^n_1,\ \hat{\Theta}_{t_n})) - L^T(\underline{y}_T;\hat{\Theta})$ is the same as that of $\zeta_\infty(t_1)^2,\ldots,\ \zeta_\infty(t_2)^2,\ldots,\ \zeta_\infty(t_n)^2$.

The uniformity of convergences in (A.4) and (A.5) and the next lemma can be proved by Theorem 9.2.1 of Gikhman and Skorokhod (1970).

Lemma 3. Let $d>0$ be a fixed number. The distribution of

$$\Psi_T(d) = \max_{t\in W^o_\delta} \eta^T_d(t,\underline{y}_T)$$

weakly converges to the distribution of

$$\Psi_\infty(d) = \max_{t\in W^o_\delta} \eta^\infty_d(t),$$

where

$$\eta^\infty_d(t) = \max_{||u||<d} u^*\Delta_\infty(t) - \frac{1}{2} u^* \Gamma(t,t)u$$

and here $\Delta_\infty(t)$, $t\in W^o_\delta$, is a zero mean Gaussian process with covariance function $\Gamma(t,s)$, $t,s \in W^o_\delta$.

We shall still need the following lemma, which can be proved by standard methods used in the theory of hypothesis testing of stochastic processes or in contiguity theory of probability measures.

Lemma 4. Assume that the sequence of numbers d_T, $T \in N$, is such that $d_T \uparrow \infty$ and $d_T T^{-1/2} \downarrow 0$ as $T \to \infty$. Then

$$\lim_{T\to\infty} P(\ \Theta \in H_\delta\ (p+1,q,r+1,s):\ d(\Theta,H_\delta^o) > d_T T^{-1/2}\quad L^T(\underline{y}_T;\Theta) - L^T(\underline{y}_T,\Theta^o) > 0 = 0$$

where $d(.,.)$ denotes the Euclidian distance between points or sets of points.

Consider an arbitrary sequence of numbers c_n, $n \in \mathbb{N}$, such that $n^{-1}c_n \downarrow 0$ and $n^{-1/2}c_n \uparrow \infty$ as $n \to \infty$. For every $n \in \mathbb{N}$ let $T_n > n$ be a time point such that for $T > T_n$ the distances of ditributions of $\Psi_T(c_n)$ and $\Psi_\infty(c_n)$ is less then $1/n$, i.e. $\rho(\Psi_T(c_n), \Psi_\infty(c_n)) < 1/n$, $T > T_n$. (Here the distance of distributions is defined as the supremum of the deviation of the distribution functions and it is denoted by $\rho(.,.)$.) Define $d_T = c_n$, if $T_n < T < T_{n+1}$ $(n=1,2,...)$.

Then clearly $d_T \to \infty$ as $T \to \infty$. Under the above conditions it follows that

$$\rho(\Psi_\infty(d_T),\Psi_\infty(\infty)) \to 0 \quad \text{and} \quad \rho(\Psi_T(d_T),\Psi_\infty(d_T)) \text{ as } T \to \infty,$$

where

$$\Psi_\infty(\infty) = \sup_{t \in W_\delta^o} \ \sup_u \{u^* \Delta_\infty(t) - \frac{1}{2} u^* \Gamma(t,t) u\}$$

By the inequality

$$\rho(\Psi_T(d_T),\Psi_\infty(d_T)) + \rho(\Psi_\infty(d_T),\Psi_\infty(\infty)) > \rho(\Psi_T(d_T),\Psi_\infty(\infty))$$

the limit relation

$$\Psi_T(d_T)) \xrightarrow{d} \Psi_\infty(\infty) \qquad \text{as } T \to \infty \tag{A.10}$$

is implied. Some easy computation shows that

$$\lim_{t,s\to\mu_i^o} k_\infty(t,s) = I, \quad \lim_{t,s\to\mu_i^o} c_\infty(t,s) = I \quad (i = 1,...,p)$$

and therefore the time domains of $\eta_\infty(t)$ and $\xi_\infty(t)$ can be extended from W_δ^O to W_δ in such a way that the extended process remains to be an almost surely continuous process on W_δ. Finally, using Lemma 4 and (A 10) the statements of Theorem 1 can be concluded.

REFERENCES

Akaike, H. (1974) A new look at statistical model identification, IEEE Trans. on Automatic Control, AC-19, 716-723.

Arató, M. (1982) Linear Stochastic Systems with Constant Coefficients, Lecture Notes in Control and Information Sciences, 45, Springer Vlg., Berlin - New York.

Bohlin, T. (1978) Maximum power validation of models without higher order fitting, Automatica, 14, 137-146.

Dunsmuir, W. (1979) A Central Limit Theorem for parameter Estimation in Stationary Vector Time Series and Its Application to Models for Signal Observed with Noise, The Annals of Statist., 7, 339-364.

Dunsmuir, W. and Hannan, E.J. (1976) Vector Linear Time Series Models, Advances in Appl. Probab., 8, 339-364.

Godfrey, L.G. (1979) Testing the adequacy of a time series model, Biometrika, 66, 67-72.

Hall, P. and C.C. Heyde (1980) Martingale Limit Theory and Its Application. Academic Press, New York.

Hannan, E.J. (1982) Testing for Autocorrelation and Akaike's Criterion, In: Essays in Statistical Science, eds. J.M. Gani and E.J. Hannan. Applied Probability Trust, Sheffield, 403-412.

Keviczky, L., J. Bokor and S. Veres (1984) Strong Consistency of ML estimators using partial fraction and elementary subsystem representation of multivariable system, Preprints of the 9th World Congress of IFAC, X. 11-16.

Van Overbeek, A.J.M. and L. Ljung (1982) On-line Structure Selection for Multivariable State-space Models, Automatica, 18, 529-543.

Poskitt, D.S. and A.R. Tremayne (1981) An Approach to Testing Linear Time Series Models, The Annals of Statistics, 9, 974-986.

Priestley, M.B. (1981) Spectral Analysis and Time Series, vol. 1-2, Academic Press, New York

Roussas, G.G. (1972) Contiguity of Probability Measures: Some Applications in Statistics, Cambridge University Press, Cambridge.

Söderström, T. (1975) Comments on "Order Assumptions of Information Matrix for Pulse Transfer Function Models", IEEE Trans. on Automatic Control, AC-20, 445-447.

Söderström,T. (1977) On model structure testing in system identification, International J. of Control, 26, 1-18.

Veres, S. (1985) Asymptotic distribution of a structural validation statistic. Preprints of the 7th IFAC Symp. on Identification and System Param. Est., York, England

Wilks, S.S (1938) Distribution of the likelihood ratio in large samples. Ann. Math. Statist. 9, 60-62.

Wilks, S.S. (1962) Mathematical Statistics, John Wiley, New York

S. Veres
Department of Probability and Statistics
Eötvös Loránd University
Budapest, Muzeum krt 6-8.
H-1088, Hungary

Proc. of the 5th Pannonian Symp. on Math. Stat., Visegrád, Hungary 1985
W. Grossmann, J. Mogyoródi, I. Vincze, W. Wertz, eds

ESTIMATING CONTAMINATION LEVEL

J.Á. VÍŠEK

Prague

A definition of contamination level is proposed and a strongly consistent estimator of it is established. An asymptotic distribution of the estimator under a null hypothesis as well as under a sequence of local alternatives is found. A numerical study of the finite sample size behaviour of the estimator is presented, too.
AMS subject classification: 62F10

1. INTRODUCTION

In the past twenty years many results belonging to the field of robust statistics were established. An idea that data had been in some way *contaminated* can be traced out in derivation of any of them. Sometimes this idea was only implicitly included in a corresponding statistical framework, at another times it was embodied by a model of contaminacy. As examples of the first group, i.e. of an implicit expression of the presence and *level of contamination* we may recall the choice of shape of ψ-function determining M-estimators (Huber (1980)), the assignment of value of α when evaluating regression quantiles (Koenker, Bassett (1978)), the selection of a type of distance for minimal distance estimators (Boos (1981)), fixing value of α for α-estimators (Vajda (1985)) or value of g for LFP-estimators (Perez (1985)), etc.. Huber's model of mixture of two populations (Huber (1965)), Rieder's model of contaminacy combining Huber's approach with total variation distance (Rieder (1977)), Huber-Strassen's model using capacities (Huber, Strassen (1973)) or models of contaminacy based on divergences as measures of distance in space of probability measures (Perez (1982), Vajda (1982)) may stay as examples of the second group. In all examples of models of contaminacy there is a parameter value of which represents our idea about the level of contamination. (This parameter may be even uncountably dimensional - see example with capacities.) Since there were estimators of this parameter only for rather special situations (e.g. a mixture of known types of distributions) statisticians have relied usually on a guess of its value based on a numerical experience gained when processing similar data. But it may be of interest to try to estimate value of the contimation level directly. Before starting such endeavour it seems reasonable to discuss a question:

"Does it pay off?".

Let us show on the three examples that assigning a wrong value to contamination level may imply considerable changes of statistical characteristics of algorithm. Let us consider simple situation when testing a hypothesis H_0: $P=P_0$ against an alternative H_1: $P=P_1$ under presence of contamination. Let us choose Rieder's model of contaminacy, i.e. we test

$$H_{0A} = \{Q: Q(A) \geq (1-\varepsilon_0)P_0(A)-\delta_0, \; Q\in\mathcal{M}, \forall A\in\mathcal{A}\}$$

against

$$H_{1A} = \{Q: Q(A) \geq (1-\varepsilon_1)P_1(A)-\delta_1, \; Q\in\mathcal{M}, \forall A\in\mathcal{A}\},$$

where $\mathcal{M}$ is the set of all probability measures over an appropriate measurable space, say $(\mathcal{X},\mathcal{A})$, and $\varepsilon_0 \geq 0$, $\varepsilon_1 \geq 0$, $\delta_0 \geq 0$, $\delta_1 \geq 0$, $0 < \varepsilon_0 + \delta_0 < 1$, $0 < \varepsilon_1 + \delta_1 < 1$. (Index "A" points out that the hypotheses were established under our assumption about level of contamination which we had expressed by assigning values ε_0, ε_1, δ_0 and δ_1 to contamination parameters.) Under some mild conditions we may find for any $\alpha > 0$ the most powerful robust test of size α. Let us denote its critical region C_A $(\subset \mathcal{X}^n)$. This region is determined by requirement that

$$\alpha = \sup\{Q^n(C_A): Q\in H_{0A}\}.$$

For the power of the test we have

$$\beta = \inf\{Q^n(C_A): Q \in H_{1A}\}.$$

If however our assumption about the contamination level was false and the true level is given by parameters $\varepsilon_0 + \tau_0$, $\varepsilon_1 + \tau_1$, $\delta_0 + \xi_0$ and $\delta_1 + \xi_1$ the actual size and power of our test is

$$\alpha_{\tau,\xi} = \sup\{Q^n(C_A): Q \in H_{0T} = \{Q: Q(A) \geq (1-\varepsilon_0-\tau_0)P_0-\delta_0-\xi_0, Q \in \mathcal{M}, \forall A \in \mathcal{A}\}\},$$

and

$$\beta_{\tau,\xi} = \inf\{Q^n(C_A): Q \in H_{1T} = \{Q: Q(A) \geq (1-\varepsilon_1-\tau_1)P_1-\delta_1-\xi_1, Q \in \mathcal{M}, \forall A \in \mathcal{A}\}\}.$$

It would have been pleasant if it had proved that for moderate valued of τ's and ξ's the differences $\alpha_{\tau,\xi}-\alpha$ and $\beta-\beta_{\tau,\xi}$ were not large. Unfortunately it is not so.

In the numerical study, a small fraction of which we present below, the following values were chosen: $\varepsilon_0 = \varepsilon_1 = .0500$, $\delta_0 = \delta_1 = .0250$ and $\tau_0 = \tau_1 = \xi_0 = \xi_1 = .0125$ (for discussion see Víšek (1983)). Then for sample size $n = 40$ and for $P_0 = N(0,1)$ and $P_1 = N(1,1)$ we obtain for

$\alpha = .191 \rightarrow \alpha_{\tau,\xi} = .296$
$\alpha = .095 \rightarrow \alpha_{\tau,\xi} = .167$
$\alpha = .051 \rightarrow \alpha_{\tau,\xi} = .097$
$\alpha = .011 \rightarrow \alpha_{\tau,\xi} = .027$.

Similarly for Weibull model, i.e. for density $f(x) = cps^{p-1}\exp\{-cx^p\}$, having taken parameters for P_0: $c=1$, $p=3$ and for P_1: $c=2$, $p=3$ (again for the sample size equal to 40) one may compute for

$\alpha = .220 \rightarrow \alpha_{\tau',\xi} = .341$
$\alpha = .117 \rightarrow \alpha_{\tau',\xi} = .205$
$\alpha = .054 \rightarrow \alpha_{\tau',\xi} = .109$
$\alpha = .014 \rightarrow \alpha_{\tau',\xi} = .033$

The Laplace distribution with density $f(x) = (2b)^{-1}\exp\{-|x-a|/b\}$ behaves in an analogous way. Taking parameters so that P_0: $a = 0$, $b = 1$ and P_1: $a = 2$, $b = 1$ we receive for

$$\alpha = .201 \rightarrow \alpha_{\tau,\xi} = .319$$
$$\alpha = .104 \rightarrow \alpha_{\tau,\xi} = .188$$
$$\alpha = .053 \rightarrow \alpha_{\tau,\xi} = .108$$
$$\alpha = .010 \rightarrow \alpha_{\tau,\xi} = .027$$

(For further results see Víšek(1983) or contact the author.) The displayed examples reveal that not large increase of contamination level (in a vague sense from 7.5% to 10%) caused that the actual size of test was nearly two times higher than assumed one. This fact hints that the answer to our question should be "Yes".

2. DEFINITION OF CONTAMINATION LEVEL

The first step in solving the problem of estimating contamination level has to be a definition of it. In the present paper let us restrict ourselves on the most simple case - Huber's model of contaminacy on the real line R.

Let $\mathcal{B}$ be the Borel σ-algebra (of subset of R) and ν a σ-finite measure on it. Let $\mathcal{P}_\nu$ be a set of all probability measures (defined on $(R,\mathcal{B})$) which are absolutely continuous with respect to ν and denote by $\mathcal{F}_\nu$ a corresponding set of distribution functions. Now, for any pair $G,F \in \mathcal{F}_\nu$ there is an $\varepsilon \in [0,1]$ and $H \in \mathcal{F}_\nu$ such that $G = (1-\varepsilon)F+\varepsilon H$. Obviously for $G \perp F$ we have $\varepsilon = 1$ and $H = G$ but for all other cases a whole family of ε's and H's are available. It leads us to give the definition of the contamination level as follows.

<u>Definition.</u> *For* $G,F \in \mathcal{F}_\nu$ *we shall understand by the contamination level of* G *with respect to* F *a value*

$$\varepsilon_{G,F} = \inf\{\varepsilon:\ g = (1-\varepsilon)f + \varepsilon h,\ H \in \mathcal{F}_\nu\},$$

where g,f *and* h *are densities of* G,F *and* H, *respectively.*

It is not difficult to show that this infimum is attained, i.e. that we may write $G=(1-\varepsilon_{G,F})F+\varepsilon_{G,F}H$, where G is assumed unknown distribution which has generated data, F a distribution by which we would like to explain data and $\varepsilon_{G,F}$ represents the contamination level, or if you want $1-\varepsilon_{G,F}$ may be considered as a level of approximation of G by F. $\varepsilon_{G,F}$ shouls be taken, in some way, into account for selection of approximate robust method. The following lemma shows that $\varepsilon_{G,F}$ is invariant to a choice of ν (in the class of all σ-finite measures equivalent with ν).

Lemma 1. *Let* $G,F\in\mathcal{F}_\nu$. *Then*

$$\varepsilon_{G,F} = \sup_{x\in R} \frac{f-g}{f}, \tag{1}$$

where again g *and* f *are densities of* G *and* F, *respectively.*

The proof (based on the only fact that h has to be non-negative) is easy and will be omitted.

Let us give an example illustrating the definition.

Example. *Consider a mixture of two normal distributions, i.e. for some* $\varepsilon_H\in(0,1)$ *put*

$$g(x) = (1-\varepsilon_H)(2\pi)^{-\frac{1}{2}}\exp\{-x^2/2\} + \varepsilon_H(2\pi)^{-\frac{1}{2}}\sigma^{-1}\exp\{-x^2/(2\sigma^2)\}.$$

Let us assume that $\sigma^2>1$. *A straightforward computation then gives that*

$$\varepsilon_{G,F} = (1-\sigma^{-1})\varepsilon_H.$$

3. ESTIMATING CONTAMINATION LEVEL

Since f in (1) is given, to estimate $\varepsilon_{G,F}$ we need only estimate g. And that is what we are going to do now. At first we introduce some necessary notation.

Let $\{X_k(\omega)\}_{k=1}^{\infty}$ be a sequence of random variables (r.v's) (defined on a probability space (Ω,C,P)) which are independent and identically distributed according to a distribution function G having (with respect to ν) a density g (which is assumed to vanish outside an interval (c,d), $-\infty \le c < d \le \infty$). Further, let $\psi = \{\psi_n(x,y)\}_{n=1}^{\infty}$ be a sequence of function defined on an interval $(a,b)^2$ with $-\infty \le a \le c < d \le b \le \infty$ and put

$$V_n = \sup_{x \in R,} \sup_{y \in R} \psi_n(x,y).$$

Finally, let us denote for any $y \in R$ and $\omega \in \Omega$ by $G_n(y,\omega)$ the empirical distribution function corresponding to the first n r.v's $X_1(\omega), X_2(\omega), \ldots, X_n(\omega)$, i.e.

$$G_n(y,\omega) = \frac{1}{n} \sum_{k=1}^{n} I_{\{X_k(\omega) \le y\}} \cdot$$

Then by a density estimator $\hat{g}_n$ of the density g we shall understand

$$\hat{g}_n(x,\omega) = g_n = \int \psi_n(x,y)\,dG_n(y,\omega) = \frac{1}{n} \sum_{k=1}^{n} \psi_n(x, X_k(\omega))$$

(see Csörgő, Révész (1981)). Moreover for any $F \in \mathcal{F}_\nu$ (having a density f) let us denote

$$\pi_f(n) = \sup_{x \in R} |E_F \hat{f}_n(x) - f(x)|.$$

Using the ideas of the proof of the Theorem 6.1.1 of Csörgő, Révész (1981) one derives the following lemma.

Lemma 2. *Let* G *and* F *belong to* $\mathcal{F}_\nu$. *Denote their densities by* g *and* f, *respectively. Further, let* $\{W_n\}_{n=1}^{\infty}$ *be a family of subsets of* R *such that* $W_n \uparrow \{f(x)>0\}$ *and*

$$[\inf_{x\in W_n} f(x)]^{-1}\max\{\pi_g(n), n^{-\frac{1}{2}}V_n \log\log n\} = o(1).$$

Then

$$\lim_{n\to\infty} \sup_{n\in W_n} \left|\frac{\hat{g}_n(x)-g(x)}{f(x)}\right| = 0 \quad \text{a.s.} \; .$$

Remark 1. It is clear that the goal of using sequence $\{W_n\}_{n=1}^{\infty}$ is to overcome some technicalities when proving (2). But it may be of interest that a proper selection of this sequence may allow us to cope with the bad tail behaviour of density estimator when applying below presented results for finite sample size situation.

Since for any $G,F\in\mathcal{F}_\nu$ the value of $\varepsilon_{G,F}\in[0,1]$, reasonable estimator $\hat{\varepsilon}_{G,F}$ has to acquire a value also from $[0,1]$. It implies that for the case $G = F$ (i.e. we have chosen the "true" distribution as explanatory model) there is no unbiased estimator of $\varepsilon_{G,F}$. Likewise for $G\perp F$.

Let us give now the theorem proving the strong consistency of an estimator of contamination level.

Theorem 1. *Let the assumption of* Lemma 2 *be fulfilled. Moreover let* $\{f_n\}_{n=1}^{\infty}$ *be a sequence of random processes such that*

$$\sup_{x\in R}|f_n(x)-f(x)| = o(\inf_{x\in W_n} f(x)) \quad \text{a.s.}$$

and put

$$\hat{\varepsilon}_{G,F} = \max\left\{0, \min\left\{\sup_{x\in W_n} \frac{f_n-\hat{g}_n}{f}, 1\right\}\right\}.$$

Then

$$\hat{\varepsilon}_{G,F} \underset{n\to\infty}{\to} \varepsilon_{G,F} \quad \text{a.s.} \; .$$

Proof: We may write

$$\sup_{x\in W_n} \frac{f_n-\hat{g}_n}{f} = \sup_{x\in W_n} \left\{\frac{f_n-f}{f} + \frac{f-g}{f} + \frac{g-\hat{g}_n}{f}\right\}.$$

Since

$$\lim_{n\to\infty} \sup_{x\in W_n} \frac{|f_n-f|}{f} = 0$$

as well as (due to Lemma 2)

$$\lim_{n\to\infty} \sup_{s\in W_n} \frac{|\hat{g}_n-g|}{f} = 0$$

the result follows.

Remark 2. It seems necessary to explain why f_n instead of f is used in numerator of estimator (although f satisfies condition (3)). The reason is following. If we use f instead of a (quantile) f_n we obtain also strongly consistent estimator but having bad behaviour for any finite sample size which is due to natural fluctuation of any estimator of density. Let us consider e.g. the situation when $G=F$, i.e. the situation when the true distribution was taken as explanatory model. An estimator with $f_n=f$ attains a.s. positive values for any finite size of sample althouth it should estimate $\varepsilon_{G,F}=0$. Substituting a quantile f_n into numerator one may shift some mass of probability to zero and improve behaviour of estimator (it naturally requieres to have a possibility to use the estimator at first on a training samples having been of the same kind as processed data - compare with adaptive estimator of location). An idea how to choose f_n is offered by Theorem 3 and 4 and by the numerical study presented at the end of paper. This study also shows that although presented theoretical results are only of asymptotical character (and hence they might be useless for practical application as many results of this kind are) they work satisfactorily for finite sample size, too.

In the rest of paper let us restrict ourselves on the kernel estimators of density. So, let $k(y)$ be a density and $\{c_n\}_{n=1}^{\infty}$ be a sequence of positive numbers converging to zero for $n\to\infty$. Then put

$$\hat{g}_n = \frac{1}{c_n n} \sum_{i=1}^{n} k(c_n^{-1}(y-x_i)).$$

Now it is necessary to cope with the task how to select f_n. Let us try to do it at first for a "null" hypothesis, i.e. when $G = F$ or in another words when $\varepsilon_{G,F} = 0$. To solve the problem we would like to have a result describing the behaviour of $\hat{\varepsilon}_{G,F}$. Such description has to be based on knowledge of fluctuation of $\hat{g}_n - g$. Study of this kind was performed in Rosenblatt (1971), where it is possible to find the following result.

Theorem 2. (Rosenblatt, p. 1828) *Let* g *be continuously differentiable and bounded away from zero on* $[0,1]$. *Assume that the bounded, integrable weight function* k *used in the estimate* $\hat{g}_n$ *is twice continuously differentiable with* $(1+x^4)|k''(x)|$ *bounded and has a Fourier transform* $\zeta(t)$ *with* $(1+t^2)\zeta(t) \in L$. *Then if* $n^{-1/24} = \mathcal{O}(c_n)$ *as* $n\to\infty$, *it follows that*

$$P\left[\max_{0\le x\le 1} \left(\frac{nc_n}{\gamma g(x)}\right)^{\frac{1}{2}} \left\{\hat{g}_n(x) - E_G\hat{g}_n\right\} \le \left\{2\log c_n^{-1}\right\}^{\frac{1}{2}} + \frac{A+z}{(2\log c_n^{-1})^{\frac{1}{2}}}\right] \to e^{-e^{-z}}$$

as $n\to\infty$, *where*

$$\gamma = \int k^2(x)dx$$

and

$$A = \log \frac{B^{\frac{1}{2}}}{2\pi}, \qquad B = -\frac{2}{\gamma}\frac{d^2}{dt^2}\left(\int k(z)k(z+t)dz\right)\Big|_{t=0}.$$

The presence of $g^{\frac{1}{2}}$ in denominator of the above result is implied by the use of results from the theory of Wiener processes and it seems that we can't easy rid off it. Another disadvantage of Rosenblatt's result is that $E_G\hat{g}_n$ instead of g

stays in numerator. It should be also taken into account that the kernel estimators are generally biased and moreover the bias is of an order of some power of c_n (not of power of n). Both these difficulties may be overcome by a transformation of data. Let us have the data $x_1, x_2, \ldots, x_n$ which are assumed to be observations of iid r.v's distributed according to some distribution function G and let us choose as an explanatory distribution F. Now we would like to find (asymptotic) distribution of $\hat{\varepsilon}_{G,F}$ under the assumption $\varepsilon_{G,F} = 0$. Then we may take instead of data $x_1, x_2, \ldots, x_n$ new data $u_1, u_2, \ldots, u_n$ such that $u_i = F(x_i)$. Under our assumption the data $u_1, u_2, \ldots, u_n$ are uniformly distributed over $[0,1]$, i.e. for transformed data we have $g(x) = 1$ for $x \in [0,1]$ and $g(x) = 0$ elsewhere. So we have $g = \sqrt{g}$ and moreover $E_G \hat{g}_n = 1$ for $x \in [0,1]$, and $E_G \hat{g}_n = 0$ elsewhere, i.e. $E_G \hat{g}_n = g$ or in other words the estimator $\hat{g}_n$ is unbiased. Now it is clear that for new data we may use following theorem.

Theorem 3. *Let for* $p \in (0,1)$

$$\tilde{z} = -\log(-\log p)$$

and put

$$b = \left(\frac{\gamma}{c_n}\right)^{\frac{1}{2}} (2\log c_n^{-1})^{\frac{1}{2}} + \frac{A + \tilde{z}}{(2\log c_n^{-1})^{\frac{1}{2}}}.$$

Finally, let $f_n(x) = 1 - bn^{-\frac{1}{2}}$ *(see* Theorem 1). *Then under assumption that* $\varepsilon_{G,F} = 0$ *we hawe*

$$P_G\{\hat{\varepsilon}_{G,F} = 0\} \underset{n\to\infty}{\to} p$$

and

$$P_G\{(2\log c_n^{-1})^{\frac{1}{2}} \left(\frac{c_n n}{\gamma}\right)^{\frac{1}{2}} \hat{\varepsilon}_{G,F} \le z - \tilde{z}\} \to e^{-e^{-z}}$$

for $z - \tilde{z} \in (0,1)$.

The proof of theorem follows in a straightforward way from Rosenblatt's result. Similarly one may obtain

Theorem 4. *Let* p *and* $\tilde{z}$ *be the same as in* Theorem 3. *Then under a sequence of local alternatives* $H_n: \varepsilon_{G,F} = \varepsilon n^{-\frac{1}{2}}$, *i.e.* $G_n = (1 - \varepsilon n^{-\frac{1}{2}})F + \varepsilon n^{-\frac{1}{2}}H$, *where* H *is a fixed d.f. from* $\tilde{\mathcal{F}}_\nu$ *and* $\varepsilon \in (0,1)$, *we have*

$$P_G\left\{(2\log c_n^{-1})^{\frac{1}{2}}\left(\frac{c_n n}{\gamma}\right)^{\frac{1}{2}}(\hat{\varepsilon}_{G,F} - \varepsilon_{G,F}) \le z - z\right\} \to e^{-e^{-z}} .$$

4. NUMERICAL STUDY

The following numerical study has been aimed to enlighten the behaviour of the estimator $\hat{\varepsilon}_{G,F}$ for finite sample size.

A little modified data published by Beran (1977, p. 460) were used and ordered. Let us denote these data $x^{(1)}, x^{(2)}, \ldots$ $\ldots, x^{(40)}$. They are normal with $\bar{x} = .1173$ and $s^2 = 1.0828$. Hence we have put

$$f(x) = (2\pi)^{-\frac{1}{2}} s^{-1} \exp\left\{- \frac{(x-\bar{x})^2}{2s^2}\right\} .$$

Then we have created a training set containing 20 samples all of them having size equal to 40. Each of these samples was built up in following way. Having employed a table of random numbers we have selected three indexes among 1,2,...,40, say i_1, i_2 and i_3. Then the observation $x^{(i_1)}, x^{(i_2)}$ and $x^{(i_3)}$ of the original data were substituted by $3x^{(i_1)}$, $3x^{(i_2)}$ and $3x^{(i_3)}$. So we have had

$$g(x) = (1 - \frac{3}{40})(2\pi)^{-\frac{1}{2}} s^{-1} \exp\left\{- \frac{(x-\bar{x})^2}{2s^2}\right\} +$$

$$+ \frac{3}{40}(2\pi)^{-\frac{1}{2}}(3s)^{-1} \exp\left\{- \frac{(x-\bar{x})^2}{2s^2}\right\}$$

and hence $\varepsilon_{G,F} = \frac{1}{40}(1 - \frac{1}{3}) = .05$ (see Example 1). Having used this training set the value of b (see Theorem 2) was established so that the mean value of $\hat{\varepsilon}_{G,F}$ over these 20 samples was equal to .05. It was found that the value of b should be 5.36925. After having acomplished this step a testing set was built up. It has included another 20 samples all of which were constructed in the same way as the samples of above described training set. Again for every samples of testing set the value of $\hat{\varepsilon}_{G,F}$ was evaluated. The values of $\hat{\varepsilon}_{G,F}$ for training as well as for testing set were gathered in a table (see Table below).

The result of numerical study supports a hope that the behaviour of $\hat{\varepsilon}_{G,F}$ could be good even for finite sample size. (Naturally, the asymptotic results of Theorem 2 and 3 only has hinted a shape of f_n - see Theorem 1.)

The mean value of $\hat{\varepsilon}_{G,F}$ over the testing set was .05079.

Table

training set			testing set	
number of sample	indexes of contaminated observations	$\hat{\varepsilon}_{G,F}$	indexes of contaminated observations	$\hat{\varepsilon}_{G,F}$
1	13,28,21	.07828	30,22,12	0
2	11,20,26	.03325	12, 8,26	.05111
3	28,40,11	0	40,29,21	.01979
4	24,22,25	.15198	13,12,40	.08201
5	29,11,12	.06856	6, 5,36	.05818
6	19,13,36	.07419	28,13,38	.07226
7	25, 1,40	.01967	24,11,34	.02639
8	11, 4, 8	.10826	12, 8, 2	.05114
9	35,36,39	.02380	37,19,27	.04115
10	23,38,19	.07419	39,13,15	.10800
11	36,26,11	.01013	14,16,30	.11569
12	32,26,13	.01698	28, 6,16	.03501
13	36,31,21	.01979	34,30,38	.03421
14	19,13,37	.07419	17, 3, 1	.00290
15	19,13,23	.08090	34,25,40	.01967
16	15,28,34	.03435	11, 4, 8	.10826
17	14,27,33	.03388	22,24,36	.06468
18	7,15,36	.03258	26, 6, 3	.03892
19	23,34, 1	.02546	11, 9,34	.06001
20	8,27,12	.05114	24,29, 1	.02653

REFERENCES

Abramowitz M., Stegun I.A. (1964): Handbook of Mathematical Functions with Formulas, Graphs and Mathematical Tables, *Dover Publications, New York*

Boos D.D. (1981): Minimum distance estimators for location and goodness of fit, *J. Amer.Statist.Soc. 76, No 375, 663-670*

Csörgő M., Révész P. (1981): Strong Approximation in Probability and Statistics, *Akadémiai Kiadó, Budapest*

Huber P.J. (1965): A robust version of the probability ratio test, *Ann. Math. Statist. 36, 1753-1758.*

Huber P.J. (1981): Robust Statistics, *John Wiley & Sons, New York*

Huber P.J., Strassen V. (1973): Minimax tests and Neyman-Pearson lemma for capacities, *Ann. Statist. 1, 251-263.*

Koenker R., Bassett G.Jr. (1978): Regression quantiles, *Econometrica 46, 33-50*

Perez A. (1982): Discrimination rate least favorable pairs of distributions for ε-contaminated statistical hypotheses or with f-divergence like neighbourhoods, *Transactions of the Ninth Prague Conference on Information Theory, Statistical Decesion Functions and Random Processes Academia, Prague, 53-66*

Perez A. (1985): LFP (Least favorable pair) approach to statistical estimating and testing using the barycenter concept, *Transactions of the Second Joint Swedish-Soviet International Workshop on Information Theory, 39-43*

Rieder H. (1977): Least favorable pairs for special capacities, *Ann. Statist. 5, 909-922*

Rosenblatt M. (1971): Curve estimates, *Ann. Math. Statist. 42, 1815-1842*

Vajda I. (1982): Theory of Information and of Statistical Decesion, *Alfa, Bratislava* (in Slovak)

Vajda I. (1985): Efficiency and robustness control via distorted maximum likelihood estimation, *Kybernetika* (to appear)

Víšek J.Á. (1983): On second order efficiency of a robust test and approximations of its error probabilities, *Kybernetika, 19, 387-407*

Jan Ámos Víšek
Czechoslovak Academy of Sciences
Institute of Information Theory
and Automation
Pod vodárenskou věží 4
182 08 Praha 8 - Libeň
Czechoslovakia

Proc. of the 5th Pannonian Symp. on Math. Stat., Visegrád, Hungary 1985
W. Grossmann, J. Mogyoródi, I. Vincze, W. Wertz, eds

LIMIT THEOREMS FOR THE LIKELIHOOD RATIO PROCESSES OF STATISTICAL PARAMETRIC MODELS

L. VOSTRIKOVA

Budapest

Using martingale methods we obtain functional limit theorems for the likelihood ratio processes of binary statistical models and general statistical parametric models. We give examples showing how one can apply these results. From these functional limit theorems we also get "predictable" sufficient conditions for asymptotic normality of maximum likelihood estimators.

AMS subject classification: 60F05, 62F12

INTRODUCTION

Suppose we are given a general statistical parametric model $\mathcal{E} = \{(\Omega,F),\ P_\theta,\ \theta\in\Theta\}$ where (Ω,F) is a measurable space and $(P_\theta)_{\theta\in\Theta}$ is a family of probability measures on (Ω,F) which depends on the parameter $\theta\in\Theta\subseteq R^m$, $m\geq 1$. Assume, our general statistical parametric model $\mathcal{E}$ is endowed with a filtration $\mathbb{F} = (F_t)_{t\geq 0}$ which is a family of nondecreasing right-continuous σ-algebras F_t, $\bigvee_{t\geq 0} F_t = F$. This additional assumption allows to associate with every $\mathcal{E}$ and fixed $\theta_0\in\Theta$ (without loss of generality assume $\theta_0 = 0$, $P_{\theta_0} \equiv P$) the likelihood ratio process $Z = (Z_t(\theta))_{t\geq 0,\theta\in\Theta}$ with

$$Z_t(\theta) = (dP_{\theta,t}/dQ_t)/(dP_t/dQ_t)$$

where Q is a dominating measure, say $(P_\theta + P)/2$, and $P_{\bullet,t} = P_\bullet|F_t$, $Q_t = Q|F_t$ are the restrictions of the corresponding measures to the σ-algebra F_t.

Among all general statistical parametric models an important role is played by the Gaussian model, i.e. by the model with the likelihood ratio process Z given by

$$Z_t(\theta) = \exp(N_t(\theta) - \frac{1}{2} < N(\theta) >_t) \tag{1}$$

where $N = (N_t(\theta))_{t\geq 0, \theta \in \Theta}$ is a continuous (a.s.) Gaussian field given on $(\bar{\Omega}, \bar{F}, \bar{P})$ and in a way that $N(\theta) = (N_t(\theta))_{t\geq 0}$ is $\bar{P}$-martingale with quadratic characteristic $\langle N(\theta)\rangle$ for every $\theta \in \Theta$, $N(0) \equiv 0$ (note that for Gaussian martingales $\langle N(\theta)\rangle_t = DN_t(\theta)$). A well known example of likelihood ratio process of such type is the case when $N_t(\theta) = (\theta, W_t)$, where $W = (W_t)_{t\geq 0}$ is a Wiener process with covariance matrix Γ.

Let $\mathcal{E}^n = \{(\Omega^n, F^n, \mathbb{F}^n), P^n_\theta, \theta \in \Theta^n\}$, $n\geq 1$, be a sequence of general statistical parametric models and $(Z^n)_{n\geq 1}$ be the corresponding sequence of likelihood ratio processes. We are interested in the conditions for a weak convergence of the distributions of Z^n to the ones of the likelihood ratio process Z of the Gaussian statistical model as $n\to\infty$. Since the likelihood ratio process Z^n is the function of (ω, t, θ), the following four types of limit theorems may be considered: i) for fixed t and θ, ii) for fixed θ with respect to t, iii) for fixed t with respect to θ, iv) with respect to (t, θ).

The limit theorems for the likelihood ratio processes of type i), i.e. for fixed t and θ, are the same as the theorems about local asymptotic normality of the likelihood ratio which can be found in the papers of Le Cam [14], [15], [16], [17] Hájek [4], Kushnir [12], Phillipou and Roussas [23], Ibragimov and Hasminskij [5], [6], Pflug [22].

The limit theorems for the likelihood ratio processes of type ii), i.e. for fixed θ with respect to t, are the functional limit theorems for the likelihood ratio processes of binary models $\mathcal{E}^n_B = \{(\Omega^n, F^n, \mathbb{F}^n), P^n_\theta, P^n\}$. The theorems of this kind have been proved for discrete time in the paper of Greenwood and Shiryayev [3] and for continuous time by Kordzahia [11].

The limit theorems of type iii), i.e. for fixed $t > 0$ with respect to θ, can be found in the papers of Prakasa Rao [24], [25], Inagaki [7], Weiss and Wolfowitz [27], Ibragimov and Hasminskij [6], Kutoyants [13].

In this paper we give functional limit theorems of type ii) and iii) for continuous time, which contain many particular cases: for example, when the measures $(P^n_\theta)_{\theta\in\Theta^n}$ correspond to diffusion type processes, counting processes, processes with independent increments, discrete-time processes.

I. LIMIT THEOREM FOR BINARY STATISTICAL MODELS

Consider the sequence of binary statistical models $\mathcal{E}^n_B = \{(\Omega^n, F^n, \mathbb{F}^n), \tilde{P}^n, P^n\}$, $n\geq 1$, where $\tilde{P}^n$ and P^n are probability measures on measurable space (Ω^n, F^n) endowed with right-continuous filtration $\mathbb{F}^n = (F^n_t)_{t\geq 0}$, $\bigvee_{t\geq 0} F^n_t = F^n$, and $Q^n = (\tilde{P}^n + P^n)/2$.

Denote by $Z^n = (Z^n_t, F^n_t)_{t\geq 0}$ the CORLOL modification of the likelihood ratio process of $\mathcal{E}^n_B$ where $Z^n_t = \tilde{\zeta}^n_t/\zeta^n_t$ with $\tilde{\zeta}^n_t = d\tilde{P}^n_t/dQ^n_t$ and $\zeta^n_t = dP^n_t/dQ^n_t$, and let $h^n = (h^n_t, F^n_t)_{t\geq 0}$ be the Hellinger process corresponding to $\mathcal{E}^n_B$ (see Liptser and Shiryayev [19], Jacod [8], Mémin and Shiryayev [20]). For every $x\in R^1$ and $t > 0$ we set $\tilde{\lambda}^n_t(x) = (1 - x/\tilde{\zeta}^n_{t-})\vee 0$, $\lambda^n_t(x) = (1+x/\zeta^n_{t-})\vee 0$ and $y^n_t(x) = (1+x/Z^n_{t-})\vee 0$ where we assume $x/0 = 0$ for every $0\leq x<\infty$, $\infty/\infty = \infty$.

Here and further on $I(\bullet)$ denotes an indicator function, ν_{Z^n,P^n} is the compensator of jump measure μ_{Z^n} of Z^n with respect to $(\mathbb{F}^n, P^n)$ and the notation $\eta^n \xrightarrow{P^n} \eta$ means that $P^n(|\eta^n - \eta|\geq\varepsilon)\to 0$ as $n\to\infty$ for every $\varepsilon > 0$.

Theorem I.1. *Let* $N = (N_t)_{t\geq 0}$ *be Gaussian martingale with quadratic characteristic* $\langle N\rangle$. *The following conditions: for every* $T > 0$, $\varepsilon > 0$

1) $Z^n_o \xrightarrow{P^n} 1$,

2) $\int_0^t \int_{R^1\setminus\{0\}} \tilde{\lambda}^n_s(x) I(\lambda^n_s(x)=0)\nu_{\zeta^n,Q^n}(ds,dx) \xrightarrow{Q^n} 0$, $\quad 0 < t \leq T$,

3) $\int_0^t \int_{R^1\setminus\{0\}} (\sqrt{Y_s^n(x)}-1)^2 I(|\sqrt{Y_s^n(x)}-1|\geq\varepsilon)\nu_{Z^n,P^n}(ds,dx) \xrightarrow{P^n} 0,\ 0<t\leq T,$

4) $h_t^n \xrightarrow{P^n} \frac{1}{8}\langle N\rangle_t,\ 0<t\leq T,$

are necessary and sufficient for a weak convergence

$$Z^n \xrightarrow{w(P^n)} \exp(N - \frac{1}{2}\langle N\rangle) \tag{2}$$

in the Skorohod space $\mathbb{D}[0,\infty)$ *with respect to* P^n.

Remark I.1. In discrete case the theorem was proved in [3]. Another formulation of this theorem (without the condition 2)) can be found in [11].

Remark I.2. The first condition of the theorem provides for the convergence of initial values of the likelihood ratio processes. The second condition is connected with local absolute continuity $\tilde{P}^n \overset{loc}{<<} P^n$ and it controls when Z^n takes infinite values with respect to the measure $\tilde{P}^n$. The third condition is the conditional Lindeberg condition and the fourth one is about convergence of variations. Note that the theorem is not true without condition 2), and this is shown by the following example.

Example I.1. Let $(\Omega,\mathcal{B},\tilde{P})$ be a probability space with $\Omega=[0,1]$, the Borel σ-field $\mathcal{B}$ and the Lebesgue measure $\tilde{P}$. Let P_n be the probability measure on $(\Omega,\mathcal{B})$ with density

$$\frac{dP_n}{d\tilde{P}} = p_n(x) = \begin{cases} 0 & \text{if } 0\leq x\leq 1/n, \\ n/(n-1) & \text{if } 1/n<x\leq 1. \end{cases}$$

Let also $\Omega^n=\Omega\times\ldots\times\Omega$, $\mathcal{B}^n=\mathcal{B}\otimes\ldots\otimes\mathcal{B}$, $\tilde{P}^n=\tilde{P}\times\ldots\times\tilde{P}$ and $P^n=P_n\times\ldots\times P_n$ be n-fold products of $(\Omega,\mathcal{B},\tilde{P})$ and P_n respectively and $F_t^n=\mathcal{B}^{[nt]}$ for $0\leq t\leq 1$, $\mathcal{B}^o=\{\emptyset,\Omega^n\}$. In this example, $Z_o^n\equiv 1$ and for sufficiently large n and $0<t\leq 1$ we have

$$\int_0^t \int_{R^1\setminus\{0\}} (\sqrt{Y_s^n(x)}-1)^2 I(|\sqrt{Y_s^n(x)}-1|\geq\varepsilon)\nu_{Z^n,P^n}(ds,dx) =$$

$$= [nt]\int_0^1 (1-\sqrt{q_n(x)})^2 I(|1-\sqrt{q_n(x)}|\geq\varepsilon)p_n(x)dx = 0,$$

$$h_t^n = \frac{1}{2}[nt]\left\{\int_0^1 (1-\sqrt{q_n(x)})^2 p_n(x)dx + \tilde{P}^n(q_n(x) = \infty)\right\}$$

where $q_n(x) = 1/p_n(x)$ for $0\leq x\leq 1$. Hence, $h_t^n \to \frac{t}{2}$ as $n\to\infty$ and the conditions 1), 3), 4) of Theorem I.1 are satisfied by $N = 2W$ where W is standard Wiener process. But $Z_1^n(x_1,\ldots,x_n) =$ $= \prod_{k=1}^n q_n(x_k) \overset{P^n\text{-a.s}}{=} (1-\frac{1}{n})^n$ and it tends to $1/e$ instead of $\exp\{2W_1 - 2\}$ as stated in Theorem I.1.

Proof of Theorem I.1. *Sufficiency*. We introduce stopping times $T^n = \inf\{t\geq 0: Z_t\leq 1/n \text{ or } Z_t\geq n\}$ with $\inf\{\emptyset\} = \infty$ and the special process $V^n = (V_t^n, F_t^n)_{t\geq 0}$ defined by

$$V_t^n = 2M_t^{nc} + 2\int_0^{t\wedge T^n}\int_{R^1\setminus\{0\}} (\sqrt{Y_s^n(x)}-1)(\mu_{Z^n}(ds,dx)-\nu_{Z^n,P^n}(ds,dx)) - $$

$$- 4b^n_{t\wedge T^n} \tag{3}$$

where M^{nc} is continuous P^n-martingale appeared in the semimartingale representation of $\frac{1}{2}\ell n\, Z^n_{t\wedge T^n}$ and $b^n = (b_t^n, F_t^n)_{t\geq 0}$ is the modified Hellinger process given by

$$b^n_{t\wedge T^n} = \frac{1}{2}\langle M^{nc}\rangle_t + \frac{1}{2}\int_0^{t\wedge T^n} (\sqrt{Y_s^n(x)} - 1)^2\nu_{Z^n,P^n}(ds,dx). \tag{4}$$

Using the results of [10], [19] and [18] we show that for every $T\geq 0$ and $\varepsilon > 0$

i) $(P_T^n) \lhd (\tilde{P}_T^n)$,

ii) $P^n(\sup_{0\le t\le T} |Z_t^n - Z^n_{t\wedge T^n}|\ge\varepsilon)\to 0,\ n\to\infty,$

iii) $P^n(\sup_{0\le t\le T} |Z^n_{t\wedge T^n} - \exp(V_t^n)|\ge\varepsilon)\to 0,\ n\to\infty,$

iv) $V^n \xrightarrow{\mathbb{D}(P^n)} N - \frac{1}{2}\langle N\rangle.$

These relations imply the needed convergence by [1].

Necessity. Using the results of [19] and [3] about contiguity, the results of [9] about convergence of quadratic variation process and also truncations with stopping times we prove that for every $T \ge 0$

i) $(2) \Rightarrow (P_T^n) \lhd \rhd (\tilde{P}_T^n),$

ii) $(2) \Rightarrow 1),\ 3),\ (2) \Rightarrow \{V^n \xrightarrow{\mathbb{D}(P^n)} N - \frac{1}{2}\langle N\rangle\},$

iii) $\{V^n \xrightarrow{\mathbb{D}(P^n)} N - \frac{1}{2}\langle N\rangle\} \Rightarrow \{[V^n,V^n]_t \xrightarrow{P^n} [N,N]_t,\ 0<t\le T\}$

iv) $\{[V^n,V^n]_t \xrightarrow{P^n} [N,N]_t,\ 0<t\le T\} \Rightarrow 4),\ 2).$ □

II. LIMIT THEOREMS FOR GENERAL STATISTICAL PARAMETRIC MODELS

Consider the sequence of general statistical parametric models $\mathcal{E}^n = \{(\Omega^n, F^n, \mathbb{F}^n),\ P_\theta^n,\ \theta\in\Theta^n\}$, $n\ge 1$. Suppose that Θ^n is convex set, $\Theta^n \subseteq \Theta^{n+1}$, $\{0\}\in\Theta^1$ and also that $F_o^n = \{\emptyset, \Omega^n\}$.

Denote by $Z^n(\theta)$ the CORLOL modification of the likelihood ratio process of the binary statistical model $\mathcal{E}_B^n = \{(\Omega^n,\ F^n,\ \mathbb{F}^n), P_\theta^n,\ P^n\}$ and let $h^n(\theta,0)$ be the corresponding Hellinger process. By $Y_t^n(x,\theta)$ we denote the analog of $Y_t^n(x)$ defined by substitution of Z_t^n by $Z_t^n(\theta)$.

Theorem II.1. *Suppose the conditions 2), 3) of Theorem I.1. with substitution of* Z^n, $Y_s^n(x)$, $\tilde{P}^n$, P^n *by* $Z^n(\theta)$, $Y_s^n(x,\theta)$, P_θ^n, P^n *respectively for every* $\theta \in \Theta^{n_0}$ *and* $T>0$. *Assume also that for every* $T>0$ *and* θ', $\theta\in\Theta^{n_0}$

$$4') \quad h_t^n(\theta',\theta) \xrightarrow{P^n} \frac{1}{8}\langle N(\theta') - N(\theta)\rangle_t, \quad 0<t\leq T.$$

Then for every $k\geq 1$ *we have*

$$(Z_{t_1}^n(\theta_1),\ldots,Z_{t_k}^n(\theta_k))^T \xrightarrow{d(P^n)} (Z_{t_1}(\theta_1),\ldots,Z_{t_k}(\theta_k))^T \tag{5}$$

where $Z_t(\theta)$ *is defined by* (1), $\xrightarrow{d(P^n)}$ *denotes the convergence in distribution and* $t_i \in [0,\infty)$ $\theta_i \in \Theta^{n_0}$, $1\leq i\leq k$.

Proof of Theorem II.1. We define the process $V^n(\theta)$ as in theorem I.1 with evident substitutions and prove that (5) is equivalent to

$$(\hat{M}_{t_1}^n(\theta_1),\ldots,\hat{M}_{t_k}^n(\theta_k))^T \xrightarrow{d(P^n)} (\tfrac{1}{2}N_{t_1}(\theta_1),\ldots,\tfrac{1}{2}N_{t_k}(\theta_k))^T \tag{6}$$

where by $\hat{M}^n(\theta)$ we denoted the martingale part of the process $V^n(\theta)$.

To prove (6) we use the Kramer-Wold method [1] together with corollary 2 of [18] and also the relations

$$\langle\hat{M}^n(\theta'), \hat{M}^n(\theta)\rangle_t - h_t^n(\theta,0) - h_t^n(\theta',0) + h_t^n(\theta',\theta) \xrightarrow{P^n} 0,$$

where $t\geq 0$ and θ, $\theta' \in \Theta^{n_0}$. The last relations in turn can be obtained using the truncations by stopping times, the modified Hellinger process $b^n(\theta,0)$ and also the Hellinger process $\hat{h}^n(\theta',\theta)$ built with the help of arbitrary dominating measure Q^n (see [8]). □

For fixed $T>0$ consider a separable with respect to θ modification $\hat{Z}_T^n = (\hat{Z}_T^n(\theta))_{\theta\in R^m}$ of the process $Z_T^n = (Z_T^n(\theta))_{\theta\in\Theta^n}$ which is continued if necessary to the whole R^m with preservation of the supremum value and the modulus of continuity. We use the notation $\varphi(y)\in\Phi_1$ for the functions $\varphi(y)$, $y\geq 0$, which monotonically decrease and are integrable on $[L,\infty)$ for sufficiently large $L>0$.

Consider the functions $\varphi_T^n(y)$, $y\geq 0$, $n\geq 1$, satisfying

$$y^{m+r-1}(\varphi_T^n(y))^{1-m/\beta}\in\Phi_1, \tag{7}$$

where $1\leq n\leq\infty$, $\varphi_T^\infty(y)=\lim_{n\to\infty}\varphi_T^n(y)$, and $r>0$ and $\beta>0$ are constants. For every nondecreasing predictable process $h=(h_t, F_t^n)_{t\geq 0}$ we set

$$\mathcal{E}(-h)_t=\begin{cases} e^{-h_t}\prod\limits_{0<s\leq t}(1-\Delta h_s)\,e^{\Delta h_s}, & \text{if } h_t<\infty,\\ 0 & \text{, if } h_t=\infty,\end{cases}$$

where $\Delta h_s = h_s - h_{s-}$.

Theorem II.2. *Suppose the conditions of Theorem* II.1. *for fixed* $T>0$. *Assume, there exist the constants* $\beta>m$, $r\geq 0$, $d\geq 0$, $D\geq 0$, $c\geq 0$, $C\geq 0$, *and the functions* $\varphi_T^n(y)$ *satisfying* (7), *such that for sufficiently large* $L>0$

1) $$\sup_n\ \sup_{|\theta|\leq L,\ |\theta'|\leq L}\frac{P_\theta^n(h_T^n(\theta',\theta)\geq DL^r|\theta'-\theta|^\beta)}{|\theta'-\theta|^{2\beta}}\leq dL^{2r}<\infty,$$

2) $$\sup_n\ \sup_{|\theta|\geq L}\frac{P^n(\mathcal{E}(-h^n(\theta,0))_T\geq C\varphi_T^n(|\theta|))}{(\varphi_T^n(|\theta|))^2}\leq c<\infty.$$

Then the distributions of $\hat{Z}_T^n$ *converge weakly in* $\mathbb{C}(R^m,P^n)$ *to the distributions of* $Z_T=(Z_T(\theta))_{\theta\in R^m}$ *defined by* (1), *as* $n\to\infty$.

Remark II.1. The condition 1) of Theorem II.2. is the one about continuity of the family of probability measures $(P_\theta^n)_{\theta\in\Theta^n}$ with respect to θ. The condition 2) is the one about asymptotic separation of the measures P_θ^n and P^n with large $|\theta|$. The two conditions 1) and 2) guarantee the tightness of the sequences of the measures, corresponding to $\hat{Z}_T^n$, in the space $\mathbb{C}(R^m,P^n)$.

Proof of Theorem II.2. First we prove that $\hat{Z}_T^n$ is continuous P^n-a.s. For this purpose we use the inequalities for the Hellinger distances and the Hellinger integrals through the Hellinger process [26], [29] and also [28]. We obtain like in [28] that for sufficiently large $L > 0$ and $h > 0$

$$P^n\Big(\sup_{|\theta-\theta'|\le h,|\theta|\le L,|\theta'|\le L} |\sqrt{\hat{Z}_T^n(\theta)}-\sqrt{\hat{Z}_T^n(\theta')}|\ge\varepsilon\Big)\le c(\varepsilon)L^m h^{\beta-m},$$

$$P^n\Big(\sup_{|\theta|\ge L}\hat{Z}_T^n(\theta)\ge\varepsilon\Big)\le c(\varepsilon)\int_{(L-h)/2}^{\infty} y^{m+r-1}[\varphi_T^n(y)]^{1-m/\beta}dy.$$

Because of Theorem II.1 we have the convergence of finite dimensional distributions of $\hat{Z}_T^n$ and using the method given in [2] and the above inequalities we prove the tightness of the probability measures corresponding to $\hat{Z}_T^n$ in $\mathbb{C}(R^m,P^n)$. Then, by Prohorov's theorem [1] we obtain the result. □

Corollary II.1. *Suppose the measures* $(P_\theta^n)_{\theta\in\Theta^n}$ *correspond to the processes with independent increments and they are (for simplicity) equivalent. Then according to* [20] *the Hellinger process* $h^n(\theta',\theta)$ *is deterministic for every* θ', $\theta\in\Theta^n$. *If for fixed* $T > 0$ *the conditions of Theorem* II.1. *are satisfied together with*

1') $$\sup_n \sup_{|\theta|\le L,|\theta'|\le L} h_T^n(\theta',\theta)/|\theta'-\theta|^\beta < DL^r < \infty,$$

2') $$\sup_n \sup_{|\theta|\ge L} \mathcal{E}(-h^n(\theta,0))_T/\varphi_T^n(|\theta|) < C < \infty,$$

then the Theorem II.2 *remains true.*

Using the formulae for calculation of the Hellinger process given in [20] and [26] we obtain the following results.

Example II.1. Let $P_\theta^n = \mu_\theta^{n,1} \times \ldots \times \mu_\theta^{n,n}$, $\theta \in \Theta^n \subseteq R^1$ be an n-fold product measure, and suppose that the measures $\mu_\theta^{n,i}$ have densities $p_\theta^{n,i}(x)$ with respect to the Lebesgue measure for all $\theta \in \Theta^n$ and $1 \le i \le n$. Assume that for a fixed $\theta \in \Theta^{n_0}$ and all $\varepsilon > 0$, $\theta' \in \Theta^{n_0}$, $0 \le t \le 1$ we have

$$1)\ \lim_{n\uparrow\infty} \sum_{i=1}^{n} \int_{-\infty}^{\infty} p_{\theta'}^{n,i}(x)\, I(p_\theta^{n,i}(x)=0)\, dx = 0,$$

$$2)\ \lim_{n\uparrow\infty} \sum_{i=1}^{n} \int_{-\infty}^{\infty} \left(\sqrt{p_{\theta'}^{n,i}(x)/p_\theta^{n,i}(x)}-1\right)^2 I\left(\left|\sqrt{p_{\theta'}^{n,i}(x)/p_\theta^{n,i}(x)}-1\right| \ge \varepsilon\right) p_\theta^{n,i}(x)\,dx = 0,$$

$$3)\ \lim_{n\uparrow\infty} \sum_{i=1}^{[nt]} \int_{-\infty}^{\infty} \left(\sqrt{p_{\theta'}^{n,i}(x)}-\sqrt{p_\theta^{n,i}(x)}\right)^2 dx = \frac{(g(\theta')-g(\theta))^2 t}{4}$$

$$4)\ \sup_{n,|\theta|\le L,|\theta'|\le L} \sum_{i=1}^{n} \int_{-\infty}^{\infty} \left(\sqrt{p_{\theta'}^{n,i}(x)}-\sqrt{p_\theta^{n,i}(x)}\right)^2 dx/|\theta'-\theta|^\beta \le DL^r < \infty,$$

$$5)\ \inf_{n,|\theta'|\ge L} \sum_{i=1}^{n} \int_{-\infty}^{\infty} \left(\sqrt{p_{\theta'}^{n,i}(x)}-\sqrt{p_\theta^{n,i}(x)}\right)^2 dx/|\theta'-\theta|^\beta \ge C > 0,$$

where $\beta > 1$, $r \ge 0$, $D \ge 0$, $\mathbb{C} > 0$ and $g(\theta)$ is a function of θ. Then the conditions of corollary II.1 are met and we get the weak convergence

$$\hat{Z}_1^n \xrightarrow{w(P_\theta^n)} (\exp(g(\theta')\xi - g^2(\theta')/2))_{\theta' \in R^1} \tag{8}$$

in the space $\mathbf{C}(R^1, P_\theta^n)$ where ξ is a standard (0,1)-normal variable.

Example II.2. Let P_θ^n, $\theta \in \Theta^n \subseteq R^1$, be the unique probability measure on (Ω, F) which corresponds to a diffusion type process satisfying the following stochastic differential equation:

$$dX_t^n = a_t^n(\theta, X^n)dt + dW_t, \quad X_o^n = 0.$$

Here $W = (W_t)_{t\geq 0}$ is a Wiener process and $a_t^n(\theta,X)$ is a non-anticipating functional with $\int_0^t [a_s^n(\theta',X)]^2 ds < \infty$ (P_θ^n-a.s.) for every $0 \leq t \leq 1$. Suppose that for a fixed $\theta \in \Theta^{n_0}$ and all $\theta' \in \Theta^{n_1}$, $0 \leq t \leq 1$ we have

1) $\int_0^t (a_s^n(\theta',X) - a_s^n(\theta,X))^2 ds \xrightarrow{P_\theta^n} (g(\theta') - g(\theta))^2$,

2) $$\sup_{n, |\theta| \leq L, |\theta'| \leq L} \frac{P_\theta^n\left(\int_0^1 (a_s^n(\theta',X) - a_s^n(\theta,X))^2 ds \geq DL^r |\theta'-\theta|^\beta\right)}{|\theta'-\theta|^{2\beta}}$$

$$\leq dL^{2r} < \infty,$$

3) $$\sup_{n, |\theta'| \geq L} \frac{P_\theta^n\left(\int_0^1 (a_s^n(\theta',X) - a_s^n(\theta,X))^2 ds \leq 4Cf^n(|\theta'-\theta|)\right)}{\exp(-4Cf^n(|\theta'-\theta|))}$$

$$\leq c < \infty,$$

where $\beta > 1$, $D \geq 0$, $d \geq 0$, $r \geq 0$, $C > 0$, $c \geq 0$ and the function $f^n(y)$ is such that $\phi_1^n(y) = \exp(-Cf^n(y))$ satisfies the condition (7). Then the conditions of Theorem II.2 are satisfied and we have (8).

III. ASYMPTOTIC NORMALITY OF MAXIMUM LIKELIHOOD ESTIMATORS

Consider the sequence of general statistical parametric models $\mathcal{E}^n = \{(\Omega^n, F^n, \mathbb{F}^n), P_\theta^n, \theta \in \Theta\}$. Let $(c_n)_{n\geq 1}$ be a non-decreasing normalizing sequence, $c_n > 0$. For fixed $\theta \in \Theta$ we set $U_\theta^n = \{u \in R^m : \theta + c_n^{-1} u \in \Theta\}$ and also $\bar{\mathcal{E}}^n = \{(\Omega^n, F^n, \mathbb{F}^n), \bar{P}_u^n, u \in U_\theta^n\}$ where $\bar{P}_u^n = P_{\theta + c_n^{-1} u}^n$. Denote by $\hat{\theta}_T^n$ the maximum likelihood estimator of the parameter θ built on the likelihood ratio process $\bar{Z}^n$ of $\bar{\mathcal{E}}^n$, i.e.

$$\hat{\theta}_T^n = c_n^{-1} \arg \sup_{u \in U_\theta^n} \bar{Z}_T^n(u) + \theta$$

where $\hat{\theta}_T^n = \infty$ if supremum with respect to u is not achieved.

Let the Gaussian field $N = (N_t(u))_{t \geq 0, u \in R^m}$ in (1) be defined by $N_t(u) = (u, V_t)$ where $V = (V_t)_{t \geq 0}$ is an m-dimensional Gaussian process with independent increments, $V_0 = 0$, $EV_t = 0$, $EV_t V_s = \min(t,s)\ \Gamma(\theta)$ with symmetric positive matrix $\Gamma(\theta)$.

Theorem III.1. *Suppose there exists the sequence* $(c_n)_{n \geq 1}$ $0 < c_n \leq c_{n+1}$, *such that for* $\bar{\mathcal{E}}^n$ *the conditions of Theorem* II.2. *are satisfied by* N *given above. Then the distributions of* $c_n(\hat{\theta}_T^n - \theta)$ *converge weakly to the normal distribution with parameters* $(0, (\Gamma^T(\theta))^{-1}/T)$ *as* $n \to \infty$.

Proof of theorem III.1. For every Borel set A we consider the functional

$$\Psi(Z) = I(\sup_{u \in A} Z(u) - \sup_{u \in A^c} Z(u) \geq 0)$$

where $Z \in \mathbb{C}(R^m)$, A^c is the complement of A. The functional $\Psi(Z)$ is continuous on the set of $Z \in \mathbb{C}(R^m)$ for which supremum with respect to u is achieved in a unique point not belonging to the boundary ∂A of A. Because the maximum of limit process Z_T is achieved in unique point $\hat{u}_T = \frac{1}{T} \Gamma^{-1}(\theta)\ V_T$ we get from [1]

$$P_\theta^n(c_n(\hat{\theta}_T^n - \theta) \in A) = E_\theta^n \Psi(\hat{Z}_T^n) \to \bar{E}\Psi(Z_T) = \bar{P}(\hat{u}_T \in A)$$

for every A such that $\bar{P}(\hat{u}_T \in \partial A) = 0$ as $n \to \infty$. □

Remark III.1. Note that in the conditions of examples II.1 and II.2 we have the results of theorem III.1 with $c_n \equiv 1$, $\Gamma(\theta) \equiv 1$.

REFERENCES

1. P. Billingsley. Convergence of probability measures, Moscow, "Nauka", 1977 (in Russian).
2. I.I. Gihman, A.V. Skorohod. Introduction to the theory of random processes, "Nauka", 1977 (in Russian).
3. S. Greenwood, A.N. Shiryayev. On the concept of contiguity and functional limit theorems under contiguous alternatives (to appear in Stochastics).
4. J. Hájek. Local asymptotic minimax and admissibility in estimation. Proc. Sixth Berkeley Symp. on Math. Statist. and Prob., 1 (1972), 175-194.
5. I.A. Ibragimov, R.Z. Hasminskij. Local asymptotic normality for not identically distributed variables. Theory Probab. Appl., 20, 2 (1975), 251-266.
6. I.A. Ibragimov, R.Z. Hasminskij. Asymptotic theory of estimations, Moscow, "Nauka", 1979 (in Russian).
7. N. Inagaki. Asymptotic relations between the likelihood estimating functions and the maximum likelihood estimator. Ann. Inst. Statist. Math., 25 (1973), 1-26.
8. J. Jacod. Processus de Hellinger, absolue continuité, contiguite. Séminaire de Probabilités, Rennes, 1983.
9. J. Jacod. Convergence en loi semimartingales et variation quadratique. Séminaire de Probabilités, XV, 1979/80. Lecture Notes in Math., 850.
10. Ju. M. Kabanov, R.Sh. Liptser, A.N. Shiryayev. Absolute continuity and singularity of locally absolutely continuous probability distributions I. Math. USSR Sbornik, 35, 5 (1975), 631-680.
11. N.E. Kordzahia. Weak convergence of likelihood ratio process. Uspehi mat. nauk, 39, 4 (1984), 167.
12. A.Ph. Kushnir. Asymptotically optimal criteria for checking the hypotheses in regression problems. Theory Probab. Appl., 13, 4 (1968), 682-700.
13. Ju.A. Kutoyants. Estimation of parameters of random processes. Yerevan, 1980 (in Russian).
14. L. Le Cam. On some asymptotic properties of maximum likelihood estimates and related Bayes estimates. Univ. of Calif. Publ. in Statist., 1 (1953), 277-330.

15. L. Le Cam. Locally asymptotically normal families of distributions. Univ. of Calif. Publ. in Statist., 3 (1960), 27-98.

16. L. Le Cam. On the assumptions used to prove asymptotic normality of maximum likelihood estimates. Ann. Math. Statist., 41, 3 (1970), 802-828.

17. L. Le Cam. Notes on asymptotic methods in statistical decision theory. Univ. de Montreal, Montreal, 1974.

18. R.Sh. Liptser, A.N. Shiryayev. A functional central limit theorem for semimartingales. Theory Probab. Appl., 25 (1980), 683-703.

19. R.Sh. Liptser, A.N. Shiryayev. On the problem of "predictable" criteria of contiguity. Probab. Theory and Math. Statist. Fourth USSR-Japan Symposium, Proceedings, 1982. Lecture Notes in Math., Springer-Verlag, 1021, 1983.

20. J. Memin, A.N. Shiryayev. Distance de Hellinger-Kakutani des lois correspondant à deux processus à accroissements indépendants. Z. für Wahr., 70 (1985), 67-90.

21. Th. Pfaff. Quick consistency of quasi maximum likelihood estimators. Ann. Statist., 10, 3 (1982), 990-1005.

22. G. Pflug. The limiting Log-Likelihood Process for Discontinuous Density Families. Z. Wahrscheinlichkeits theorie verw. Gebiete, 64 (1983), 15-35.

23. A. Phillipou, G. Roussas. Asymptotic distribution of the likelihood function in the independent not identically distributed case. Ann. Statist., 1, 3 (1973), 454-471.

24. Prakasa Rao B.L.S. Estimation of the location of the cups of a continuous density. Ann. Math. Statist., 39, 1 (1968), 76-87.

25. Prakasa Rao B.L.S. Estimation for distributions with monotone failure rate. Ann. Math. Statist., 41, 2 (1970), 507-519.

26. E. Valkeila, L. Vostrikova. An integral representation for the Hellinger distance (to appear in Mathematica Scandinavica).

27. L. Weiss, J. Wolfowitz. Maximum Probability Estimators and Related Topics. Lecture Notes in Mathematics, 421, 1974, Springer-Verlag, Berlin.

28. L. Vostrikova. On Criteria for c_n-Consistency of Estimators. Stochastics, 11 (1984), 265-290.

29. L. Vostrikova. Hellinger distances, Hellinger integrals and Hellinger processes. International Seminar on Mathematical Modelling and Stability of Stochastic Models. Abstracts of Communications, Varna, 1985.

L.VOSTRIKOVA
Computing Center of
Eötvös Loránd University
Budapest, Hungary
Bogdánfy u. 10/b. 1117

Proc. of the 5th Pannonian Symp. on Math. Stat., Visegrád, Hungary 1985
W. Grossmann, J. Mogyoródi, I. Vincze, W. Wertz, eds

ESTIMATION IN LINEAR MODEL WITH CENSORED DATA

P. VOLF

Prague

The contribution deals with the nonparametric estimation of regression function, provided the data are randomly right-censored. The simpliest form of K-estimator is derived and its consistency is proved.
AMS subject classification: 62G05

1. INTRODUCTION

Let $Y_1, Y_2, \dots, Y_N$ be random variables satisfying the model

$$Y_i = r(x_i) + \varepsilon_i, \quad i = 1, \dots, N, \tag{1}$$

where x_i are known values of regressor, in R_1 (in R_k more generally), ε_i are i.i.d.r.v.-s with zero mean and continuous distribution function F, r(.) is a real function expressing the dependence EY_i on x_i.

We cannot observe the realizations of Y-s directly, instead we observe

$$T_i = \min(Y_i, V_i) \quad \text{and} \quad \delta_i = I[Y_i \le V_i],$$

where I[A] denotes the indicator of set A, $V_1, \dots, V_N$ are random variables independent mutually and of $\varepsilon_1, \dots, \varepsilon_N$. They may depend on x_i, we suppose them to fulfil a linear model

$$V_i = v(x_i) + e_i, \tag{2}$$

with continuous distribution function G of e_i.

Our aim is to estimate the value $r(x)$ for fixed x. The estimator will be computed from the observations corresponding to x_i from some neighbourhood of x. Let us denote such neighbourhood as $O_d(x) = \{z \in R_1 : |x-z| \le d\}$.

A1: Fix a sequence of positive d_N such that $\lim_{N\to\infty} N \cdot d_N = \infty$ and $\lim_{N\to\infty} d_N = 0$.

A2: Let us imagine that x_i are the realizations of random variable $\mathcal{X}$ in $(R_1, \mathcal{B}_1)$, with density function $h(.)$ continuous in some neighbourhood of x, $h(x)>0$.

It is well known from the theory of nonparametric density estimation that these two assumptions are sufficient for mean square convergence of simple K-estimator: Denote by M_N the number of points x_i in $O_{d_N}(x)$, then $M_N / 2N \cdot d_N \to h(x)$. Simultaneously, $M_N \to \infty$ at least in probability (see for instance in Gasser, Rosenblatt, 1979).

2. ESTIMATION

Define the estimator of $r(x)$ by the following expression:

$$r_N(x) = \sum_{i=1}^{N} T_i \cdot \Delta F^Y_{N,x}(T_i) \cdot I[\{x_i \in O_{dN}(x)\} \wedge \{|T_i| \le A_N\}],$$

where A_N is a real sequence tending to infinity, $F^Y_{N,x}$ is product limit estimator (PLE, Kaplan and Meier, 1958) of the distribution function of Y-s provided the corresponding x-s are in $O_{d_N}(x)$: Let $T_{k1} \le \ldots \le T_{kM}$ be all results with x_i in $O_{d_N}(x)$, ranked, then

$$F^Y_{N,x}(t) = 1 - \prod_{j=1}^{M} \left(\frac{M-j}{M-j+1}\right)^{\delta_{kj} \cdot I[T_{kj} \le t]} .$$

This function has jumps $\Delta F^Y_{N,x}$ at points T_{kj}, $\delta_{kj} = 1$. It is an useful convention to put $\delta_{kM} = 1$. Evidently, there M stands for M_N and indices kj depend on N, too.

Theorem. Under following assumptions $r_N(x)$ is probability-consistent estimator of $r(x)$:

B1: $d_N = C \cdot N^{-\beta}$, $\beta \in \left(\frac{2}{3}, 1\right]$, $C > 0$.

B2: A2 holds.

B3: Distribution functions F and G are Lipschitz-continuous.

B4: Functions r and v are Lipschitz-continuous, too, in some neighbourhood of x. Denote the constants -measures of continuity as ρ and ν respectively.

B5: Let $\mathcal{T}(F) = \sup\{t: F(t) < 1\}$, $T < \mathcal{T}(F)$. There exist $a, b \in (0,1]$ that

$$a[1-F(t)]^b \leq 1-G[t+r(x)-v(x)] \quad \text{for all} \quad t \in [T, \mathcal{T}(F)).$$

B6: A_N are such that

$$\lim_{N\to\infty} A_N = \infty, \quad \lim_{N\to\infty} A_N \cdot \left[\frac{\log(N \cdot d_N)}{(N \cdot d_N)^a}\right]^{\frac{1}{2+b}} = 0.$$

We shall proceed with proof after several lemmas are stated. Instead of B5 we can use another assumption stated by Rejtő (1982) to be sufficient for strong uniform consistency of PLE. The degree of this a.s. uniform consistency is given by mentioned paper as $[\log M \,/\, M^a]^{\frac{1}{2+b}}$, M being extent of sample, tending to infinity.

Lemma 1. Let $U_1^M \leq U_2^M \leq \ldots \leq U_M^M$ be order statistics from M independent copies of r.v. U distributed uniformly on (0,1). Put $U_0^M = 0$. Then for $\gamma > 2$, $\alpha \in (0,1)$:

I. $\lim_{M\to\infty} \Pr\left\{\min_i (U_i^M - U_{i-1}^M) > M^{-\gamma}\right\} = 1.$

II. $\lim_{M\to\infty} \Pr\left\{\max_i (U_i^M - U_{i-1}^M) < M^{-\alpha}\right\} = 1.$

Proof: Denote $\Delta_i^M = U_i^M - U_{i-1}^M$.

I. $\Pr\left\{\min_i \Delta_i^M > M^{-\gamma}\right\} = 1 - \Pr\left\{\Delta_i^M < M^{-\gamma} \text{ for some } i\right\} \geq 1 - \sum_{i=1}^{M} \Pr\left\{\Delta_i^M < M^{-\gamma}\right\}$,

we intend to show that the limit of the sum is zero.

Evidently, all Δ_i^M-s have the same distribution as Δ_1^M. After denoting $z = 1/M$ and using l'Hospital's rule,

$$\lim_{M\to\infty} M\cdot\Pr\left\{\Delta_1^M < M^{-\gamma}\right\} = \lim_{M\to\infty} M\cdot\left[1 - (1-M^{-\gamma})^M\right] = \lim_{z\to 0+} \frac{1}{z}\left[1-(1-z^\gamma)^{1/z}\right] =$$

$$= \lim_{z\to 0+} (1-z^\gamma)^{1/z}\left[\frac{\gamma\cdot z^{\gamma-1}}{z(1-z^\gamma)} + \frac{\ln(1-z^\gamma)}{z^2}\right] = 0.$$

II. By the same way,

$$\Pr\left\{\max_i \Delta_i^M < M^{-\alpha}\right\} \geq 1-M\cdot\Pr\left\{\Delta_1^M > M^{-\alpha}\right\} = 1-M\cdot(1-M^{-\alpha})^M,$$

and

$$\lim_{M\to\infty} M(1-M^{-\alpha})^M = \lim_{M\to\infty} M\left[(1-M^{-\alpha})^{M^\alpha}\right]^{M^{1-\alpha}} = 0,$$

because the term in square brackets tends to e^{-1}.

<u>Corollary.</u> Let $\tau_1^M,\ldots,\tau_M^M$ represent the order statistics from i.i.d.r.v.-s with distribution function ϕ. Suppose that ϕ is Lipschitz-continuous with constant φ. Then

$$\lim_{M\to\infty} \Pr\left\{\min_i(\tau_i^M - \tau_{i-1}^M) > M^{-\gamma},\ \gamma > 2\right\} = 1 \quad (\tau_o^M = \inf\{t: \phi(t)>0\}).$$

<u>Proof:</u> As for various $s < t$ $\phi(t) - \phi(s) \leq (t-s)\cdot\varphi$, if $\gamma > 2$,

$$\Pr\left\{\min_i(\tau_i^M - \tau_{i-1}^M)>M^{-\gamma}\right\} \geq \Pr\left\{\min_i[\phi(\tau_i^M) - \phi(\tau_{i-1}^M)] > \varphi\cdot M^{-\gamma}\right\} \to 1,$$

as it follows from Lemma 1, part I.

<u>Lemma 2.</u> Denote $Z_1^M \leq \ldots \leq Z_M^M$ order statistics from M independent copies of r.v. Z having continuous distribution function F, finite mean μ. Let F_M^z be a uniformly consistent estimator of F : $\sup_z|F_M^z(z) - F(z)| = \mathcal{O}(c_M)$ a.s., where c_M is a sequence of positive real numbers tending to zero. Then for a real sequence A_M such that $A_M\to\infty$, $A_M c_M \to 0$, $A_M\cdot M^{-\alpha} \to 0$ for some $\alpha \in (0,1)$, $Z_o^M = -\infty$:

$$\underset{M\to\infty}{\text{P-lim}} \sum_{i=1}^{M} Z_i^M \cdot \left\{F_M^z(Z_i^M) - F_M^z(Z_{i-1}^M)\right\} \cdot I\left[|Z_i^M| \le A_M\right] = \mu .$$

<u>Proof:</u> Denote $I_A = \int_{-A}^{A} t\, dF(t)$, mean may be defined as

$$\mu = I_\infty = \lim_{M\to\infty} I_{A_M} . \tag{3}$$

Further denote $\Delta_i = F(Z_i^M) - F(Z_{i-1}^M)$, $\hat{\Delta}_i = F_M^z(Z_i^M) - F_M^z(Z_{i-1}^M)$, define $F(Z_o^M) = F_M^z(Z_o^M) = 0$, all summations are over indexes i with $|Z_i^M| \le A_M$. Then

$$|\Sigma Z_i^M \hat{\Delta}_i - \mu| \le |\mu - I_{A_M}| + |I_{A_M} - \Sigma Z_i^M \Delta_i| + |\Sigma Z_i^M (\hat{\Delta}_i - \Delta_i)| . \tag{4}$$

The first term of the right side tends to zero, it follows from (3). The second term yields

$$\left| \int_{-A_M}^{A_M} t\, dF(t) - \Sigma Z_i^M \Delta_i \right| \le \omega_M + \Sigma \left| \int_{Z_{i-1}^M}^{Z_\ell^M} t\, dF(t) - Z_i^M \Delta_i \right| \le$$

$$\le \omega_M + \Sigma (Z_i^M - Z_{i-1}^M)\Delta_i \le \omega_M + 3A_M \cdot \max \Delta_i .$$

There ω_M stands for $\int_{Z_k^M}^{A_M} t dF(t) + \int_{-A_M}^{Z_1^M} t dF(t)$, where $k = \max\{i : Z_i^M \le A_M\}$ and $l = \min\{i : Z_{i+1}^M \ge -A_M\}$. P-convergence of ω_M to zero is evident. From Lemma 1, part II and from the supposition $A_M \cdot M^{-\alpha} \to 0$ the P- convergence of $A_M \cdot \max \Delta_i$ results.

As

$$\Sigma Z_i^M (\hat{\Delta}_i - \Delta_i) = \mathcal{O}[c_M A_M] ,$$

P-limit of the third term of (4) is zero, too.

3. PROOF OF THEOREM

Let us denote $\tau_i = \min\{\varepsilon_i, e_i + v(x) - r(x)\}$, $|T_i - \tau_i - r(x)| \le$ $\le (\rho+\nu)\cdot d_N$ for indices i such that $x_i \in O_{d_N}(x)$. All following summations are over indices i for which x_i is in $O_{d_N}(x)$, $|T_i| \le A_N$. Then

$$|r_N(x) - \Sigma\tau_i \Delta F^{Y}_{N,x}(T_i) - r(x)\cdot\Sigma F^{Y}_{N,x}(T_i)| \le (\rho+\nu)d_N. \qquad (5)$$

Let us imagine that $F_{N,x}$ is PLE of F from the sample $\{\tau_i\}$. Assumption B5 enable us to use the result of Rejtő(1982):

$$\sup_t |F_{N,x}(t) - F(t)| = \mathcal{O}\left[\left\{\frac{\log(Nd_N)}{(Nd_N)^a}\right\}^{\frac{1}{2+b}}\right] \text{ a.s.}$$

Now, assumption B6 together with Lemma 2 yield

$$\underset{N\to\infty}{\text{P-lim}}\ \Sigma\tau_i\cdot\Delta F_{N,x}(\tau_i) = 0 \quad \text{and} \quad \underset{N\to\infty}{\text{P-lim}}\ \Sigma\Delta F_{N,x}(\tau_i) = 1. \qquad (6)$$

As we can see from (5), P-consistency of our estimator will be proved, if asymptotically, for the indexes $i: x_i \in O_{d_N}(x)$:

I. τ_i and T_i have the same rank for the same i,

II. $\delta_i^{\tau} = \delta_i$, where $\delta_i^{\tau} = I[\varepsilon_i \le e_i + v(x) - r(x)]$,

I. Lipschitz-continuity of distribution function of τ_i is obvious and from the corollary of Lemma 1 we get

$$\lim_{N\to\infty} \Pr\left\{\min_{i\ne j} \frac{|\tau_i - \tau_j|}{M_N^{-\gamma}} > K\right\} = 1 \text{ for every real K and } \gamma > 2.$$

Rewrite

$$d_N = C\cdot N^{-\beta} = C\cdot\left(\frac{N^{1-\beta}}{M_N} M_N\right)^{\frac{-\beta}{1-\beta}},$$

denote $\gamma = \beta/(1-\beta) > 2$, as soon as $\beta > 2/3$. As $N^{1-\beta}/M_N \xrightarrow{P} 2h(x)$,

$$\lim_{N\to\infty} \Pr\left\{\min_{i\neq j} \frac{|\tau_i-\tau_j|}{d_N} > K\right\} = 1,$$

too. Part I. is proved. The distance between T_i and $\tau_i + r(x)$ is comparable with d_N and it tends to be considerably less than the distance between τ_i and the closest τ_j.

II. Let us denote $Z_i = |\varepsilon_i - e_i - v(x) + r(x)|$, its distribution function

$$\phi(z) = 1 - \int_{-\infty}^{\infty} [1-F(z+s)+F(-z+s)]dG[s-v(x)+r(x)]$$

is Lipschitz continuous, too. By the same method as above we can point out that

$$\lim_{N\to\infty} \Pr\left\{\min_{i} Z_i > K\cdot d_N\right\} = 1.$$

The distance between censorig and censored variable tends to be considerably larger than the distance between T_i and $\tau_i+r(x)$, part II. is proved, too. And the proof of Theorem is completed.

REFERENCES

Gasser Th., Rosenblatt M.-editors(1979) Smoothing Techniques for Curve Estimation, Lecture Notes in Math. V. 757, Springer-Verlag.

Kaplan E.L., Meier P.(1958) Nonparametric estimation from incomplete observations, J. Amer. Stat. Assoc. 53, 457.

Rejtő L.(1982) On the fixed censoring model and consequences for the stochastic case, in Trans. 9th Prague Conf. on Inf. Theory..., Vol. B. Academia, Prague.

Susarla V., Van Ryzin J.(1980) Large sample theory for an estimator of the mean survival time from censored samples, Ann. Statist. 8, No5.

Petr Volf
Czechoslovak Academy of Sciences
Institute of Information Theory
and Automation
Pod vodárenskou věží 4
182 08 Praha 8, Czechoslovakia

Proc. of the 5th Pannonian Symp. on Math. Stat., Visegrád, Hungary 1985
W. Grossmann, J. Mogyoródi, I. Vincze, W. Wertz, eds

LOCAL ASYMPTOTIC NORMALITY WITH RATES

W. WEFELMEYER

Cologne

We describe necessary and sufficient conditions for degenerate convergence with rates of standardized sums of independent random variables. These degenerate convergence conditions are used to define differentiability with rates of one-parameter families of probability measures. In the i.i.d. case, this differentiability concept gives necessary and sufficient conditions for the log-likelihood ratio to admit a first order stochastic expansion with rates.
AMS subject classification: 62A10.

0. INTRODUCTION

Let P_t, $t \in [0,\varepsilon)$, be a path of mutually absolutely continuous probability measures on a measurable space $(X,\mathcal{A})$. Fix $P = P_0$, and let f_t denote a density of P_t with respect to P. Given $u > 0$ and n i.i.d. observations $\underline{x} = (x_1,\dots,x_n)$, we denote the *log-likelihood ratio* by

$$\Lambda_{n,u}(\underline{x}) := \sum_{\nu=1}^{n} \log f_{n^{-1/2}u}(x_\nu).$$

For a measurable function $g\colon X \to \mathbb{R}$ with *expectation* $P(g) = 0$ we write the *standardized sum* as

$$\tilde{g}(\underline{x}) := n^{-1/2} \sum_{\nu=1}^{n} g(x_\nu).$$

In Section 2 we describe a differentiability condition on f_t which is necessary and sufficient for the following particular form of *local asymptotic normality with rates*: For every bounded sequence $u_n > 0$, $n \in \mathbb{N}$,

$$P^n \{|\Lambda_{n,u_n} - u_n\tilde{g} + \frac{1}{2}u_n^2 P(g^2)| > cn^{-a/2}\} = o(n^{-b/2}), \quad c>0. \tag{0.1}$$

The proof is based on a degenerate convergence criterion, given in Section 1.

Stochastic expansions in the sense of (0.1) for the log-likelihood ratio are of interest in several statistical applications.

For $b = a$, relation (0.1) can be used to obtain a normal approximation of order $o(n^{-a/2})$ for the distribution of the log-likelihood ratio: Let $a \in [0,1)$, and assume that relation (0.1) holds with $b = a$. If $P(g) = 0$ and $P(|g|^{2+a}) < \infty$, then uniformly for $s \in \mathbb{R}$,

$$P^n\{\Lambda_{n,u} \le s\} = N(-\tfrac{1}{2}u^2P(g^2), u^2P(g^2))(-\infty, s] + o(n^{-a/2}).$$

(See, e.g., Bhattacharya and Rao, 1976, p. 184, Corollary 18.3.)

For $b = 2a$, relation (0.1) is assumed by Bickel et al. (1981) to prove that a test statistic is efficient of order $o(n^{-a})$ if it approximates the log-likelihood ratio up to $o(n^{-a/2})$. In this sense, "first order efficiency implies second order efficiency".

Similar characterizations of higher order stochastic expansions of the log-likelihood ratio $\Lambda_{n^{-1/2}u}$ can be given in terms of higher order differentiability of f_t. Such stochastic expansions lead in particular to second order bounds for the power of tests and for the concentration of estimators. (See Pfanzagl and Wefelmeyer, 1985, p. 165, Theorem 6.2.2, and p. 254, Theorem 8.2.16.)

1. THE DEGENERATE CONVERGENCE CRITERION

We introduce the following degenerate convergence condition.

1.1. <u>Definition</u>. Let $b \geq 0$. A path of measurable functions $r_t: X \to \mathbb{R}$, $t \in [0,\varepsilon)$, fulfills (at $t = 0$) the *degenerate convergence condition* DCC_b if

(i) $P\{|r_t| > \varepsilon t^{-1}\} = o(t^{2+b})$ for every $\varepsilon > 0$,

(ii) $P\left(r_t 1_{\{|r_t| \leq t^{-1}\}}\right) = o(t)$,

(iii) $P\left(r_t^2 1_{\{|r_t| \leq t^{-1}\}}\right) = o(t^o)$.

For $b = 0$, this is the usual degenerate convergence condition for $\tilde{r}_{n^{-1/2}u}$.

The following Theorem 1.2 presents necessary and sufficient conditions for degenerate convergence with rates of standardized sums of independent random variables. The proof of the necessity part (i) is closely related to the corresponding part of Theorem 4 in Baum and Katz (1965, p. 113), which characterizes convergence rates in the weak law of large numbers. The technique is due to Erdős (1949).

1.2. <u>Theorem</u>. *Let* $b \geq 0$, *and let* $r_t: X \to \mathbb{R}$, $t \in [0,\varepsilon)$, *be a path of measurable functions. Then the following assertions hold.*

(*i*) *If, for every sequence* $u_n \to 1$,

$$P^n\{|\tilde{r}_{n^{-1/2}u_n}| > c\} = o(n^{-b/2}),\ c > 0,$$

then r_t, $t \in [0,\varepsilon)$, *fulfills* DCC_b.

(*ii*) *Conversely, if* $t^{-\delta} r_t$, $t \in [0,\varepsilon)$, *fulfills* DCC_b *for some* $\delta > 0$, *then* (1.3) *holds for every bounded sequence* $u_n > 0$, $n \in \mathbb{N}$.

The proof is deferred to Section 3.

If $b = 0$, then equivalence holds in (i). This is the usual degenerate convergence criterion (see, e.g., Loève, 1977, p. 329, Corollary). We have introduced the sequences u_n, $n \in \mathbb{N}$, to obtain a degenerate convergence criterion for $t \in [0,\varepsilon)$, not only for $t \in \{n^{-1/2}: n \in \mathbb{N}\}$. See in this connection Lemma 3.1.

For $b > 0$ we cannot expect equivalence in (i). If P is the standard normal distribution and $r_n(x) = \varepsilon_n x$ with $\varepsilon_n \to 0$, then DCC_b is fulfilled for arbitrary $b \geq 0$. But

$$P^n\{|\tilde{r}_n| > \varepsilon\} = 2\Phi(-\varepsilon/\varepsilon_n),$$

which is of order $o(n^{-b/2})$ for $b > 0$ only if $\varepsilon_n = o((\log n)^{-1/2})$.

2. STOCHASTIC EXPANSIONS OF LOG-LIKELIHOOD RATIOS

As an application of the degenerate convergence criterion (Theorem 1.2) we obtain necessary and sufficient conditions for the log-likelihood ratio to admit a stochastic expansion with rates (Theorem 2.2). These conditions are expressed in terms of differentiability of f_t with respect to degenerate convergence in the sense of Definition 1.1.

2.1. Definition. Let $a, b \geq 0$, and let $g: X \to \mathbb{R}$ be a measurable function. The path P_t, $t \in [0,\varepsilon)$, is DCC_b-*differentiable* (at $t = 0$) *at a rate* $o(t^a)$ with *derivative* g if

$$f_t = 1 + tg + t^{1+a} r_t$$

with r_t, $t \in [0,\varepsilon)$, fulfilling DCC_b.

2.2. Theorem. *Let* $a \in [0,1)$ *and* $b \geq a$. *Assume that* $P(g) = 0$ *and* $P(|g|^{2(2+b)/(2-a)}) < \infty$. *Then the following assertions hold.*

(*i*) *If, for every sequence* $u_n \to 1$,

$$P^n\{|\Lambda_{n,u_n} - u_n\tilde{g} + \tfrac{1}{2}u_n^2 P(g^2)| > cn^{-a/2}\} = o(n^{-b/2}),\ c > 0, \quad (2.3)$$

then P_t, $t \in [0,\varepsilon)$, *is* DCC_b*-differentiable at a rate* $o(t^a)$.

(*ii*) *Conversely, if* P_t, $t \in [0,\varepsilon)$, *is* DCC_b*-differentiable at a rate* $o(t^{a+\delta})$ *for some* $\delta > 0$, *then* (2.3) *holds for every bounded sequence* $u_n > 0$, $n \in \mathbb{N}$.

The extension of this result to higher order stochastic expansions is straightforward.

If $a = b = 0$, then equivalence holds in (i). In this case, (2.3) is known to be equivalent to Hellinger differentiability (at $t = 0$) of P_t, $t \in [0,\varepsilon)$. Hence Hellinger differentiability is equivalent to DCC_0-differentiability at a rate $o(t^0)$. This can also be proved directly. (See Pfanzagl and Wefelmeyer, 1985, p. 27.)

3. LEMMAS AND PROOFS

3.1. Lemma. *Let* $h: [0,\varepsilon) \to \mathbb{R}$, *and let* $t_n > 0$, $n \in \mathbb{N}$, *be a sequence with* $t_n \to 0$ *and* $t_n^{-1} - t_{n-1}^{-1}$, $n \in \mathbb{N}$, *bounded. Then* $h(t) \to 0$ *for* $t \to 0$ *if and only if* $h(t_n u_n) \to 0$ *whenever* $u_n \to 1$.

For a proof see Pfanzagl and Wefelmeyer (1985, p. 24, Lemma 1.2.10).

A proof of the following Bernstein-type inequality may be found in Bennett (1962).

3.2. Lemma. *Let* $g: X \to \mathbb{R}$ *be measurable with* $P(g) = 0$ *and* $|g| \leq n^{1/2}c$. *Then*

$$P^n\{\tilde{g} > \varepsilon\} \leq \exp\left[-\frac{\varepsilon^2/2}{P(g^2) + \varepsilon c/3}\right].$$

The following lemma shows that DCC_b-differentiability is preserved under smooth transformations of f_t. It will be applied for the transformation $H(z) = \log(1+z)$. The proof is straightforward (see Pfanzagl and Wefelmeyer, 1985, p. 30, Lemma 1.3.4).

3.3. Lemma. *Let* $a \in [0,1)$ *and* $b \ge a$. *Assume that* $P(g) = 0$ *and* $P(|g|^{2(2+b)/(2-a)}) < \infty$. *Let* $H: \mathbb{R} \to \mathbb{R}$ *be twice differentiable in a neighborhood of* 0, *with* $H(0) = 0$, $H'(0) = 1$, *and second derivative* H'' *Lipschitz at* 0. *For* $t \in \mathbb{R}$ *let* r_t *and* s_t *be related by*

$$H(tg+t^{1+a}r_t) = tg + \frac{1}{2}t^2H''(0)P(g^2) + t^{1+a}s_t .$$

Then r_t, $t \in [0,\varepsilon)$, *fulfills* DCC_b *if and only if* s_t, $t \in [0,\varepsilon)$, *fulfills* DCC_b.

Proof of Theorem 1.2. (i) Conditions DCC_0 follow from Lemma 3.1 and the well known degenerate convergence criterion without rates (see, e.g., Loève, 1977, p. 329, Corollary 1).

To prove DCC_b(i), let $s_t(x,y) := r_t(x)-r_t(y)$ denote the symmetrization of r_t. Define $Q := P\times P$ and

$$A_{n\nu} := \{(\underline{x},\underline{y}): s_{n^{-1/2}u_n}(x_\nu,y_\nu) > n^{1/2}\varepsilon\},$$

$$B_{n\nu} := \{(\underline{x},\underline{y}): \sum_{\substack{\mu=1\\ \mu\neq\nu}}^{n} s_{n^{-1/2}u_n}(x_\mu,y_\mu) \ge 0\}.$$

We have

$$Q^n\{\tilde{s}_{n^{-1/2}u_n} > \varepsilon\} \ge Q^n(\bigcup_{\nu=1}^{n} A_{n\nu} \cap B_{n\nu})$$

$$\ge nQ^n(A_{n1} \cap B_{n1} \cap (\bigcup_{\mu=2}^{n} A_{n\mu})^c)$$

$$\ge nQ^n(A_{n1} \cap B_{n1}) - n\sum_{\mu=2}^{n} Q^n(A_{n1} \cap A_{n\mu})$$

$$= nQ^n(A_{n1})(\frac{1}{2} - (n-1)Q^n(A_{n1}))$$

$$\ge \frac{1}{4}nQ^n(A_{n1})$$

for n sufficiently large, since by $DCC_0(i)$,

$$Q^n(A_{n1}) = Q(s_{n^{-1/2}u_n} > n^{1/2}\varepsilon\}$$

$$= \frac{1}{2}Q\{|s_{n^{-1/2}u_n}| > n^{1/2}\varepsilon\} \le P\{|r_{n^{-1/2}u_n}| > \frac{1}{2}n^{1/2}\varepsilon\}$$

$$= o(n^{-1}).$$

Relation $DCC_0(i)$ implies that there exist $\varepsilon_n \downarrow 0$ such that the median m_n of $P*r_{n^{-1/2}u_n}$ fulfills $|m_n| \le n^{1/2}\varepsilon_n$. Hence for n sufficiently large,

$$Q^n(A_{n1}) = \frac{1}{2}Q\{|s_{n^{-1/2}u_n}| > n^{1/2}\varepsilon\}$$

$$\ge \frac{1}{4}P\{|r_{n^{-1/2}u_n} - m_n| > n^{1/2}\varepsilon\}$$

$$\ge \frac{1}{4}P\{|r_{n^{-1/2}u_n}| > 2n^{1/2}\varepsilon\}.$$

Relations (1.3), (3.4), (3.5) imply

$$P\{|r_{n^{-1/2}u_n}| > 2n^{1/2}\varepsilon\} = o(n^{-1-b/2}).$$

Hence $DCC_b(i)$ follows with Lemma 3.1.

(ii) By assumption, the truncated remainder term

$$\bar{r}_t := r_t 1_{\{|r_t| \le t^{-1+\delta}\}}$$

fulfills

$$|\bar{r}_t| \le t^{-1+\delta},\ P(\bar{r}_t) = o(t^{1+\delta}),\ P(\bar{r}_t^2) = o(t^{2\delta}).$$

Hence $f_t := t^{-\delta}(\bar{r}_t - P(\bar{r}_t))$ fulfills, for t sufficiently small,

$$|f_t| \le 2t^{-1}, \; P(f_t) = 0, \; P(f_t^2) = o(t^0).$$

Therefore, Lemma 3.2 and condition DCC_b(i) imply for $u_n > 0$, $n \in \mathbb{N}$, bounded,

$$P^n\{|\tilde{r}_{n^{-1/2}u_n}| > \varepsilon\}$$

$$= P^n\{|\tilde{\bar{r}}_{n^{-1/2}u_n}| > \varepsilon\} + nP\{r_{n^{-1/2}u_n} \neq \bar{r}_{n^{-1/2}u_n}\}$$

$$\le P^n\{|\tilde{f}_{n^{-1/2}u_n}| > \tfrac{1}{2}n^{\delta/2}u_n^{-\delta}\}$$

$$+ nP\{|r_{n^{-1/2}u_n}| > n^{(1-\delta)/2}u_n^{-1+\delta}\}$$

$$= o(n^{-b/2}).$$

<u>Proof of Theorem 2.2</u>. (i) Let s_t be defined by

$$\log f_t = tg - \tfrac{1}{2}t^2P(g^2) + t^{1+a}s_t.$$

From (2.3),

$$P^n\{|\tilde{s}_{n^{-1/2}u_n}| > c\} = o(n^{-b/2}), \quad c > 0.$$

Theorem 1.2(i) implies that s_t, $t \in [0,\varepsilon)$, fulfills DCC_b. Hence P_t, $t \in [0,\varepsilon)$, is DCC_b-differentiable by Lemma 3.3.

(ii) By Lemma 3.3,

$$\log f_t = tg - \tfrac{1}{2}t^2P(g^2) + t^{1+a}s_t$$

with $t^{-\delta}s_t$, $t \;\; [0,\varepsilon)$, fulfilling DCC_b. Theorem 2.1(ii) implies (3.6) for every bounded sequence $u_n > 0$, $n \in \mathbb{N}$. Hence

$$P^n\{|\Lambda_{n,u_n} - u_n\tilde{g} + \frac{1}{2}u_n^2 P(g^2)| > cn^{-a/2}\} = o(n^{-b/2}), \quad c > 0.$$

REFERENCES

Baum, L.E., and Katz, M. (1965). Convergence rates in the law of large numbers. *Trans. Amer. Math. Soc.* 120 108-123.

Bennett, G. (1962). Probability inequalities for the sum of independent random variables. *J. Amer. Statist. Assoc.* 57 33-45.

Bhattacharya, R.N., and Rao, R.R. (1976). *Normal Approximation and Asymptotic Expansions*, Wiley, New York.

Bickel, P.J., Chibisov, D.M., and van Zwet, W.R. (1981). On efficiency of first and second order. *Internat. Statist. Rev.* 49 169-175.

Erdős, P. (1949). On a theorem of Hsu and Robbins. *Ann. Math. Statist.* 20 286-291.

Loève, M. (1977). *Probability Theory*. Vol. 1. 4th ed. Springer-Verlag, New York.

Pfanzagl, J. With the Assistance of W. Wefelmeyer (1985). *Asymptotic Expansions for General Statistical Models.* Lecture Notes in Statistics 31, Springer-Verlag, Berlin.

Wolfgang Wefelmeyer
Mathematical Institute
University of Cologne
Weyertal 86
D-5000 Cologne 41

SUBJECT INDEX

estimation theory

fields

random variables

stochastic processes